C000271095

The Foundations of British Maritime Ascendancy

British power and global expansion between 1755 and 1815 have mainly been attributed to the fiscal-military state and the achievements of the Royal navy at sea. Roger Morriss here sheds new light on the broader range of developments in the infrastructure of the state needed to extend British power at sea and overseas. He demonstrates how developments in culture, experience and control in central government affected the supply of ships, manpower, food, transport and ordnance as well as the support of the army, permitting the maintenance of armed forces of unprecedented size and their projection to distant stations. He reveals how the British state, although dependent on the private sector, built a partnership with it based on trust, ethics and the law. This book argues that Britain's military bureaucracy, traditionally regarded as inferior to the fighting services, was in fact the keystone of the nation's maritime ascendancy.

ROGER MORRISS is Senior Lecturer in the Department of History, University of Exeter, and General Editor of the Navy Records Society. His previous publications include *Naval Power and British Culture, 1760–1850: Public Trust and Government Ideology* (2004).

Cambridge Military Histories

Edited by

HEW STRACHAN, Chichele Professor of the History of War, University of Oxford, and Fellow of All Souls College, Oxford

GEOFFREY WAWRO, Major General Olinto Mark Barsanti Professor of Military History, and Director, Center for the Study of Military History, University of North Texas

The aim of this new series is to publish outstanding works of research on warfare throughout the ages and throughout the world. Books in the series will take a broad approach to military history, examining war in all its military, strategic, political and economic aspects. The series is intended to complement Studies in the Social and Cultural History of Modern Warfare by focusing on the 'hard' military history of armies, tactics, strategy and warfare. Books in the series will consist mainly of single-author works – academically vigorous and groundbreaking – which will be accessible to both academics and the interested general reader.

A list of titles in the series can be found at:

www.cambridge.org/militaryhistories

The Foundations of British Maritime Ascendancy

Resources, Logistics and the State, 1755–1815

Roger Morriss

CAMBRIDGE
UNIVERSITY PRESS

CAMBRIDGE
UNIVERSITY PRESS

The Edinburgh Building, Cambridge CB2 8RU, UK

Published in the United States of America by Cambridge University Press, New York

Cambridge University Press is part of the University of Cambridge.

It furthers the University's mission by disseminating knowledge in the pursuit of education, learning and research at the highest international levels of excellence.

www.cambridge.org
Information on this title: www.cambridge.org/9781107670136

First published 2011
First paperback edition 2013

A catalogue record for this publication is available from the British Library

Library of Congress Cataloguing in Publication data
Morriss, Roger.
The foundations of British maritime ascendancy : resources, logistics and the State, 1755–1815 / Roger Morriss.
   p.   cm. – (Cambridge Military histories)
Includes bibliographical references and index.
ISBN 978-0-521-76809-2 (hardback)
1. Great Britain – History, Naval – 18th century.   2. Great Britain – History, Naval – 19th century.   3. Sea-power – Great Britain – History – 18th century.
4. Sea-power – Great Britain – History – 19th century.   5. Great Britain – Politics and government – 18th century.   6. Great Britain – Politics and government – 19th century.   I. Title.   II. Series.
DA87.M67   2010
359.00941′09033 – dc22   2010035498

ISBN 978-0-521-76809-2 Hardback
ISBN 978-1-107-67013-6 Paperback

Contents

Tables

Preface

Britain emerged from the wars between 1755 and 1815 with the world's greatest overseas empire; with the reputation of a great military power; and with trade and industry that culminated in the world's first industrial revolution. Britain's maritime ascendancy in the period 1755–1815 has been vital to the formation of the British national identity. Much has been written about its achievement. Contemporary mercantilists pointed out the importance of colonies and of seaborne trade for an island state. Nineteenth-century naval officers claimed a 'natural naval superiority over... continental neighbours whose habits and feelings are drawn more particularly to land operations'.[1] More recently, great importance has been attached to Britain's financial system at a time when, to contemporaries, 'security, trade, empire and military power really mattered'.[2] Colonies, naval power and money were all visible and vital manifestations of power. This book looks for the nexus of that power in the organisation of the state and the culture of its servants.

In seeking that central source of Britain's power, this book examines 'the logistics' of British maritime ascendancy. This word is now commonly used with regard to the provision of a chain of supply. Before the mid twentieth century, it was little used. Neither 'logistics' nor the word 'supply' appear in the index of C. Oman's book on *Wellington's Army*, written before the First World War. Indeed, indicative of the contemporary order of interest, Oman's chapter on the army Commissariat comes after those on uniforms and weapons. But, almost in his final sentence, Oman admits: 'That the Peninsular War was successfully maintained... was surely, at bottom, the work of the much maligned commissaries'.[3] Now that order of priority is reversed. Military historians speak of power projection, the complement to which is support by supply. In *Rethinking*

---

[1] Sir George Cockburn to Sir Thomas Cochrane, 15 Sept. 1850, NLS 2291, fo. 176.

[2] P. K. O'Brien, 'Political preconditions for the Industrial Revolution' in *The Industrial Revolution and British Society*, ed. P. K. O'Brien and R. Quinault (Cambridge, 1993), 124–55.

[3] C. Oman, *Wellington's Army 1809–1814* (London, 1913), 319.

Military History, Jeremy Black acknowledges 'the powers best able to wage war were those who got close to a synthesis of military organisation and political/administrative capacity'.[4]

In the eighteenth century, 'national characters and manners were closely integrated into economic diagnoses'. At the beginning of the present century, Emma Rothschild re-focussed 'on matters of *geist* or *esprit* as the source of Britain's success as a global maritime and mercantile power'. She emphasised the importance of protection, communication and trade by sea to British ways of thinking.[5] Here that maritime character of British thinking is reinforced with an awareness that the political as well as the geographical environment was important to maritime power. Britain's relationship with the sea shaped the nature of the state as well as its people. It also shaped an economy in which long-distance trade was cheap and convenient compared to that carried out by land. The combination of maritime economy and state in turn fostered branches of supply that were capable of maintaining Britain's armed forces wherever they were located throughout the world.

This capability of the British state lay partly in the experience that accrued through half a century of overseas operations; partly in its effective union with the private sector; and partly in the development of an efficient administrative infrastructure. The relationship between the state and wider society was vital. For not only did the state draw motivation, resources and ethics from society, the state in turn affected the way in which society developed through spending and war. The relationship developed a unique capability in logistical matters nurtured in the bureaucracy of the British state.

Much here remains for analysis. Each aspect of state operation has had its historian, but the data available for each still require a team of statisticians. This book scratches the surface and suggests trends of development. For support in its research and writing, I must thank a range of colleagues, both past and present. As always, I am grateful to Roger Knight with whom interests have been shared for nearly forty years. More recently, Michael Duffy and Jeremy Black have generously given moral support and encouragement, while the managers of the History Department at Exeter University have provided the time to write, without which nothing could have been done. Stephen End read parts of the manuscript while Gareth Cole and John Day provided insights, for which I shall always be grateful. The staffs of the National Archives of

---

[4] J. Black, *Rethinking Military History* (London, 2004), 163–4.
[5] E. Rothschild, 'The English Kopf' in *The Political Economy of British Historical Experience 1688–1914*, ed. P. K. O'Brien and D. Winch (Oxford, 2002), 31–60.

the United Kingdom, the National Maritime Museum and Royal Naval Museum have my thanks for the indispensable service they provide. A range of other librarians and archivists remain unnamed but remembered. The Rhode Island Historical Society kindly gave permission for quotation from the papers of Christopher Champlin. Michael Watson, Leigh Mueller and Cambridge University Press have been unfailingly helpful. For errors of fact and interpretation, the writer of course remains responsible and he awaits their illumination with interest.

Glossary of British weights, measures, casks and money values

British measure		Abbreviation used here	Metric equivalent
Avoirdupois Weights			
1 dram		dram	1.77 grammes
1 ounce	(16 drams)	oz	28.35 grammes
1 pound	(16 ounces)	lb	0.45 kilograms
1 quarter	(28 pounds)	qtr	12.7 kilograms
1 hundred-weight	(112 pounds)	cwt	50.8 kilograms
1 ton	(20 hundredweights)	ton	1.02 tonnes
Linear Measures			
1 inch		in	25.39 millimetres
1 foot	(12 inches)	ft	0.30 metres
1 yard	(3 feet)	yard	0.91 metres
1 mile	(1,760 yards)	mile	1.61 kilometres
Measures of Capacity			
1 pint		pint	0.57 litres
1 gallon	(8 pints)	gall	4.55 litres
1 peck	(2 gallons)	peck	9.09 litres
1 bushel	(8 gallons)	bush	3.64 dekalitres
1 tun	(252 gallons)	tun	1,146.6 litres

Casks
1 firkin   (9 gallons)
1 half-barrel (2 firkins, 18 gallons)
1 barrel   (2 half-barrels, 36 gallons)
1 tierce   (35 gallons, one third of a pipe, 42 gallons by twentieth century)

1 puncheon (varied, 72-120 gallons, becoming standardised at
            72 gallons)
1 butt   (varied, 105-40 gallons, becoming standardised at
            108 gallons)
1 pipe   (105 gallons)

Money Values

1 farthing		$\frac{1}{4}$d
1 half-penny	(2 farthings)	$\frac{1}{2}$d
1 penny	(2 half-pennies)	d
1 shilling	(12 pence)	s
1 pound	(20 shillings)	£

Abbreviations

BIHR	*Bulletin of the Institute of Historical Research*
BL	British Library
British Naval Documents	*British Naval Documents 1204–1960*, ed. J. B. Hattendorf, R. J. B. Knight, A. W. H. Pearsall, N. A. M. Rodger and G. Till
CR	*House of Commons Reports*
DNB	*Dictionary of National Biography*
DRO	Devon Record Office
EconHR	*Economic History Review*
EHR	*English Historical Review*
HJ	*Historical Journal*
HR	*Historical Research*
HRO	Hampshire Record Office
IJMH	*International Journal of Maritime History*
JBS	*Journal of British Studies*
JMH	*Journal of Military History*
JSAHR	*Journal of the Society for Army Historical Research*
LC	Library of Congress
MM	*The Mariner's Mirror*
NAS	National Archives of Scotland
NLS	National Library of Scotland
NMM	National Maritime Museum
NRS	Navy Records Society
OHBE	*Oxford History of the British Empire*, ed. N. Canny, P. J. Marshall and R. W. Winks
PP	*Parliamentary Papers*
RNM	Royal Naval Museum

TNA	The National Archives of the United Kingdom
WLC	William L. Clements Library, University of Michigan
WRP	William R. Perkins Library, Duke University

Introduction

Until the twentieth century, Britain's maritime history – then largely naval history – was written to inspire pride and emulation in its readers. It has given rise to a triumphal view of the expansion of Britain's maritime empire, naval power and economic wealth, the heroes of which have been statesmen, military officers and gentlemen capitalists. This largely class-bound history has fostered an understanding of Britain's status in the maritime world that has tended to ignore, indeed disdain, clerical skills, labour, transportation and supply. After all, these were the work of working men and servants. Almost at the bottom of a hierarchy of explanation for Britain's maritime ascendancy has been bureaucracy. Throughout British history the back-room bureaucrats have often been scapegoats for the blame of its military leaders. The consequence has been disregard for, if not deliberate derision of, the organisation which has always been necessary to ventures beyond Britain's shores.

This book turns this scale of values on its head. It attaches great importance to the bureaucratic culture which evolved under the aegis of the state during the eighteenth century. By 'culture' here is meant simply a way of thinking and performing tasks – in this case, ones necessary to the state. Naturally, both were shaped by historical legacy and by public opinion, which in Britain had a particular capacity for influence through the political structure of the state. Often overlooked, but fundamental to these environmental influences, was Britain's maritime nature. Here the inner working of the British state is correlated to its role in the maritime world. It is an approach which is offered in complement to existing explanations. For these, being drawn from imperial, military and economic history, have their strengths as well as limitations as sources of explanation for Britain's maritime ascendancy between 1755 and 1815.

Imperial history has been struggling to achieve unity.[1] Until the late twentieth century, it was split by the loss of the American colonies into

---

[1] A. Webster, *The Debate on the Rise of the British Empire* (Manchester, 2006), 68–116.

the 'first British empire' of self-governing Anglo-Saxon communities,[2] and a second in the east consisting of indigenous peoples ruled by colonial governors.[3] That division has only recently been bridged by focus on continuities of culture and identity.[4] Since then Peter Marshall has enhanced unity by observing how war, conquest and loss transformed ideas about the nature, ethics and organisation of empire.[5] At the same time historians dealing with the acquisition of territories in both east and west have identified among contemporary statesmen, both before and after the beginning of the American War of Independence in 1775, a common reluctance to take on new responsibilities without real necessity.

Before 1775 popular support for the protection of colonies was jingoistic.[6] Yet state policy, as Daniel Baugh has argued, was pragmatic. Trade was preferred to property: 'Possession, settlement, governance and territorial defence entailed needless and unwise costs [and were avoided] so long as trade could be carried on otherwise.' Territories acquired in wartime were generally employed as bartering counters, returned at peace to recoup losses elsewhere. They were taken to deprive an enemy of trade revenues or of privateer bases and rarely to add to the existing empire or for the purpose of creating a naval base.[7] It was a policy that re-emerged during the Napoleonic War.

Pragmatism persisted after 1775 with good reason. Acquisition by the East India Company of the right to collect the revenues of Bengal, Bihar and Orissa after the battles of 1757 and 1764 created uncertainty about how to manage the company. To control costs and corruption, the India Act of 1773 gave the state oversight of the company's accounts. That of 1784 created a Governor General, the board of Control and vice-presidencies in India. That of 1813 placed the company's territories under the control of the Crown. But, as Hew Bowen has shown, trade profits and tax revenues were still largely eaten up by the costs of maintaining military forces in India. Empire remained a financial liability,

---

[2] P. J. Marshall, 'The First British Empire' in *OHBE* (5 vols., Oxford, 1999), V, 43–53, discussing J. R. Seeley, *The Expansion of England* (1883) and V. T. Harlow, *The Founding of the Second British Empire* (1952).

[3] C. A. Bayly, 'The Second British Empire' in *OHBE*, V, 44–72.

[4] P. J. Marshall, 'Introduction' to vol. II, *The Eighteenth Century*, in *OHBE*: 1–27.

[5] P. J. Marshall, *The Making and Unmaking of Empires: Britain, India and America c. 1750–1783* (Oxford, 2005), 25–44.

[6] J. Brewer, 'The eighteenth-century British state: contexts and issues' and K. Wilson, 'Empire of virtue' in *An Imperial State at War: Britain from 1689 to 1815*, ed. L. Stone (London, 1994), 52–71, 128–64.

[7] D. A. Baugh, 'Maritime strength and Atlantic commerce. The uses of "a grand marine empire"' in *An Imperial State at War*, ed. Stone, 185–223.

paid for from the proceeds of trade – moreover, that to China rather than India.[8]

If imperial history has found uniting themes, military history has remained divided into that of the navy and of the army, although several historians have written about amphibious operations.[9] For the period between 1755 and 1815, the navy has been regarded as the more important to British maritime ascendancy. After all, even before the Seven Years' War, it had demonstrated its power to erode the shipping and navies of Britain's enemies. It denied use of the sea to the latter, while providing protection to Britain's own trade, and did much to safeguard a maritime economy that provided a significant proportion – although not as much as might be assumed – of the revenues and loan capital necessary to Britain's military operations.[10] Then, and subsequently, the navy's role has appeared vital to Britain's financial capability.[11]

But naval power had its limitations. Victories made little impact on the continental dominance of France.[12] Naval power was effective against states with oceanic and coastal trades but not on those without them.[13] Blockade distorted economies but did not destroy them.[14] To exhaust continental powers demanded allies willing to take them on, the payment of subsidies on a huge scale,[15] and the military involvement of Britain on the continent. Sea power thus had to be complemented by land power, a factor which transformed state investment in the armed forces during wartime, and made the finance and supply of the army as important to Britain's ascendancy as the maintenance of a navy. After all, bases,

---

[8] H. V. Bowen, *The Business of Empire: The East India Company and Imperial Britain, 1756–1833* (Cambridge, 2006), 3–19, 222–45.

[9] See, for example, R. Harding, 'Sailors and gentlemen of parade: some professional and technical problems concerning the conduct of combined operations in the eighteenth century'; P. Mackesy, 'Problems of an amphibious power: Britain against France, 1793–1815'; and D. Syrett, 'The methodology of British amphibious operations during the Seven Years' and American Wars' – all in *Naval History 1680–1850*, ed. R. Harding (Burlington, VT, and Aldershot, 2006), 127–47, 117–26, 309–20, respectively.

[10] *The Influence of History on Mahan*, ed. J. B. Hattendorf (Newport, RI, 1991), discussing A. T. Mahan, *The Influence of Sea Power upon History 1660–1783* (1890).

[11] P. Kennedy, *The Rise and Fall of British Naval Mastery* (London, 1976), 97–147, and *The Rise and Fall of the Great Powers: Economic Change and Military Conflict from 1500 to 2000* (London, 1989), 167–9.

[12] E. Ingram, 'Illusions of victory: the Nile, Copenhagen and Trafalgar revisited', *Military Affairs* 48(1984), 140–3.

[13] R. Harding, *The Evolution of the Sailing Navy, 1509–1815* (Basingstoke, 1995), 120, 140.

[14] F. Crouzet, 'Wars, blockade and economic change in Europe 1792–1815', *Journal of Economic History* 24(1964), 567–88.

[15] J. M. Sherwig, *Guineas and Gunpowder: British Foreign Aid in the Wars with France 1793–1815* (Cambridge, MA, 1969), 345–6.

colonies and strategic territory could only be held by land forces, which in amphibious operations were the point of the spear.

Important though they might appear, the relationships of the navy to empire and to economic growth in the maritime sphere were, and still are, matters of faith. In 1964 Gerald Graham could assert unequivocally that without 'command of the sea there would have been no British empire'.[16] Although he acknowledges the many conditional factors, the title of Nicholas Rodger's most recent volume in his new naval history maintains that view.[17] Yet in 1999 Barry Gough claimed 'the general linkage of navy to empire continues to escape historians, perhaps because the task [of establishing connections] is such a daunting one'.[18] The relationship between naval power and merchant shipping is also questionable. The American advocate of sea power, A. T. Mahan, argued that nations 'advanced to power at sea' through 'service of their [merchant] ships'.[19] But the relationship did not always operate in reverse. Ralph Davis noticed that during wartime before 1783 naval vessels 'were always too few to be fully effective'.[20] It was a problem that persisted especially for the coastal trade and in spite of the Convoy Act of 1798 which made convoy compulsory for most ocean-going vessels.[21]

As a source of explanation, British economic history has suffered as much from division as imperial and military history. Its maritime component has been graced by just a few distinguished scholars. Yet, as an island state, Britain needed shipping for the import of naval stores, food and industrial raw materials, for trade, and for the transportation of its armed forces with the supplies they needed. About a tenth of Britain's ocean-going shipping was under hire to the state by the end of the eighteenth century. State employment aided the growth of shipping which in turn contributed to economic expansion, for ship managers were able to cheapen transport costs for their customers, facilitating capital

---

[16] G. S. Graham, *The Politics of Naval Supremacy: Studies in British Maritime Ascendancy* (Cambridge, 1965), 3.

[17] N. A. M. Rodger, *The Command of the Ocean: A Naval History of Britain 1649–1815* (London, 2004); see also his 'Sea-power and Empire, 1688–1793' in *OHBE*, II, 169–83; *British Naval Documents*, 326–31.

[18] B. M. Gough, 'The Royal Navy and empire' in *OHBE*, V, 327–41.

[19] A. T. Mahan, *The Influence of Sea Power upon History 1660–1783* (Boston, MA, 1890, repr. 1965), 53.

[20] R. Davis, *The Rise of the English Shipping Industry in the Seventeenth and Eighteenth Centuries* (Newton Abbot, 1962, repr. 1972), 315–37.

[21] See, for example, the complaint of losses to privateers suffered by contractors shipping provisions from Ireland to London, who claimed that the English Channel from Scilly to the Forelands was infested with French privateers: NMM, ADM. BP/30B, 15 Nov. 1810.

accumulation and investment elsewhere in the economy.[22] Already global in operation by 1755, with distinct regions of trade and specialisation, the shipping industry flourished despite the recurrence of war, partly under the Navigation Laws that still endeavoured to preserve colonial trade for English ships.[23]

The flexibility, economies and military importance of shipping for an island state have been under-acknowledged. So too has its importance in carrying the trade that contributed indirect taxation to the financing of wars in the eighteenth century. To John Brewer in *The Sinews of Power*, shipping was no more than a victim of hostilities. For Cain and Hopkins, it was only important after 1850 as a link between 'gentlemanly capitalism' practised in the City of London and that in the outposts of empire.[24] Brewer terminated his examination of the 'fiscal-military state' in 1783, while Cain and Hopkins were 'notably thin on everything before the 1790s'.[25] Patrick O'Brien's essays demonstrate the growth of Britain's financial capability throughout the period of hostilities. Yet even his focus on the fiscal state, in which trade revenues played a major part, tends to ignore the importance for Britain of shipping.[26]

Ironically, since publication of *The Sinews of Power*, the preoccupation with finance as the mainspring of state power has given rise to studies of the financial arrangements of other eighteenth-century states and reduced the apparent exceptionality of Britain's methods of raising money, if not the scale and stamina of that capability.[27] Less distinguished by fiscal-military arrangements, the question of what gave Britain the ability to become the dominant power at sea remains open. In this study the maritime nature of the British economy assumes great importance. But even more important was the bureaucratic culture of the British state.

This book examines that culture in operation, in the supply of the military forces that Britain projected throughout the world between 1755

---

[22] S. Ville, 'The growth of specialization in English shipowning, 1750–1850', *EconHR* 46, 4(1993), 702–22, and 'Total factor productivity in the English shipping industry: the north-east coal trade, 1700–1850', *EconHR* 39, 3(1986), 355–70.

[23] R. Davis, *Rise of the English Shipping Industry*, 300–14; S. Ville, *English Shipowning during the Industrial Revolution: Michael Henley and Son, London Shipowners, 1770–1830* (Manchester, 1987), 59, 95, 129.

[24] J. Brewer, *The Sinews of Power: War, Money and the English State 1688–1783* (London, 1989), 197–8; P. J. Cain and A. G. Hopkins, *British Imperialism: Innovation and Expansion 1688–1914* (Harlow, 1993), 170–1, 179–80.

[25] B. Lenman, *Britain's Colonial Wars 1688–1783* (Harlow, 2001), 4.

[26] Most notable is P. K. O'Brien, 'The political economy of British taxation, 1660–1815', *EconHR* 2nd ser. 41, 1(1988), 1–32.

[27] P. K. O'Brien, 'Fiscal exceptionalism: Great Britain and its European rivals from Civil War to triumph at Trafalgar and Waterloo' in *The Political Economy of British Historical Experience 1688–1914*, ed. D. Winch and P. K. O'Brien (Oxford, 2002), 245–65.

and 1815. Chapters deal successively with the supply of strategic ideas, finance, the navy, ordnance, manpower, food, shipping and the organisation serving land forces overseas. The common factors in the examination of each are the resources, organisation, methods and innovations with which the servants of the state met the needs of Britain's armed forces around the world. For much of the time, these servants included merchants and ship owners, acting under contract, as well as the central commissioners, their clerks and agents in Britain and abroad. How the private sector was employed and trained was an important aspect of each branch of state service. This book is thus particularly concerned with the attitudes and practices that made for efficiency of performance, and for the smooth coordination of the public and private sectors.

For this, the role of the state was critical in providing a framework for the regulation of international and domestic relations, social order and commercial business.[28] It provided a legal framework which was subject to changing attitudes and meant that, as Peter Marshall suggested, government gradually adapted to movements in contemporary ethics as well as administrative and organisational ideas. These ethics and ideas supplied the cement that provided cohesion between the state and its servants, especially those in the private sector. It was their cooperation that made for economic strength and was the foundation of military initiatives and expansion. But, as this book will show, it was the thinking, organisation and administrative ability of the bureaucracy at the heart of the state, founded on an expanding maritime economy and financial capability, which made Britain the dominant power at sea between 1755 and 1815.

---

[28] P. K. O'Brien, 'Central government and the economy, 1688–1815' in *The Economic History of Britain since 1700*, ed. R. Floud and D. McCloskey (2nd edn, Cambridge, 1994), 205–41.

1   The British state in evolution

During the second half of the eighteenth century the power of the British state grew. It drew that strength from within from those who wielded power locally and in commerce. Driven by war, the state grew and developed efficient forms of managing innovation and change. Ideas about the efficiency of organisation shifted, with a view to the production of greater motivation among its servants. Management of the state's resources was placed in the hands of men open to new thinking, and ready to consult those with expert knowledge. With competition, resources were always in short supply, but policy towards them was equally concerned with the removal of obstacles to existing sources as with enlargement of the resource base. State law reflected this reduction of restrictions, but was balanced by the growing framework in law and policy of equity between the state and its servants in the private sector, whether small-scale contractors or great commercial companies. Ideas, management, policy and law all shaped the logistics of state supply. They made for a state which could summon, control, organise and provide resources for its armed forces throughout the globe.

The British state

The structure, culture and capabilities of the British state developed during the second half of the eighteenth century. Its growing power was reflected in the perceptions of its leading economic critics. Writing on the eve of the American Revolution, Adam Smith assigned to it a small role, limiting its duties to defence, administration of justice and the performance of certain public works. The state, he proposed, should simply create a milieu suited to the uninhibited conduct of private enterprise which could be conducted more efficiently by private entrepreneurs than by public bureaucrats.[1] Writing four decades later, after experience of

---

[1] J. J. Spengler, 'Adam Smith's theory of economic growth, parts I–II' in *Adam Smith: Critical Assessments*, ed. J. Cunningham Wood (3 vols., Beckenham, Kent, 1983), III, 110–31.

population growth, urbanisation, industrial and agricultural expansion, David Ricardo realised the state could also affect capital accumulation, investment and employment, and was critical to the well-being of its subjects.[2]

The sources of its power

Historians writing about the British state have tended to reflect this change in view of contemporaries. Writing of the mid eighteenth century, Stephen Conway located the strength of the state in its partnership with private and local interests. It relied on the assistance of local Justices of the Peace and on the tax-raising capability of Parliament. He emphasised the 'overlapping and competing jurisdictions' of government departments and the relative autonomy of privateers, the regiments raised by noblemen and the army of the East India Company. On the one hand, he suggests the dependence of the state on contractors and financiers was a weakness. On the other, he argues that the ability to pay and equip the armed forces through contractors on terms set by the state was a strength in which there was growing confidence, a product of repeated wars, growing expenditure, experience and efficiency.[3]

Conway's explanation for growing confidence in the state's strength holds true for the period of the Revolutionary and Napoleonic Wars. By the time of Ricardo that strength was used to regulate industry, commerce and social problems. Historians recognise the state's power and the attitudes which made for reform.[4] But the sources of that power and those attitudes in developments during the periods of hostilities prior to 1815 tend to have been ignored.[5]

Yet these wartime developments took palpable form in the growth of bureaucracy. Even by later standards, that bureaucracy was not small. At the very beginning of the eighteenth century, employees of the state have been estimated at 12,000, with 114 commissioners sitting on 18 different

---

[2] J. H. Hollander, 'The work and influence of Ricardo', and D. V. Ramana, 'Ricardo's environment', in *David Ricardo. Critical Assessments*, ed. J. Cunningham Wood (4 vols., Beckenham, Kent, 1985), I, 42–5, 196–208, respectively.

[3] S. Conway, *War, State and Society in Mid-Eighteenth Century Britain and Ireland* (Oxford, 2006), 31–55.

[4] U. Henriques, 'Jeremy Bentham and the machinery of social reform' in *British Government and Administration: Studies presented to S. B. Chrimes*, ed. H. Hearder and H. R. Loyn (Cardiff, 1974), 169–86.

[5] See, for example, E. J. Evans, *The Forging of the Modern State: Early Industrial Britain* (Harlow, 1983; 3rd edn, 2001).

boards.[6] By 1797 the figure had reached 16,267, and by 1815 24,598.[7] Some central offices remained small. In 1793 the Treasury, Home Office and Foreign Office contained only 17, 19 and 19 personnel respectively. But the Navy Office in 1792 had a staff of 98 established with 10 extra clerks;[8] and by 1813 it had grown to 151 established with 80 extra clerks – a total of 231. The Victualling Office grew from 65 in 1787 to 105 staff by 1805.[9] Meanwhile the Admiralty Office, with 45 personnel in 1797, grew to 65 in 1815.[10] The departments that grew most dealt with the state's finances and military requirements. Just those dealing with finance, by 1755, numbered 6,484 employees, and by 1782/3 this had risen to 8,292.[11] Meanwhile the Royal dockyards – which were enlarged in wartime and reduced in peace – grew from around 8,100 employees about 1745 to more than 15,000 in 1815.[12]

The scale of the fiscal and military departments reflected the workload they bore. By 1755 the British state possessed a financial system capable of rivalling that of any other European power and it had the most powerful navy in Europe. Its West Indian and American colonies were part of an Atlantic economy that was served by a merchant fleet that rivalled that of the Dutch. After 1755 the state built on these assets. During the Seven Years' War, Britain was manifestly the most dynamic state in Europe. The loss of her American colonies in 1783 temporarily diminished that standing but not the latent power and dynamism. Britain's expansion in India, south-east Asia, the South Pacific and the West Indies between 1755 and 1815 was a product of this underlying strength.

War built state power. A 'cycle of war and state formation' has long been recognised.[13] 'War became the great flywheel for the whole political enterprise of the modern state . . . the constant rivalry among the powers . . . produced an unheard-of exertion of energy, especially military and financial energy'.[14] For maritime powers, the growth of naval power was

---

[6] Brewer (quoting G. Holmes), 'The eighteenth-century British state'.

[7] P. Harling and P. Mandler, 'From "fiscal-military" state to laissez-faire state, 1760–1850', *JBS* 32(1993), 44–70.

[8] C. Emsley, *British Society and the French Wars 1793–1815* (London, 1979), 9.

[9] R. Knight, 'Politics and trust in victualling the navy, 1793–1815', *MM* 94(2008), 133–49. But see also NMM, ADM. DP/27, 21 Feb. 1807.

[10] NMM, ADM. BP/28A, 25 Mar. 1828; ADM. DP/201B, n.d.

[11] Brewer, *The Sinews of Power*, 66.

[12] D. A. Baugh, *British Naval Administration in the Age of Walpole* (Princeton, NJ, 1965), 264; R. Morriss, *The Royal Dockyards during the Revolutionary and Napoleonic Wars* (Leicester, 1983), 106.

[13] B. D. Porter, *War and the Rise of the State: The Military Foundations of Modern Politics* (New York, 1974), 36–9, 58–9, 72–121.

[14] T. Ertman (quoting O. Hintze), 'The *Sinews of Power* and European state-building theory' in *An Imperial State at War*, ed. Stone, 33–51.

as important to state building as armies were to land-bound states.[15] The way Britain grew appears not to have been different from other states.[16] Agreements between the governing power and interest groups facilitated the extraction of revenues, usage of manpower, protection of trade and investment in military forces.[17] Jan Glete argues that political transformation went hand in hand with military transformation and that resource problems were normally solved by resort to the commercial market. Glete refers particularly to early modern states but his observations apply equally to state development in Britain in the eighteenth century.[18]

State bureaucracy served as the linchpin of military force and the commercial market. Britain's bureaucracy already had a shape and character formed by the experience of the seventeenth century.[19] Basic structures were in place and tested during the War of William III, the first of the series of wars with France and its allies.[20] A growing workload, increasing numbers of state employees and knowledge of political arithmetic improved bureaucratic efficiency by 1715.[21] Repeated wars after 1739 enhanced this efficiency. By the time of the Napoleonic War, Britain's bureaucracy would conduct the business of war with all the authority and regulation it would later apply to the social problems of the mid nineteenth century.

Fundamental to the power of the state was its financial system. That too developed, as did the attitudes to the state's system of war funding. Before 1793 the national debt was a cause of concern central to parliamentary politics. Indeed before 1765 Parliament routinely limited military expenditure. However, after 1793, the national debt gradually lost its horror. The state set time limits on short-term debt and increased the rate at which the navy was funded. At the same time the introduction

---

[15] J. Glete, *Navies and Nations: Warships, Navies and State-building in Europe and America, 1500–1860* (2 vols., Stockholm, 1992).

[16] Ertman, 'The *Sinews of Power* and European state-building theory'.

[17] J. Glete, *Warfare at Sea, 1500–1650: Maritime Conflicts and the Transformation of Europe* (London, 2000), 60–75, 186–7.

[18] J. Glete, *War and the State in Early Modern Europe: Spain, the Dutch Republic and Sweden as Fiscal-Military States, 1500–1660* (Abingdon, 2002), 43–66, 213–17.

[19] For a summary examination of British naval administration at its crucial formative stage, see Rodger, *The Command of the Ocean*, 95–111.

[20] See, for example, 'A scheme for governing the business of Victualling the Navy, of Comissary General of the Land Forces and of Commissioner for Transports under one Management', 1 October 1686; and 'Some few Reasons for managing the Victualling by a Commission rather than by Contract', May 1698; both TNA, T.48/89, fos.167–8, 275–7.

[21] J. H. Plumb, *The Growth of Political Stability in England* (London, 1967, repr. 1980), 11–12, 116–22.

of income tax increased the revenues available to the state for funding debt. Needless to say, however, expenditure grew the greater, supported by an economy still expanding despite the difficulties imposed on overseas trade by the war.[22] Remarkably, although state expenditure and debt continued to grow, by 1815 Patrick Colquhoun could regard the nation's debt as an asset rather than a liability.

Meanwhile, Parliament had not the time, qualification or knowledge to regulate state budgets. Finance committees kept it informed but, for day-to-day management of the state's finances, the government came to rely almost completely on the state's bureaucracy. Increasingly, and especially during the Napoleonic War, the military departments worked under the eye of the Treasury. Even so, the proportion of state expenditure devoted to the supply of war gave the military spending departments considerable influence in Britain's economy.

The military infrastructure

The bureaucracy principally responsible for military expenditure was the War Office, and the Admiralty Office, which respectively represented the needs of the army and navy. The Secretary for War was advised by the Commander-in-Chief who relied upon subordinates – the Adjutant General, Quartermaster General, military secretary, and field officers – to secure information and implement decisions. The War Office secured funding for the army and obtained sanction to order the issue of arms and give marching orders. Likewise the First Lord of the Admiralty was advised by the board of Admiralty which obtained information and carried out decisions through its subordinate Navy, Victualling and Sick and Hurt Boards. The latter existed only until 1806 when, with its office staff, it was absorbed by the Transport Board that was established in 1794. The armaments needed by both the army and the navy were supplied by the Ordnance Board which worked to the directions of the Master General of the Ordnance.

During the second half of the eighteenth century, this patchwork of departments was subject to a number of criticisms – for failures of supply, for the length of time decisions took, and for the incompetence of its officials.[23] The origin of these criticisms was invariably the military men conducting operations at sea or in the field who were unaware of the

---

[22] I. R.Christie, *Wars and Revolutions: Britain 1760–1815* (London, 1982), 296.

[23] R. Glover, *Peninsular Preparation: The Reform of the British Army 1795–1809* (Cambridge, 1988), 14; D. Syrett, *Shipping and the American War 1775–83: A Study of British Transport Organization* (London, 1970), 2; N. Baker, *Government and Contractors: The British Treasury and War Supplies 1775–1783* (London, 1971), vi.

problems, especially of communication, encountered by the departments they relied upon for supply.[24] The result has been history written from the point of view of the operational officer, who often had little sympathy for or understanding of the work of the government departments.[25] Their reputation has, moreover, suffered from the moral judgement of historians who condemned relations between government officials and private contractors.[26]

This was regrettable since the military departments of government were subject to innovations, reform and rising expectations after 1755 which progressively made for greater efficiency. Some of these innovations especially affected the projection of British power overseas. The Commissariat, organised by the Treasury and responsible for supply of the army on campaign, developed in the Seven Years' War and, by the time of the Napoleonic War, replicated the functions of the British state in microcosm. Between 1779 and 1793, responsibility for the shipment of food supplies to the army overseas was shifted from the Treasury to the Navy Board, and then to the Victualling Board. The duty of hiring ships to serve as transports was shifted from individual departments to the Transport Board in 1794. Henceforward, any department of government that needed shipping to transport men or materials around the world simply applied to the Transport Board.

The board of Ordnance has particularly suffered from criticism for conservatism and inefficiency. The criticism lasted until the end of the eighteenth century. Yet that department saw the introduction of officials specifically aimed at raising standards. From 1780, there was an Inspector of Artillery and from 1789 an Inspector of Gunpowder Manufactories. A significant attempt was also made to improve the operational performance of the civil departments of the navy with the appointment at the Admiralty in 1796 of an Inspector General of Naval Works, an officer authorised to make recommendations for improvements in methods of management as well as physical infrastructure. All these offices were linked to important developments in, for example, dock capacity, cannon design and the strength of gunpowder. Equally important for the projection of power by sea was the establishment at the Admiralty in 1795 of the office of Hydrographer of the Navy. The man appointed, Alexander Dalrymple from the East India Company, was to spend his first five years organising his office. But from the early nineteenth century he was able

---

[24] R. J. B. Knight, 'Civilians and the navy, 1660–1832' in *Sea Studies: Essays Presented to Basil Greenhill*, ed. P. G. W. Annis (Greenwich, 1983), 63–70.

[25] C. D. Hall, *British Strategy in the Napoleonic War 1803–15* (Manchester, 1992), xii.

[26] R. G. Albion, *Forests and Seapower: The Timber Problem of the Royal Navy, 1652–1862* (Cambridge, MA, 1926), 53–4.

to start supplying charts to ships, up-dating old ones, and developing a chart production process.

These innovations raise the question of whether Britain's armed forces in the late eighteenth century participated in change tantamount to a military revolution. The proposed timing and nature of such a development have altered over the last half-century. It was first suggested by Michael Roberts in 1955 from a study of Swedish military operations in the period 1560–1660 when portable firearms were introduced and gave rise to the training of troops, new tactics, more complex strategies, more permanent forces and the need to provide for them. Twenty years later, from a study of siege armies and their supply problems, Geoffrey Parker broadened the period, extending it to the middle of the eighteenth century and the growth in numbers of purpose-built warships, naval tactics and fleet size.[27] In 1991 Jeremy Black reacted against the length and continuity of this period and looked for armed forces that actually altered, rather than reflected, the society in which they existed.[28] More recently, he has stressed the influence of long-term incremental change and the equal importance of ideological pull as well as innovatory push factors.[29]

Black's suggestions raise important questions, of the extent to which military change in Britain affected British society, and how civil and military cultures interact. Wars extended military experience and deepened national identity.[30] Repeated, protracted and demanding hostilities unquestionably affected British society. Conversely, contemporary social attitudes affected military performance. In the organisation of government, especially the managerial structures of supply, ideology was crucial to efficiency, based on the ideas, beliefs, standards and values of its managers and workers. It shaped the habitual behaviour of bureaucrats and became institutionalised in their organisation.[31]

Examined singly, the foundation of offices like those of the Inspectors of Gunpowder Manufactories and of Naval Works tell us a limited amount about the culture of the British state. But examined collectively,

---

[27] G. Parker, *The Military Revolution: Military Innovation and the Rise of the West, 1500–1800* (Cambridge, 1988; 2nd edn, 1996).

[28] J. Black, *A Military Revolution? Military Change and European Society 1550–1800* (Basingstoke, 1991), 93–6.

[29] J. Black, 'Was there a military revolution in early modern Europe?' *History Today* 58, 7(2008), 34–41.

[30] H.V. Bowen, *War and British Society, 1688–1815* (Cambridge, 1998), 40–55; L. Colley, *Britons: Forging the Nation 1707–1837* (New Haven and London, 1992), 5, 9, 167–88, 285–300.

[31] P. Bagby, *Culture and History: Prolegomena to the Contemporary Study of Civilisation* (Berkeley and Los Angeles, 1958), 72–7; J. Black, 'Military organisations and military change in historical perspective', *JMH* 62(1998), 871–92.

they tell us a great deal. Indeed, the evolving structure of British military government between 1775 and 1815 documents a culture that was in step with contemporary thinking in society at large, probably in some respects farther ahead. For war was a forcing house for organisational change, at the heart of which lay ideological change.

Bureaucratic organisation and expertise

The ideology of motivation

The ideology of the state's servants has been little noticed in the history of empire, naval power and shipping. Yet it was vital to the motivation of the state's servants and to the enlargement of the resources available to the state. In the past the establishment of Britain's maritime ascendancy has been most closely associated with its sea-going naval service where the distinction of individuals was a vital part of the promotion process. Private enterprise too operated on the basis of individual responsibility. By contrast, the British government establishment in the eighteenth century operated on a different ideological basis. Obliged to check against fraud and abuse, it practised collective responsibility. A vital part of the regeneration of the state's military bureaucracy towards the end of the eighteenth century was the creation of greater consistency between these three areas of state activity.

Collective responsibility in civil administration arose in the seventeenth century, or even before, when instructions were issued to the board and local officials responsible for the state's assets.[32] The commissioners of the navy, later known as the Navy Board, were directed to keep their business 'digested in books' so that they might 'better inspect the transactions of each . . . relating one to the other' and 'to trace one another in their distinct and separate duties'. At least three members were to sign their letters and orders to subordinates. Similarly in the dockyards, the principal officers were required to sign all letters to the Navy Board of common concern in triplicate as a check upon the veracity of their contents. As different naval departments were equipped with instructions, so the principle of collective responsibility was extended, and it survived. In 1801 one yard officer could report the 'grand principle of government in the dockyards' was the check yard officers kept on one another.[33]

---

[32] R. Morriss, *Naval Power and British Culture, 1760–1850: Public Trust and Government Ideology* (Aldershot, 2004), 64–7.

[33] *The Economy of His Majesty's Navy Office* (London, 1717), 20–1; TNA, ADM. 1/5126, 10 March 1801.

However, as the burden of business mounted in the management
of departments of government, the collective checks broke down. It
remained the theoretical check against fraud and abuse. Yet in practice, by the second half of the eighteenth century, commissioners and
officers had no the time to check the transactions of their colleagues. In
1767 the board of Admiralty led by Edward Hawke observed that 'the
present disorder and irregularities in the yards have gradually increased
to their present height from a neglect and relaxation of that discipline
which it is the duty of the officers of the said yards, according to their
instructions and rules of the navy, to maintain'.[34] An aggravation was
the dispersal of standing orders and warrants, issued at different times,
through the papers of each office, where, in any case, different circumstances determined different interpretations. In essence, the systems of
management were in need of a thorough overhaul.

By then, to supplement wage and salary rates established in the seventeenth century, a private economy of recompense had grown up. Fees
were charged on transactions, even with public employees – to collect for
example the paper copy of a commission – and also with members of the
public, most notably contractors. Rates were charged at a percentage of
the value of the contracts. This distorted motivation in the public offices.
Officers and clerks aspired to perform the tasks that were rewarded by
receipt of the highest fees, which could far exceed the amount of their
official salaries. Naturally, once in place, officials remained as long as
possible in to order receive these fees. Premiums were paid on first entry
and on every promotion in office to recompense and provide a lump sum
for employees retiring from the department. Place holders thus naturally
wanted to make sure they obtained in fees value for the money they
had invested in premiums. There was usually no fraud involved in these
private transactions. Fee rates were posted in most offices. But the total
amounts of fee earnings were kept private. And, though the integrity of
officials was defended, suspicion of fraud always existed, to break out in
allegations at times of stress, disorder and failure.[35]

Defeat in America and public opinion shortly before 1780 gave rise
to the movement for economical reform of government.[36] One of its
first achievements was the establishment in 1785 of the Commission
on fees, gratuities, perquisites and emoluments in the public offices. It
examined only ten of the twenty-four departments slated for inspection.
Nevertheless, those ten included the naval departments. Its principal

---

[34] NMM, ADM. A/2596, 23 June 1767.
[35] Morriss, *Naval Power and British Culture*, 73–8, 107–15, 137–40.
[36] E. Hellmuth, 'Why does corruption matter? Reforms and reform movements in Britain
and Germany in the second half of the eighteenth century' in *Reform in Great Britain
and Germany 1750–1850*, ed. T. C. W. Blanning and P. Wende (Oxford, 1999), 6–23.

recommendation for a prohibition on the receipt of unofficial payments by state officials was effected in the late 1790s – in the Navy Office in 1796, in the Admiralty and Victualling Office in 1799–1800, and in the dockyards in 1801.[37]

However, the abolition of the profit motive in civil officials left an ideological vacancy in their motivational thinking. It was filled, significantly, by the principle of individual responsibility, advocated by Sir Samuel Bentham who was appointed to the post of Inspector General of Naval Works in 1796. Samuel was the brother of Jeremy Bentham, the moral philosopher and writer on jurisprudence who, in 1793–5, was working on his *Manual of Political Economy*.[38] Jeremy and Samuel collaborated closely. While Jeremy was the theorist, Samuel was the practical executer. In the naval departments, the latter worked to establish individual responsibility among office-holders whom Jeremy would later describe as 'single-seated functionaries'. Other key features of Samuel's schemes were education, classification and central control, especially of finance, which became important elements in the agenda of *utilitarian* administrative reformers.[39] However, as a radical and out-spoken advocate in a post attached to the Admiralty, Samuel made enemies.[40] In 1807 he lost his influence at the Admiralty and was placed at the Navy Board until 1812. He nevertheless made converts who were rising to positions of power in naval bureaucracy, while his ideas gained support from wider movements in public opinion.

The strength of opinion during a long and expensive war should not be under-estimated. One victim was Henry Dundas, Viscount Melville, who failed, while Treasurer of the Navy during the 1780s, to supervise adequately his Paymaster of the Navy who (against a regulation established by Dundas himself) profited from public money charged to his care. Impeached and tried in the House of Lords, his career was ruined.[41] The charges against him arose from the tenth report of the Commissioners for enquiring into irregularities, frauds and abuses practised in the naval departments. This Commission of naval enquiry had been obtained in 1802 by St Vincent, First Lord of the Admiralty until 1804. It produced

---

[37] J. Breihan, 'William Pitt and the Commission on Fees, 1785–1801', *HJ* 24(1984), 59–81.

[38] In 1801–4, Jeremy also wrote his 'Institute of Political Economy': D. Lieberman, 'Economy and polity in Bentham's science of legislation' in *Economy, Polity and Society. British Intellectual History 1750–1950*, ed. S. Collini, R. Whatmore and B. Young (Cambridge, 2000), 107–34.

[39] See Morriss, *Naval Power and British Culture*, part II.

[40] R. Morriss, 'Samuel Bentham and the management of the royal dockyards, 1796–1807', *BIHR* 54(1981), 226–40.

[41] H. Furber, *Henry Dundas, 1st Viscount Melville, 1742–1811* (London, 1931), 160–5; M. Fry, *The Dundas Despotism* (Edinburgh, 1992), 263–75.

fourteen reports by 1805 when it was replaced by the Commission for revising and digesting the civil affairs of the navy, which produced another fourteen reports between 1805 and 1809. Meanwhile, the management of the army was subject to similar investigation and revision. The Commission of military enquiry, appointed in 1805, issued nineteen reports by 1812.

These inquiries and revisions of instructions aimed to clarify the duties of all naval and military officials. Although boards survived in the naval departments until 1832, the instructions aimed to ensure officials were individually responsible for the execution of their duties. It was no coincidence that by 1807 Samuel's brother, Jeremy Bentham, was employing the idea of 'single-seated functionaries' in his Constitutional Code, which, like the commissions of enquiry and revision, attempted to analyse and define duties in fine detail. Although the recommendations of these commissions were only implemented in the second half of the Napoleonic War, the values they represented began to permeate the military departments during the 1790s.[42] They thus began a silent revolution soon after the French Revolution.

Hence the military departments were subject to reform along lines consistent with what became known as utilitarian principles far in advance of the reform of social institutions in the nineteenth century.[43] The idea of individual responsibility and the detailed prescription of duties created a new foundation for the conduct of civil business. In appearance, in 1815, military services retained the features they possessed in the mid eighteenth century, but their civil branches existed in a very different administrative environment. Now the pursuit of private perquisites was punished and duty to the state was as important as in the military branches. Co-incidental with the introduction of utilitarian ideas, the reach of the state had changed dramatically. From an Atlantic empire, the British state had grown to one encompassing the antipodes as well as the Indian sub-continent. As officials were despatched to these new colonies, the new ideology went with them, harmonising motivational belief in the civil, military and imperial branches of state operation.[44]

Management and expertise

The policy of appointing individually responsible officers to new positions of vital importance – as Inspectors of Artillery, of Gunpowder

---

[42] Morriss, *Naval Power and British Culture*, 147–65.

[43] See J. Dinwiddy, *Bentham* (Oxford, 1989), 90–108; Henriques, 'Jeremy Bentham and the machinery of social reform', 169–201.

[44] E. T. Stokes, 'Bureaucracy and ideology: Britain and India in the nineteenth century', *Transactions of the Royal Historical Society*, 5th series, 30(1980), 131–56.

Manufactories, of Naval Works – had a remarkable impact on the performance of the British military infrastructure partly because the men appointed were of great character and ability, and partly because they were masters of their specialisation. The appointments of Thomas Blomefield and William Congreve to the Ordnance departments in the wake of industrial change dramatically affected the quality of British guns and powder.

The Inspector General of Naval Works could not be master of all the trades under his purview but he therefore had working under his individual authority a staff of eleven, including an architect/engineer, a mechanist, a chemist and a draughtsman. There is probably some truth in the observation of Captain Thomas Hamilton: 'General Bentham is a clever, knowing man; "amongst the blind the one-eyed are kings"; his mechanical science goes no further than what is common among ingenious artists.' But, his great gift was an imagination that saw what was possible both in practical construction and in management. In contrast to administrators pre-occupied with the progress of day-to-day business, pragmatically making the best of existing facilities, he possessed striking lateral thought. In 1795 Sir Charles Middleton commended him as 'undoubtedly a man of first-rate abilities and of great experience in practical mechanics'.[45]

Middleton's opinions are central to the way in which British naval administration developed before 1815. Having gone to sea in 1741 and served throughout the Seven Years' War, he was Comptroller of the Navy Board for twelve years from 1778, served at the Admiralty between 1794 and 1795, and became First Lord of the Admiralty at the time of Trafalgar.[46] Between 1785 and 1787 he fed suggestions for improvements in the management of the navy to the Commissioners on fees, gratuities, perquisites and emoluments in the public offices. At that time, he himself started to collate and revise standing orders issued by the Navy Board to the dockyards since the beginning of the century. Appropriately, between 1805 and 1809 he became Chairman of the Commission for revising and digesting the civil affairs of the navy.[47] In 1805 he was made Baron Barham. A friend of William Pitt, the Younger, he had a formidable

[45] *Selections from the Correspondence of Admiral J. Markham during 1801–4 and 1806–7*, ed. Sir C. Markham (NRS, 1904), 342; *The Private Papers of George, Second Earl Spencer, 1794–1801*, ed. J. S. Corbett and H. W. Richmond (4 vols., NRS, 1913–24), I, 46.

[46] *Letters and Papers of Charles, Lord Barham 1758–1813*, ed. J. Knox Laughton (3 vols., NRS, 1907–11), prints some of Middleton's papers now in the NMM. For biographies, see J. E. Talbott, *The Pen and Ink Sailor: Charles Middleton and the King's Navy, 1778–1813* (London, 1998), and *New DNB*.

[47] R. Morriss, 'Charles Middleton, Lord Barham' in *Precursors of Nelson: British Admirals of the Eighteenth Century*, ed. P. Le Feure and R. Harding (London, 2000), 301–23.

appetite for work and has been termed a bureaucratic imperialist. During the American War of Independence, despite a daunting workload managing the dockyards and all the business of the Navy Office, he accepted organisation of the loading and despatch of food transports to the British army in North America and garrisons elsewhere. The experience lay behind the rationalisation of food and transport supply in 1793–4.

Yet, for all this experience, men like Middleton were still dependent on experts in subordinate offices and by the quayside. In 1789 he was taking advice from George Teer, agent for transports in the River Thames, on how to send troops to Australia. In 1805–6, he was consulting John Payne, clerk in the office of the Secretary to the Navy Board, on the rate of decay of the British fleet.[48] Then only twenty-six, Payne had entered the Navy Office in 1795 at the age of fifteen and developed a clear grasp of the logistics of building, repairing and maintaining warships. By 1811, having worked on the recommendations of the Commissioners of naval revision for the dockyards, he established a system of measurement, quantifying and paying for piecework on the spot within the week. It permitted weekly payment of the workforce and the simultaneous budgeting of the wages bill. A spin-off was the ability to calculate the rate at which all six dockyards could refit, repair and build warships.[49]

The work of men like Blomefield, Congreve, Bentham, Middleton, Teer and Payne link the defence of empire and fleet operations to the minutiae of finance, chemistry and labour in the dockyards. Their common concern was the conversion of resources into instruments of maritime power. Men like them worked in, and for, all the military supply departments of the British state. The Navy, Victualling and Transport Offices each had their phalanx of agents working at the ports and overseas. Some, in departments like the army Commissariat demanding dynamic organisation, performed remarkable achievements of supply. Through their everyday business, the state preserved and institutionalised their experience. Through subordinates and successors selected on merit, especially after the abolition of perquisites, their knowledge accrued cumulatively to the benefit of all those involved in operations at sea and overseas.

The knowledge available to the state was given a geographical dimension as, for territorial security and trade protection, naval squadrons and army garrisons grew more numerous around the globe. In 1755 the navy

---

[48] See NMM, MID.10/4, 4 pages, last page missing, but headed 'Private, My Dear Sir, Since my conversation with you Thursday evening I have been turning to several papers I have by me which tend to elucidate the present state of the Navy.'
[49] NMM, ADM. BP/32B, 27 May, 27 June 1812; see also Morriss, *Royal Dockyards*, 205–8.

had five overseas refitting yards – that on Minorca being lost in 1756. By 1814 it had ten, each with a naval commissioner, and another seven depots managed by naval storekeepers. These yards and depots provided maintenance materials and facilities. The commissioners were invariably naval captains and, by the second half of the Napoleonic War, took many decisions on their own responsibility, guided by comprehensive instructions issued by the Commissioners of naval revision.[50] The officers of the Commissariat, catering for garrisons in distant locations and for troops on campaign, had also to take many decisions on their own responsibility. Surprisingly, in view of historians' preference for citing criticisms, the Duke of Wellington was complimentary about them.

These officials, operating at the periphery of the British state, were themselves often supplied by transports despatched by government to carry naval stores, munitions or victuals to its depots overseas. But, more usually, they were supplied by contractors, merchants who engaged to ship and supply stores to a particular depot for profit. Sometimes, as to the new colony of New South Wales, state store ships supplied the garrison and nascent colony in tandem with merchants developing their capability of making the lengthy voyage. At the same time, experience of managing merchants as contractors developed in the different naval boards and at the Treasury. This experience, at the interface of public and private sectors, married state and private enterprise, and provided for the growth of stability in this relationship.

Since the mid eighteenth century, the British navy had been deliberately commissioned 'to make discoveries of countries hitherto unknown'.[51] But that knowledge was of no more than academic use until the servants of the state and their commercial contractors permitted military forces to survive and protect British interests in distant locations. The distribution of garrisons, commissioners and storekeepers by 1815 provided the practical knowledge and resources necessary to the geographical enlargement of the state.

Resources and state policies

The resource base

The geographical enlargement of the state made a greater part of the world's resources accessible to it and to commercial entrepreneurs who

---

[50] I am grateful to John Day, University of Exeter, for this information.
[51] R. Drayton, 'Knowledge and empire' in *OHBE*, II, 231–52; G. Williams, '"To make discoveries of countries hitherto unknown": the Admiralty and Pacific exploration in the eighteenth century', *MM* 82(1996), 14–27.

included military contractors. The latter supplied most commodities employed by the state, be it money, warships, guns, powder, food or shipping. In theory, the state might seem to possess a position in which it could command a monopoly of any particular resource. In reality, however, it competed for it. For money and naval stores, the competition was international. For manpower, food and shipping, the competition was principally domestic. This did not make the competition any less intense, for Britain's expanding economy demanded all the resources it could get.

For manpower, the state had to compete with the private sector. For skilled seamen, the navy competed with merchant shipping; for soldiers, the army competed with civil occupations. Although agriculture was releasing labour, industry was expanding its demand for manpower. Indeed in the mid eighteenth century, the army may have suffered greater recruitment problems than the navy, which had the advantage of employing a stick as well as a carrot to obtain its manpower.[52]

Initially, the navy improved the terms of recompense for state service.[53] But seamen remained reluctant to enter even though provision was made for the support of relatives and for the seamen themselves in old age. The Spithead mutiny in 1797 at last gave rise to higher rates of pay, but these were still dwarfed by earnings in private trade. These difficulties forced the navy to employ impressment. The naval Impress Service developed from the 1740s to recruit seamen on shore. During the mid eighteenth century, its net spread across Britain. In 1794–5 its work was temporarily supplemented by the Quota Acts but the supply of men was still deficient, and not helped by the rate of desertion, which remained constant throughout this period. In common with rates on shore, mortality was reduced, but that reduction appeared the greater partly because the discharge of invalids increased.

For food, the state had to compete with the growing demand from population and towns in England. The landed interest would like to have monopolised this growing demand, as it did by achieving the exclusion of imported cattle products before 1758 and corn after 1815. But by the middle of the eighteenth century there were shortages of beef, pork and butter and, by the end of the century, shortages of wheat, oats and beans.[54] Deficiencies were highlighted by the occurrence of food riots in the southern counties of England. During the last decade of the eighteenth century, food prices more than doubled. Between 1808 and 1810 only large-scale imports helped to prevent the public price of wheat from rising another 50 per cent.

[52] P. J. Marshall, *The Making and Unmaking of Empires*, 61.
[53] See pp. 232–4.   [54] See pp. 273–6.

For shipping, the state had to enter the market for the hire of cargo space or ships either for the duration of a voyage or for a certain period. During the eighteenth century there was a doubling in the number of ships serving as coasters, constituting six-fold growth in tonnage. Between 1755 and 1815 the total British merchant fleet approximately doubled in size. This enlarged the supply of shipping available for hire during wartime but the take-up of the state only reduced the supply available to the mercantile community and enhanced charter prices. The state did not fare well before 1794 when there were four boards competing for shipping, each using different methods of measurement and different terms of employment, permitting ship owners to choose the department they preferred.[55] Insufficient central control also permitted commanders-in-chief overseas to divert ships to local purposes, reducing the availability of ships in Britain.

State policies towards resources

What could the British state do to make the employment, purchase or hire of these resources cheaper and easier? Its consistent policy was to find new sources of supply either within the British Isles, within the empire, or from beyond, and to reduce the competition for those resources.

These policies are most evident in the case of the state's manpower. The army recruited everywhere in the British Isles. The 1745 rebellion bred distrust of the highland Scots but only a decade later their regiments formed a significant part of the army. Despite its colonial status, Ireland also provided a high proportion of the army: 16 per cent at the time of the American War of Independence. After 1801, well over 30 per cent of some English regiments were Irish. In 1797–8 the navy recruited 8.3 per cent of its 120,000 seamen from Ireland.[56] Colonies more distant from Britain also supplied soldiers and sailors. In the Seven Years' War, North America, India and Jamaica all provided regiments, either for the regular army or for the East India Company. Native Indians as well as Europeans were recruited in America. The growth of the British empire to 18 million 'Europeans', 42 million 'free persons of colour' and more than a million slaves by 1815 enlarged the pool for recruitment. By that time, India contained 20,000 British soldiers, but a 'sepoy' army seven times that size.[57]

---

[55] See pp. 327–9.
[56] C. J. Doorne, 'Mutiny and sedition in the Home commands of the Royal Navy 1793–1803', unpub. University of London Ph.D. thesis, 1997.
[57] P. Colquhoun, *Treatise on the Wealth, Power and Resources of the British Empire* (London, 1815), 47.

Non-colonial foreign territories which supplied troops included the dependency of Hanover and independent states of Hesse, Brunswick, Calabria, Piedmont, Greece, Switzerland, even France. Slaves and free black men were recruited into both the army and the navy. Ship crews were distinctly cosmopolitan.[58] From 1793 the proportion of non-British nationalities recruited into merchant ships, and thus available for transfer to naval vessels, was enlarged by the suspension of the clause in the Navigation Laws requiring four-fifths of crews to be British.

In the case of food supply, resources were similarly enlarged. English agriculture grew in productivity about 50 per cent in the eighteenth century, but England's population doubled in the same period. The demands of the state also grew, a factor in 1758–62 when the Cattle Acts of 1666–7 were suspended, then repealed, opening English markets to Irish cattle products. Thereafter the contribution of the Irish rural economy to the supply of England and Britain's armed forces grew, and there was no shortage of beef, pork and butter, as there was of cereals at the end of the century. The British government in 1793 reduced competition between the army and the navy for the available grain by placing responsibility for their common supply with the Victualling Board. Shortly after, with the necessity of supplying British forces in the West Indies, those clauses of the Navigation Laws blocking the import of foreign goods into the British West Indies were suspended.

Most ships employed as state transports were chartered from British brokers and owners. To an island state, shipping was indispensable and not easily hired beyond the River Thames, though some later came from other ports. To reduce the competition for shipping, in 1794 the Transport Board was established to serve all departments of government. It hired ships more quickly than previously and provided fleets of transports that served all the subsequent expeditions and operations overseas. By the end of the War of 1812, the Transport Board had nearly 1,000 ships under hire, which was 5 per cent of the ships and 11 per cent of the tonnage of the British merchant fleet. Despite a well-regulated system of agents, transports were sometimes detained overseas and the board still suffered occasional shortages of shipping. Nevertheless the competition for hired ships that had existed prior to 1794 was rationalised.[59]

---

[58] P. J. Marshall, *The Making and Unmaking of Empires*, 60–8; N. A. M. Rodger, *The Wooden World: An Anatomy of the Georgian Navy* (London, 1986), 156–9; R. Holmes, *Redcoat: The British Soldier in the Age of Horse and Musket* (London, 2001), 48–56.

[59] S. Palmer, *Politics, Shipping and the Repeal of the Navigation Laws* (Manchester, 1990), 40–3; Ville, *English Shipowning during the Industrial Revolution*, 96; see also chapter 8, 'The service at the time of the War of 1812'.

The demands of the state for manpower, food supplies and shipping between 1755 and 1815 gave rise to two main responses. One was the enlargement of the state's sources of supply within Britain, its colonies and beyond. The other was the removal of regulations and competition that inhibited supply.[60] Deregulation, increasing access, was a step towards free trade. Adam Smith in 1776 accurately predicted the passing of the mercantilist age, and the demands of the state during wartime enhanced this process.[61] 'Laissez-faire' proved to be 'the only practical strategy for governments to pursue'.[62]

State law and public policy

Hitherto, most attention has been paid to the state's laws relating to trade and navigation between the colonies and mother country, and to the Admiralty's prize laws.[63] Little has been given to the evolution of grandly named 'Laws of Nations' or anything that protected or fostered domestic resources. Of these, those relating to seamen are best known, but laws applied to all other human and physical resources valuable to maritime power. Especially important for the eradication of fraud and the development of trust in commercial relationships was the development of equity law.

Equity law and commercial practice

The legal framework for commercial transactions was vital to relations between the public and the private sectors. It determined the tenor of relations between the state and its civil employees, contractors and client corporations. One element in the strength of the British state lay in its ability to maintain fruitful partnerships, and developments in the law fostered that capability. The law, as a framework for business, became all the more important as distances between markets, the scale of shipments and time between transactions increased.

During the eighteenth century, statute law was growing in volume but was prolix and virtually incomprehensible to non-lawyers. Common law

---

[60] For the complementary context, see K. H. O'Rourke, 'The worldwide economic impact of the French Revolutionary and Napoleonic Wars, 1793–1815', *Journal of Global History* 1(2006), 123–49.

[61] R. Unger, 'Warships, cargo ships and Adam Smith: trade and government in the eighteenth century', *MM* 92(2006), 41–59.

[62] O'Brien, 'Political preconditions for the Industrial Revolution'.

[63] For the latter, see R. Hill, *The Prizes of War: The Naval Prize System in the Napoleonic Wars 1793–1815* (Stroud, Glos., 1998).

was more understandable and, being based on precedent, was evolving
with the values and practices of society. However, it was complex, technical and had to be disentangled from disparate cases. Parliament and the
courts steadily amended and re-interpreted these traditional forms of law,
adding at its margins. Religion instilled morality through contemporary
attitudes. Yet the formal process of legal practice was slow, expensive and
avoided by most businessmen faced with cases of fraud, debt or breach
of contract. In general, they thus preferred their own informal, cheaper
and quicker systems of arbitration conducted among their peers.[64]

More useful than statute, and replacing common law in commercial
cases, was equity law. Governed by logical rules and bound by precedent, it looked beneath the surface of transactions to discern the real
intentions of the parties concerned. During the second half of the eighteenth century it developed a body of precedents, and gradually achieved
a virtual monopoly of fraud cases.[65] It made an important contribution to the financial development of the state through, for example, the
thinking of the Commissioners for examining the public accounts of the
kingdom, appointed in 1780, whose chairman was an equity lawyer.[66]
It also served to shape the thinking of board commissioners of government departments required to regulate the conduct of contractors. By
the early nineteenth century, the London courts accorded primacy to
promises and intentions.

The aim of achieving equity for the state was perhaps most evident
in dealings between the solicitor acting for the Victualling Board and its
suppliers. Until the early 1780s the latter had claimed the sanction of
custom to perpetuate the substitution of one type of food for another,
even though their financial values were different. In a test case, a charge
of fraud was brought against one merchant partnership in the Exchequer
court in 1784, but the Victualling Board had to accept arbitration by
the merchants' peers and the case went against the state. However, the
board then recast all its legal documentation to ensure contractors kept
to the letter of their contracts and their deliveries did justice to the state.
On this new foundation, during the 1790s the Victualling Board developed a good working relationship with most contractors, which persisted
through the Napoleonic War.[67]

[64] O'Brien, 'Central government and the economy, 1688–1815'.
[65] L. A. Sheridan, *Fraud in Equity: A Study in English and Irish Law* (London, 1957), xi–xv,
13–15.
[66] J. Torrance, 'Social class and bureaucratic innovation: the Commissioners for examining
the public accounts, 1780–1787', *Past and Present* 78(1978), 56–81.
[67] For more detail of this case, see chapter 3, 'Contractors and their management',
pp. 109–13; also Morriss, *Naval Power and British Culture*, 115–21.

Equity in state policy

In other areas, too, the state promoted the principle of equity. This was evident in relations between the state and its employees. The state has had a bad press for its treatment of seamen. But in justice and their subsistence they were treated on exactly the same basis as officers. Procedures in courts martial were identical for seamen and officers: cross-examinations equally probing and verdicts exactly scrupulous.[68] Similarly, in the supply of food, when crews were put on short allowance, the purser was 'strictly charged not to supply any officer at whole allowance, whilst the rest of the company are at short, but all are to be equal in point of victualling'. He was also directed to be careful 'that the men be punctually paid' their deficiencies in money.[69]

Public opinion backed the state in its pursuit of equity in principle, as shown in its relations with the East India Company. In return for protection by state forces, the company was expected to contribute to the costs of the defence of its territories and trade. This became legally enforceable after Lord North's Regulating Act of 1773 which primarily reformed the company's electoral system but also, among other things, required that regular accounts of company business be made available to Parliament and the Treasury.[70] Until 1781 the Admiralty was able to obtain cargo space on board outward-bound company ships for the shipment of naval stores to the Indian sub-continent. Until that time the company always had sufficient shipping tonnage to fulfil its own commitments and provide the navy with free cargo space. However, in March 1781, because it was transporting troops to India, the company had to decline the carriage of 450 tons of naval stores on account of a lack of tonnage; in April, it could not take 40 tons of sheet copper; and, in June, was forced to leave some 2,000 tons of goods behind. That month the company informed the Admiralty that it could take no more naval, victualling or other stores that season.[71]

The government immediately obtained the Act of 21 Geo. III ca. 65, 'for the Company's supplying His Majesty's Fleet in the East Indies with stores, government repaying them the first cost'. In June 1782, the Navy Board and the company took advantage of a clause in the Act permitting them to settle regulations 'as they should think fit', deciding the Navy Board should supply the stores and the company pay the freight.[72] The

---

[68] For their procedutres, see *Naval Courts Martial, 1793–1815*, ed. J. Byrn (NRS, 2009), 1–146.
[69] *Regulations and Instructions relating to His Majesty's Service at Sea* (London, 1787), 'Of the Provisions', article II, p. 61.
[70] Bowen, *The Business of Empire*, 70–2.   [71] Syrett, *Shipping and the American War*, 72–3.
[72] NMM, ADM. BP/3, 11 June 1782.

company attempted to agree the same with the Victualling Board, arguing it was obliged only to supply those provisions that could be procured in India. Yet the Attorney General was not convinced of the company's case, claiming 'the words of the 18th section [of the Act] are extremely strong that the company "shall provide and supply all the victualling to be delivered in the East Indies necessary for His Majesty's ships"'. He maintained that 'the whole burden falls on the Company and they are to provide and supply and to carry to India what cannot be provided on the spot'.[73]

In 1782–3 the cost and freight of naval provisions purchased in, or conveyed to, the East Indies amounted to more than half a million pounds. During the French Revolutionary and Napoleonic Wars the costs of victualling the fleet in the East Indies were far greater. Those for just the squadron in the Red Sea in 1800–1 were over a quarter of a million pounds.[74] Naturally, the company was careful to keep accounts of its own expenses in other public services like the carriage of troops, which were used to liquidate its naval debts. After Pitt's India Act in 1784 and the creation of Indian vice-presidencies, many of their bills were paid by vice-presidential treasuries.[75] The state and its greatest commercial company were thus obliged to work in financial partnership, an interdependence which was not just necessary but based on a public sense of equitability supported by legal opinion.

The gradual deconstruction of the East India Company's monopoly of trade into and beyond the Indian Ocean had a similar basis. From 1788, individuals were allowed to ship freight in East India ships, and from 1793 the company had to make 3,000 tons of shipping available for private trade. In 1813 the company lost its monopoly of trade to India, inaugurating a new era of open competition beyond the Cape of Good Hope. By then the proceeds of trade with India were far outgrown by those with China, with which the company retained a monopoly of trade until 1833.[76]

The direction and projection of the state

At a cost to its monopolistic chartered companies, the pursuit of equity in law and principle strengthened relations between the public and the private sectors. The partnership reinforced the state's form of consensus

[73] W. R. P., Papers of Lloyd Kenyon, 1st Baron Kenyon, case, 18 Oct. 1782.
[74] NMM, ADM. DP/4, 24 Mar. 1784; ADM. DP/23, 8 Sept. 1803.
[75] BL, OIOC, L/AG/1/6/24, pp. 406–7, cited by H. Bowen in workshop on 'The Contractor State' at Greenwich Maritime Institute, 13 April 2007; NMM, ADM. DP/34A, 11, 13 Aug. 1814; ADM. DP/36B, 18 Oct. 1816.
[76] Bowen, *The Business of Empire*, 2, 11, 252–4.

government, at the heart of which responsibilities in the Cabinet were gradually distinguished between foreign relations, colonial affairs and the direction of war. The clarification of responsibilities permitted, by the last decade of the century, the expression of clear views on the potential direction of state war policy. At the same time, the duties of the departments of government responsible for logistical support to the armed forces were rationalised. Indeed, with the appointment of a new Transport Board, the work of projecting British forces overseas and supporting them was simplified.

The restructuring of the Cabinet

In theory this state was directed by the King in Cabinet. George II, the last monarch to lead the British army on the battlefield, remained until his death a key influence on strategic decisions. But the King in 1756 was a limited monarch who acknowledged Parliament as the arbiter of its influence. In 1688 the Glorious Revolution had ensured that Britain would no longer succumb to autocratic, non-parliamentary government when it replaced James II by William of Orange. The revolution established a monarchy that was in essence contractual. The Bill of Rights of 1689 embodied a statement of political expectations with which William and Mary and succeeding monarchs were expected to comply. The protestant succession was spelt out and formed a basis for the Act of Settlement in 1701.[77]

Derived from ideas generated by the civil war four decades earlier,[78] the financial arrangements of the new constitution provided for the monarch to live, after 1689, not simply off his or her own inherited resources, but from a 'civil list' for household and civil government offices voted by Parliament. The latter became responsible for military and naval expenditure too. Though owing their first allegiance to the monarch, the armed forces thus became dependent for their funding on opinion within Parliament. Arrangements allowed for the development of new methods of funding war which placed parliament between government and the new Bank of England. Confidence in the new system of funding would promote the growth of banking, of the economy and of the capability of raising money to wage war.[79]

---

[77] R. Pares, *Limited Monarchy in Great Britain in the Eighteenth Century* (History Association, 1957); G. Holmes, *The Making of a Great Power: Late Stuart and Early Georgian Britain* (Harlow, Essex, 1993), 212–20.

[78] O'Brien, 'Fiscal exceptionalism'.

[79] *Ibid.*, 266–7; P. G. M. Dickson, *The Financial Revolution in England: A Study in the Development of Public Credit, 1688–1756* (London, 1967), 3–13.

From Parliament, a Prime Minister emerged to lead other royal councillors who composed the Cabinet council. During the early eighteenth century this contained officers of the state, the court, the church, the law, the armed services and other peers or statesmen summoned to add weight and experience. But by the mid eighteenth century this great gathering had become ceremonial and was replaced as a working body by an 'efficient' Cabinet composed of principal office-holders – the Secretaries of State, First Lord of the Treasury, Chancellor of the Exchequer and President of the Council. But it was an elastic body, and heads of other government departments attended according to need. Throughout the second half of the eighteenth century, it was this 'efficient' Cabinet which, as it became more formal, methodical and regular, made British policy.[80]

Cabinet decisions were in theory made collectively. Political history has emphasised the importance of personality in the influence wielded by office-holders in the Cabinet. But departmental responsibilities carried great weight. Between 1715 and 1835, when the Prime Minister was often the First Lord of the Treasury, sometimes also Chancellor of the Exchequer, his knowledge of revenues and expenditure reinforced his primacy.[81] But the Prime Minister could hold another office, for example that of a Secretary of State, as did William Pitt, Earl of Chatham. Then, however, the Duke of Newcastle managed the Treasury which, with its patronage in the revenue departments, was vital in influencing constituency votes for members of the Commons who supported the King's ministers.[82]

Secretaries of State controlled foreign and domestic policy. Until 1782 there were Secretaries for the north and for the south; after 1782 they became instead responsible for foreign affairs and home or internal affairs. A third Secretary's post, the Secretary at War, was created during the Seven Years' War and survived until the end of the French Revolutionary War. It gave its name to the War Office which handled the financing of the army, directed the issue of arms to troops, and gave marching orders to the troops in Britain.[83] Between 1768 and 1782 there was a

---

[80] L. Namier, 'The end of the nominal Cabinet' in *Crossroads of Power: Essays on Eighteenth Century England* (London, 1962), 118–23; I. R. Christie, 'The Cabinet in the reign of George III, to 1790' in *Myth and Reality in Late Eighteenth Century British Politics*, ed. I. R. Christie (London, 1970), 55–108.

[81] J. C. Sainty, *Treasury Officials 1660–1870* (London, 1972), 16–17.

[82] D. Kynaston, *The Secretary of State* (Lavenham, Suffolk, 1978), 136–44.

[83] For the development of the War Office, see H. C. Tomlinson, *Guns and Government: The Ordnance Office under the later Stuarts* (London, 1979), 209; *An Eighteenth Century Secretary at War: The Papers of William, Viscount Barrington*, ed. T. Hayter (Army Records Society, London, 1988).

fourth Secretary of State, for the American colonies.[84] During the War of American Independence, the post was held by Lord George Germain who was conveniently blamed for failures of coordination in troop and transport movements to America.[85]

This fourth Secretary's post lapsed in 1782. It was succeeded, however, in 1794 by the office of Secretary of State for War and Colonies. The post was to have a vital influence on the direction of British war policy until 1815. The new post had responsibility both for planning and coordinating the overseas expeditions of the army and for defence of the colonies.[86] Now the post appears critical to the transformation of the British state, for in both the French Revolutionary War and the Napoleonic War the vision of this office-holder shaped British war policy.[87]

Direction and coordination

Between 1794 and 1801 the post was held by Henry Dundas.[88] Between 1784 and 1801 he was also Commissioner, then President, of the board of Control for India. Although his view diametrically opposed those of the Foreign Secretary, until the very end of the Revolutionary War Dundas was for avoiding involvement in Europe while pursuing British interests overseas, attacking France's colonies in the East and West Indies. In July 1793 he claimed that British success in these regions was 'of infinite moment, both in the view of humbling the power of France, and with the view of enlarging our national wealth and security'. The West Indies were 'the first point to make perfectly certain'. Two years later, he was still arguing that 'complete success in the West Indies' was 'essential to the interests and I will add to the contentment of this country': 'by success in the West Indies alone you can be enabled to dictate the terms of peace'.[89] Such was the power of his convictions, in 1801 he planned and carried out the campaign to remove Napoleon's troops from Egypt against the opinion of Pitt, the Prime Minister, and of George III.[90]

---

[84] J. C. Sainty, *Officials of the Secretaries of State 1660–1782* (London, 1973), 1–2.

[85] For the coordination of transports with troop embarkations, see NMM, MID. 1/106 and MID. 2/20, Middleton's correspondence with the Secretary at War and his secretary.

[86] R. Glover, *Britain at Bay: Defence against Bonaparte* (London, 1973), 30, 33.

[87] Hall, *British Strategy in the Napoleonic War*, 74–81.

[88] Henry Dundas, first Viscount Melville (1742–1811), Privy Counsellor and Treasurer of the Navy, 1782–3, 1784–1800; Home Secretary 1791–4; President of the board of Control for India, 1793–1801; Secretary for War and Colonies, 1794–1801; First Lord of the Admiralty, 1804–5.

[89] Quoted in M. Duffy, *Soldiers, Sugar and Seapower: The British Expeditions to the West Indies and the War against Revolutionary France* (Oxford, 1987), 25, 161.

[90] *DNB*; Fry, *The Dundas Despotism*, 225. Dundas planned the expedition to Egypt with General Abercromby – old friend, political ally, and relation of one year following the marriage of the latter's son to Dundas's favourite daughter.

In July 1805 Lord Castlereagh became Secretary of State for War and Colonies. Like Dundas before him, in 1805–6 he too combined his Secretary's post with that of President of the board of Control for India. Returning to the Secretary's post in 1807–9, he supported the war in the Spanish Peninsula and the command of Arthur Wellesley, Lord Wellington.[91] Between 1812 and 1822, Castlereagh was Foreign Secretary. In the opinion of Richard Glover, as Secretary for War and Colonies, Castlereagh 'did more than any other man since the Seven Years War to enable Britain to fulfil the role of a great power in a continental war . . . his strategy . . . first paved the way to victory, and then enabled him to negotiate from strength when, as Foreign Secretary, he helped decide the future of Europe at the . . . Congress of Vienna in 1814–15'.[92]

Complementary to the creation of this new Secretary's post was growth in the Admiralty's coordinating role. The political status and power of the Admiralty had started to grow in the middle of the eighteenth century.[93] This can be partly attributed to its growing patronage, though that of the Sea Service was defended on grounds of the expertise required in its officers. More relevant was the growing importance of the navy in the wars of the period to Britain's defence and expansion overseas. Although sometimes a distinguished naval officer, increasingly during the second half of the eighteenth century First Lords were civil politicians or statesmen supported at the Admiralty Board by at least two naval officers serving as commissioners.

The First Lord and Board of Admiralty were supported in turn by subordinate boards responsible for the logistics of the navy: the Navy, Victualling, and Sick and Hurt Boards.[94] The addition of a new Transport Board in 1794 was a critical supplement to these offices. At the same time the addition of responsibility for feeding the army overseas to the duties of the naval Victualling Board rationalised logistical support to the armed forces. Meanwhile overseas, the army Commissariat had become a virtual department of government, accompanying the army wherever it went on amphibious operations, on campaign in Europe or on garrison duty throughout the world. The work of all the naval boards was complemented by the Ordnance Board and by the Treasury Board, the latter being responsible not only for sustaining state finance but for maintaining the army overseas through the Commissariat. Under the direction and management of the Secretaries of State, the board of Admiralty and the Treasury, these logistical departments served to project the British

[91] J. W. Derry, *Castlereagh* (London, 1976), 116–19.
[92] Glover, *Britain at Bay*, 147–50.
[93] N. A. M. Rodger, *The Admiralty* (Lavenham, Suffolk, 1979), 48, 60–1, 87.
[94] J. C. Sainty, *Admiralty Officials 1660–1870* (London, 1975), 23–7; J. M. Collinge, *Navy Board Officials 1660–1832* (London, 1978), 18–25.

state and to maintain its control of both sea and territory throughout the world.

During the second half of the eighteenth century, restructuring at Cabinet and managerial board level thus improved the direction of state policy and the coordination of its resources. The regular incidence of war enhanced the power of the state which benefited especially from its partnership with contractors, relations with whom were facilitated at the end of the century by the employment of a policy of equity in contract management and by the development of equity law in legal proceedings. Both made for an ethos of fair dealing. The state suffered from competition for resources but, when possible, sought new sources of supply and removed obstacles to the employment of existing sources. In this, it benefited from the experience of able managers and the expertise of specialists who were motivated at the end of the century by a shift from collective to individual responsibility. Efficiency in the performance of duties was enhanced at the beginning of the nineteenth century by the comprehensive revision of instructions for the civil branches of both armed forces. The motivation and efficiency of the bureaucracy gave the British state a special capability that built on its existing financial power and dynamic private sector. The following chapters look at the resources available to that bureaucracy and the organisation of those branches of government which were important to Britain's maritime ascendancy. As this derived from defensive and offensive operations, the product of ideas and experience tested over time, it is to these resources and their effects that this book now turns.

2   Defence and expansion

By 1755 Britain already possessed an overseas empire, an expanding maritime economy and the strongest navy in Europe. Strategic policy would ensure that, over the next sixty years, despite the loss of colonies in America, Britain's maritime power would grow. The dominant position achieved by 1815 was the product of long-term success in maintaining the security of the British Isles and of its territorial interests overseas. The implementation of this policy demanded the maintenance of armed forces and of logistical support for them, which in turn required resources. Defence of the British Isles, its trade and colonies was complemented by a consistent concern to prevent any European power or alliance from achieving hegemony on the continent. This demanded the use of land forces which were also used in amphibious operations overseas and for garrisoning territories controlled by Britain. All these demands existed in British strategic thought from the sixteenth century but they were not all addressed together until the early 1700s, and were not answered by a coordinated policy until the mid eighteenth century.[1] Thereafter, state expansion overseas, building upon effective defence in home waters, gave rise to a system of supply permitting the global disposition of both the army and the navy.

The shaping of Britain's defence policy

Geography and defence

The most constant factor in British policy was the effect of geography. The Mediterranean and Baltic inland seas fostered early maritime powers which employed galley technology. The location of the British Isles to the west of the European land mass, itself a series of inlets and peninsulas

---

[1] W. S. Maltby, 'The origins of a global strategy: England from 1558 to 1713' in *The Making of Strategy: Rulers, States and War*, ed. W. Murray, M. Knox and A. Bernstein (Cambridge, 1994), 151–77.

favouring maritime activity, gave Britain a position that commanded lines of sea communication and trade between the north and south of Europe. With the development of long-distance global trade and the three-masted ship, the British were able to increase their command of the sea routes focussing on Western Europe. During the seventeenth century, the Dutch developed ascendancy before the English. But their geographical disadvantages – most notably their proximity to the French and inescapable involvement in continental warfare – told against them. Partly by default of the Dutch, the English emerged as the dominant maritime power in Western Europe.[2]

The replacement of the Dutch by the French as the principal power which threatened English interests shifted the grounds but did not diminish the necessity for a strategic response to hostile powers across the English Channel and North Sea.[3] The simple proximity of Britain to the continent dictated the necessity for naval defence. Already in the mid seventeenth century the hostility of the continental powers to the differences in attitude of the regicide English republic to monarchy, religion and trading policy had given rise to a great surge in the size of the English navy. The threat from France after 1674 gave rise to further naval building and the development of a nascent system of managing, preserving and supplying warships.[4]

Against the French, power had to be contested in the North Atlantic as well as the English Channel and the North Sea. Like the Dutch republic, France was well positioned to invade England and after 1688 gave shelter to the Stuart court, preserving until the 1760s the fear of a Jacobite rising timed to coincide with a French invasion.[5] The consolidation of the British state was a first response. Wales had been welded to England by the Act of Union in 1536; Scotland followed in 1707, and Ireland in 1801.[6] With an army that rarely fell below a quarter of a million men, and was sometimes double that number, France had also to be countered by allies on the continent of Europe. The necessity to defend the homeland of the Hanoverian monarchy of Britain after 1714 affected the choice of those allies, and partly accounts for the cultivation of friendly relations

---

[2] Glete, *Warfare at Sea, 1500–1650*, 186–7.

[3] D. Davies, 'The birth of the imperial navy? Aspects of English naval strategy *c.* 1650–90' in *Parameters of British Naval Power 1650–1850*, ed. M. Duffy (Exeter, 1992), 14–38.

[4] M. Duffy, 'The foundations of British naval power' in *The Military Revolution and the State 1500–1800*, ed. M. Duffy (Exeter, 1980), 49–90.

[5] I. R. Christie, *Stress and Stability in Late Eighteenth Century Britain: Reflections on the British Avoidance of Revolution* (Oxford, 1984), 37–8.

[6] C. A. Bayly, *Imperial Meridian: The British Empire and the World 1780–1830* (Harlow, 1989), 81.

with Prussia and Russia, both growing European powers in the eighteenth century.[7]

Coalitions with, and subsidies to, continental powers complemented the work performed by the British navy at sea.[8] Without these allies to engage and absorb the resources of Britain's continental enemies and of France in particular, the power of those enemies would have been channelled into their naval campaigns. Fortunately, only during the American War of Independence were these enemies free to concentrate on a war at sea. Yet, even then, Britain's resources were sufficient, when summoned, to wage a resilient and ultimately successful defensive naval compaign. The imperatives arising from these continental threats impressed all realistic politicians in mid-eighteenth-century Britain and produced strategic responses – including a commitment to continental affairs – that invariably transcended parliamentary divisions.[9]

British strategy demanded military land forces, as well as a navy, that could be used against, and in alliance with, continental powers. However, after the English civil war, Commonwealth and Restoration, a constant feature of British public opinion was distrust of a standing army. Moreover, the War of William III, 1689–97, and the War of Spanish Succession, 1702–13, demonstrated the cost of long-term, large-scale military commitments to the continent.[10] Henceforward, British strategy would favour the funding of continental allies and the limitation of British land forces to relatively small-scale expeditions around the fringes of Europe. Only in the Napoleonic War would a large-scale army again be employed on the continent, by which time the costs and logistical challenges were judged to be manageable.

Even then, however, the navy remained vital to the defence of the British Isles. A growing fleet, and a growing merchant navy that provided transports, permitted Britain to command and exploit the external, sea-borne lines of communication around Europe. By sea, occasional allies such as Portugal, Spain, Savoy, Austria and Prussia could be supported at far less cost than the continental scale of warfare demanded of France, even after 1733 when that power had the assistance of Spain in a Bourbon

[7] J. Black, 'British naval power and international commitments: political and strategic problems, 1688–1770' in *Parameters of British Naval Power 1650–1850*, ed. Duffy, 39–59; J. Black, *A System of Ambition? British Foreign Policy 1660–1793* (Stroud, 2000), 32–45.
[8] Kennedy, *The Rise and Fall of British Naval Mastery*, 101, 104, 136.
[9] N. A. M. Rodger, 'The continental commitment in the eighteenth century' in *War, Strategy and International Politics: Essays in Honour of Sir Michael Howard*, ed. L. Freedman, P. Hayes and R. O'Neill (Oxford, 1992), 39–55.
[10] J. Hattendorf, *England in the War of the Spanish Succession: A Study of the English View and Conduct of Grand Strategy, 1702–1712* (New York and London, 1987).

alliance. Large-scale land warfare was exorbitantly expensive and a consistent feature of the Anglo-French wars was the bankruptcy of France and her inability to fund the expansion, or even proper maintenance, of her navy after a few years of warfare. Damaging naval battles – La Hougue in 1692, Toulon in 1744, Lagos Bay and Quiberon Bay in 1759 – left the French navy in need of repair and new construction. France was regularly weakened and rendered vulnerable at sea. Even so, at the beginning of every war, she remained the principal maritime threat.

Apart from in the Channel and North Sea, France had to be countered in the Mediterranean, from where the Toulon fleet was able to join the Brest fleet, or *vice versa*. The Mediterranean fleet permitted French power to extend eastward against the Italian states or even, as in 1798, into Egypt and the Middle East. From the seventeenth century, England worked to established bases in the Mediterranean. Before 1688 naval experience was gained from trade-protection convoys and cruises against the North African corsair states.[11] After 1688 the experience of William III added to England's capability, revealed in 1694–5 when a fleet wintered in the Mediterranean with the support of stores and facilities established at Cadiz.[12] The seizure of Gibraltar in 1704, and of Minorca in 1708, consolidated Britain's position in the Mediterranean. Both possessed strategic locations and, although neither before 1815 was equipped with a dock, they provided stores and facilities for small-scale maintenance of ships.[13]

While a squadron or fleet in the Mediterranean became a constant feature of Britain's wartime strategy (see table 2.1), its armed forces entered the Baltic for shorter periods. There were at least twenty naval expeditions to the Baltic between 1658 and 1813.[14] By this means intermittent allies like Russia and Prussia were supported. Even more important, local powers that threatened the supply of naval stores from the Baltic were confronted. In places the waters were shallow, the northern waters froze in winter, and there was a danger the Danes or Swedes would close The

---

[11] S. R. Hornstein, *The Restoration Navy and English Foreign Trade 1674–1688* (Aldershot, 1991), 99–154.

[12] J. Ehrman, 'William III and the emergence of a Mediterranean naval policy, 1692–4', *HJ* 9(1949), 269–92; J. Ehrman, *The Navy in the War of William III, 1689–1697* (Cambridge, 1953, 1975), 540–1, 575; J. S. Corbett, *England in the Mediterranean, 1603–1714* (2 vols., London, 1904), II, 443–9.

[13] J. D. Davies, 'Gibraltar in naval strategy c. 1600–1783', and K. Breen, 'Gibraltar: pivot of naval strategy in 1781', *Transactions of the Naval Dockyards Society* 2(2006), 10–18, 47–54, respectively; D.W. Donaldson, 'Port Mahon, Minorca: the preferred naval base for the English fleet in the Mediterranean in the seventeenth century', *MM* 88(2002), 423–36.

[14] From Duffy, 'The foundations of British naval power', 61.

Table 2.1 *The deployment of the British fleet, 1755–1813*[a]

(Stations, squadrons and functions categorised according to contemporary terminology – which changed throughout the period.)

	1 Dec. 1755	1 Dec. 1760	1 Jan. 1777	1 Jan. 1782	1 Jan. 1800	1 Jan. 1810	1 July. 1813
European waters							
Western Squadron	–	52	–	–	75	–	–
Channel fleet	–	–	–	19	–	32	33
Ireland	–	–	–	–	12	13	15
Downes Squadron	–	21	–	22	–	33	18
Jersey & Guernsey	–	–	–	–	–	12	8
N.Sea Squadron	–	–	–	–	50	–	–
Texel & Scheldt	–	–	–	–	–	67	30
Portugal	–	–	–	–	–	15	21
Baltic	–	–	–	–	23	25	45
Mediterranean	6	29	5	·5	61	82	91
West Indies							
Jamaica	9	16	13	22	45	33	17
Leeward Isles	5	18	11	48	22	69	39
North America							
N. American Coast	7	17	82	77	–	36	60
Newfoundland	2	2	9	7	–	5	13
Nova Scotia	6	–	–	–	15	–	–
East Indies	6	21	6	12	18	32	19
Coast of Africa	–	7	–	–	–	–	–
Cape of Good Hope	–	–	–	–	14	15	6
South America	–	–	–	–	–	10	14
British Isles							
Convoys & cruisers	58	50	–	74	47	–	–
Foreign cruisers	–	–	8	–	–	–	–
Cruisers coasts of GB	–	–	14	–	–	–	–
Convoys and particular services	–	–	–	–	–	20	31
At Home Ports	38	42	62	112	226	66	61
Troop ships	–	–	–	–	–	–	18
Stationary ships	–	–	–	–	–	78	38
Unappropriated	–	–	–	–	–	81	47
General Abstract	137	275	210	398	585	722	624

[a] From Admiralty List Books, TNA, ADM. 8/30, 35, 53, 58, 79, 99, 100. It should be noted that these were 'working documents' and that the clerks responsible for them did not always revise the General Abstract.

Sound and prevent the easy withdrawal of the British fleet. The presence maintained by Admiral Saumarez between 1808 and 1812 was thus exceptional, but by that means he was able to counter French control of the southern shore.[15]

More constant were squadrons stationed to protect colonies in the West Indies and on the coast of North America.[16] To support them, naval bases were established at Port Royal, Jamaica, from 1721; at Antigua in the Leeward Islands from 1728; and at Halifax, Nova Scotia, from 1758.[17] Until the American Revolution, Boston too served British ships on the east coast of America. From the mid eighteenth century, a naval squadron supported the East India Company on the coast of India where, fortunately, the navy was able to use company facilities at Bombay and Madras. Bombay even possessed docking facilities and built ships.[18] These local bases were indispensable to military forces operating overseas. They provided storage for supplies, facilities for refitting ships, and shore refreshment for crews and troops who had sometimes been confined to ships for weeks on end. They received stores from Britain but also drew upon local supplies of materials and provisions. Towns like Boston, Bombay and Madras acted as commercial centres for contractors, especially those engaged in the supply of foodstuffs.[19]

Trade and colonies

The contractors who were employed to help maintain naval yards and squadrons overseas were part of the infrastructure of colonial development. The great trades in sugar from the Caribbean, tobacco from the Chesapeake region, and furs from Canada were thriving before local naval bases were developed, and the imperial economy grew steadily during the eighteenth century – despite the dip in trade between Britain and America between 1776 and 1785.[20] The escalating scale of British oceanic shipping assisted the growth of this transatlantic imperial economy.[21]

---

[15] J. Black, 'Naval power and British foreign policy in the age of Pitt the Elder' in *The British Navy and the Use of Naval Power in the Eighteenth Century*, ed. J. Black and P. Woodfine (Leicester, 1988), 91–107.

[16] P. Webb, 'British squadrons in North American waters, 1783–1793', *The Northern Mariner* 5(1995), 19–34.

[17] J. Gwyn, *Ashore and Afloat: The British Navy and the Halifax Naval Yard before 1820* (Ottawa, 2004), 3; Morriss, *Royal Dockyards*, 4.

[18] R. A. Wadia, *The Bombay Dockyard and the Wadia Master Shipbuilders* (Bombay, 1955), 31–9.

[19] R. Harding, *Seapower and Naval Warfare 1650–1830* (London, 1999), 217–18.

[20] J. M. Price, 'The imperial economy, 1700–1776' in *OHBE*, II, 78–104.

[21] Davis, *Rise of the English Shipping Industry*, 40–1.

Contractors provided supplies overseas on networks of connection and credit that were well established by the mid eighteenth century.[22]

The colonies were important in their own right for timber and naval stores and were cultivated by the Councils for Trade and for Plantations from 1660. The two councils were united from 1672 and became better known after 1696 as the board of Trade and Plantations.[23] Colonial trade and shipping flourished under the Navigation Laws to the point where the colonies by the third quarter of the eighteenth century were ready – in the view of Adam Smith and some American colonists – for freedom from regulation.[24] At the same time, however, they were already seen as part of a global economy which was indispensable to the financing of British conflict with France, especially if the latter was supported by Spain.[25]

With trade indispensable to survival, it had to be defended. The obligations of the state in this respect had been established during the War of William III. In 1694 the Admiralty was obliged to state the number of ships it would devote to protecting trade. In 1708, during the subsequent War of Spanish Succession, by the Convoy and Cruisers Act, forty-three naval vessels had to be reserved to serve as convoy escorts, the operations of which had to be reported annually to Parliament.[26] The need to provide escorts for convoys brought about a structural change in the navy, with an increase in smaller fourth, fifth and sixth rate ships and sloops.[27] By the last half of the eighteenth century, convoy experience had instituted a body of Admiralty regulations governing the organisation of shipping under protection at sea. Commercial groups like the West India merchants of London recognised the efficacy of the convoy system.[28] Indeed, after a century of experience, in 1798 Lloyd's insurance company and the Admiralty produced a Convoy Act which made convoys compulsory for all ships engaged in foreign trade, with the exception of fast-sailing licensed ships, Hudson's Bay and East India Company ships, and vessels making for Ireland.[29]

---

[22] P. Mathias, 'Risk, credit and kinship in early modern enterprise' and K. Morgan, 'Business networks in the British export trade to North America, 1750–1800' in *The Early Modern Atlantic Economy*, ed. J. J. McCusker and K. Morgan (Cambridge, 2000), 15–35 and 36–62, respectively.

[23] J. C. Sainty, *Officials of the Board of Trade 1660–1870* (London, 1974), 1–7.

[24] Unger, 'Warships, cargo ships and Adam Smith'.

[25] P. J. Marshall, *The Making and Unmaking of Empires*, 81–2.

[26] P. Crowhurst, *The Defence of British Trade 1689–1815* (Folkestone, 1977), 55.

[27] Harding, *Seapower and Naval Warfare 1650–1830*, 290–1, derived from Glete, *Navies and Nations*, II.

[28] *British Naval Documents*, 401–3, 444–8.

[29] Crowhurst, *Defence of British Trade*, 71.

Until that time, some insurance policies had included the necessity to sail in convoy. Merchants and ship owners at the ports around Britain applied to the Admiralty for convoy protection. At the Admiralty, the routing requirements of ships were recorded in a register of applications and, at the main anchorages, fleet commanders were directed to supply vessels to serve as escorts. From 1794, convoys departed 'on the first fair wind after' regular monthly dates. Warships sailing for foreign stations were often used as escorts; otherwise small warships were employed, convoy being rated as a particular station in the Admiralty's Disposition Books. Some escorts thus accompanied merchant ships virtually to their destinations, while others only provided protection 'clear of the track of the enemy's cruisers'.[30]

After 1798 convoy lists were sent to Lloyds with details of the vessels under escort. By the Napoleonic War, these convoys were huge. In July 1808, 138 vessels departed for the Mediterranean under escort; in August that year 184 went out to the West Indies in one convoy, while another containing 153 merchant ships sailed directly for Jamaica. In 1814 the number brought back from Jamaica was 201.[31] For the navy, the great advantage of convoy was that, in relation to the number of escorts, the number of ships under protection could grow exponentially.[32] This had wide ramifications for sea trade and shipping. For after 1748 the number of foreign-going English merchant ships continued to grow even during wartime.[33] Indeed, it was the distant trades to America, the West Indies and south-east Asia in which shipping most expanded, a growth which demanded an equivalent expansion in trade to the eastern Baltic for the materials used in the shipping industry.[34]

While the British state worked to preserve its own colonies and trade, it also worked to weaken her enemies by taking their colonies and shipping. The policy dated back to the capture of Jamaica from Spain in 1655 and of New Amsterdam from the United Provinces in 1665. However, overseas expeditions were difficult to organise. New Englanders

---

[30] TNA, ADM. 7/60; see, for example, 1 Jan. and 5 Dec. 1794.

[31] TNA, ADM. 7/782, 796, registers of convoys 1793–6, 1811–12; ADM. 7/64, convoy lists sent to Lloyds, 1808–15; ADM.7/65, lists of ships and vessels under convoy, 1810.

[32] D. W. Waters, 'Seamen, scientists, historians and strategy', *The British Journal for the History of Science* 13(1980), 207–9, printed as appendix B to the introduction of *The Defeat of the Enemy Attack on Shipping, 1939–1945*, ed. E. J. Grove (NRS, 1997), xl–xlii.

[33] Growth probably ceased during the American War of Independence, which deprived Britain of American-built ships and gave rise to great losses from capture. But numbers between 1775 and 1783 are uncertain. Ralph Davis ended his survey of the English shipping industry in 1775, and the registration of British shipping only began in 1787.

[34] Davis, *Rise of the English Shipping Industry*, 40; B. R. Mitchell and P. Deane, *Abstract of British Historical Statistics* (Cambridge, 1962), 312.

attacked French Quebec in 1690 and 1711, but both expeditions failed because of, among other things, insufficient manpower, food supplies and support from Britain.[35] With demands for war in Europe at this time, resources for such operations could not easily be spared. The New Englanders did better in 1745. With some British naval support, they took the French fortress of Louisburg on Cape Breton Island at the mouth of the St Lawrence. Other British expeditions cost so many lives they could not be termed successes.[36] Their failure derived from a variety of factors, including deficient experience of the logistics of long-distance amphibious operations and inadequate intelligence of the enemy waters, territories and naval movements, which caused apprehension and tension between the commanders.[37] Even so, during the Wars of Jenkins's Ear and of Austrian Succession in the 1740s public opinion favoured expeditions against French and Spanish colonies, strengthened by an inherited sense of protestant virtue and antipathy towards the catholic states.[38]

Overseas experience, improved intelligence, direction and public opinion were to come together during the Seven Years' War after 1756. The logistical lessons of operations overseas in the previous war had been learned.[39] The knowledge accompanied a growing navy, a readiness to invest in amphibious operations, and strategy at sea in European waters that would provide Britain with the opportunity to mount long-distance expeditions in what became the first world-scale war. Among other places, the French were attacked at Louisburg, Quebec and Montreal in Canada, and the Spanish at Havana in Cuba and at Manila in the Philippines. All five assaults were successful, and their success arose from an understanding of both the logistics and the physical challenges involved.

The making of state policy

Success in the Seven Years' War owed a great deal to the experience of its military leaders. The experience of the First Lord of the Admiralty was critical to the development of an effective defensive strategy in home waters, which served as a shield for overseas expeditions. Execution of that defensive strategy depended on logistical support. It also benefited

---

[35] R. Harding, 'The expeditions to Quebec 1690 and 1711: the evolution of trans-Atlantic amphibious power' in *Guerres maritimes 1688–1713* (Vincennes, 1996), 197–212.

[36] R. Harding, *Amphibious Warfare in the Eighteenth Century: The British Expedition to the West Indies 1740–1742* (Woodbridge, 1991).

[37] Harding, 'Sailors and gentlemen of parade'.

[38] Baugh, 'Maritime strength and Atlantic commerce'.

[39] C. Buchet, 'The Royal Navy and the Caribbean, 1689–1763', *MM* 80(1994), 30–44.

from a supply of information or intelligence about the intentions and forces of the enemy. By the time of the Napoleonic War, the First Lord of the Admiralty would think of the arrangements to defend Britain and her empire as a system which united command at sea, the growth of the navy and the supply of naval stores.

Experience and vision

In January 1755, when the French prepared to counter the British expedition to America, the two military officers who were called to the Cabinet to provide advice had experience dating back to the beginning of the century. Representing the army was Sir John Ligonier, who was then seventy-five and had been active under Marlborough during the War of Spanish Succession. He was distinguished by his strategic and administrative abilities as well as for his experience in the field. Between 1757 and 1766 he would serve as Commander-in-Chief of the army, doubling as Master General of the Ordnance in the crucial period 1759–63.[40]

Representing the navy was George, Lord Anson, aged fifty-six. A volunteer in 1712, a lieutenant from 1716, Anson had achieved wealth and fame from his circumnavigation of 1740–4. The scale of mortality upon the voyage had been as extraordinary as the value of the captured Spanish silver. Upon his return in 1744 Anson had become an Admiralty Board commissioner under the Duke of Bedford and Lord Sandwich, in which role he also served at sea during the last two years of the War of Austrian Succession. In 1748 when Sandwich replaced Bedford as First Lord of the Admiralty, Anson remained at the board and in 1751 became himself First Lord. Anson held this post, with only a seven-month absence in 1756–7, until he died in June 1762.[41]

At the Admiralty, Anson had few assistants. Even in 1760, at the height of the Seven Years' War, the Admiralty comprised only seven commissioners, two secretaries, seven established and twelve extra clerks, the latter serving just the length of the hostilities.[42] However, thirteen years at the Admiralty gave Anson the opportunity to make numerous reforms, fundamentally affecting the efficiency of both the sea personnel and the material of the navy. He instituted the sea officers' 'additional'

---

[40] J. S. Corbett, *England in the Seven Years' War* (2 vols., London, 1907; repr. 1992), I, 33–4; C. Barnett, *Britain and her Army 1509–1970: A Military, Political and Social Survey* (London, 1970), 213.

[41] Sainty, *Admiralty Officials*, 107; N. A. M. Rodger, 'George, Lord Anson 1697–1762' in *Precursors of Nelson*, ed. Le Fevre and Harding, 176–99.

[42] Sainty, *Admiralty Officials*, 101–2.

fighting instructions, revised the articles of war, and re-organised the marines.[43] He dealt with dockyard administration, naval architecture, ordnance, stores: indeed, 'the whole system of supply was placed on a new and more satisfactory basis'.[44]

During the Seven Years' War, his most noteworthy achievement was to develop the use of a western squadron of the Channel fleet to command the western approaches to the English Channel. Others before him had mooted the idea. But the scheme needed ships in sufficient numbers to permit reliefs and officers that understood its strategic importance. The location of the squadron was critical to effectiveness. As Admiral Vernon observed in 1745, a large squadron 'posted within the Channel between the Lizard and the coast of France' would 'leave all Ireland, the western coasts of this island and even the Bristol Channel and all our East and West Indian trade expected home, open to them [the French] to do what they please'.

Whereas a western squadron formed as strong as we can make it . . . and got speedily out into the Soundings, might face their united force, cover both Great Britain and Ireland and be in condition to pursue them wherever they went and be at hand to secure the safe return of our homeward bound trade from the East and West Indies.[45]

In 1746, having command of the squadrons in the English Channel, Anson had formed from them a single 'western squadron' which had much success in catching French warships and convoys that attempted to cross the Atlantic.[46] During the Seven Years' War, Anson reconstituted the western squadron and, to keep it on station, arranged a supply of fresh food for it.[47]

Described as 'the linchpin' of British naval strategy, the 'essential tool for naval supremacy', the western squadron contributed to each of Britain's maritime achievements during the Seven Years' War. Its principal task was the blockade of the French port of Brest. During the French Revolutionary and Napoleonic Wars, this job was done by the greater part of the Channel fleet and proved crucial to the naval control of France. As the Secretary for War and Colonies then observed, 'unless the great

---

[43] R. Middleton, 'Naval administration in the age of Pitt and Anson, 1755–1763' in *The British Navy and the Use of Naval Power in the Eighteenth Century*, ed. Black and Woodfine, 109–27.

[44] Corbett, *England in the Seven Years' War*, I, 35.

[45] *The Vernon Papers*, ed. B. McL. Ranft (NRS, 1958), 459.

[46] Rodger, 'George, Lord Anson', 182.

[47] S. F. Gradish, *The Manning of the British Navy during the Seven Years War* (London, 1980), 165–7, 209–10.

fleet of Brest is kept in its proper subjection by the great fleet of England all subordinate operations must be nugatory'.[48]

Information and system

Decisions about defence were best made with sound intelligence. War leaders were inundated with financial and strategic data and, during the second half of the eighteenth century, the quantity tended to grow. From sightings, naval observations and agents, the secretary of the Admiralty collated incoming information relating to the movements of the enemy. By the time of the French Revolutionary War, the First Lord of the Admiralty was able to make such intelligence available to other members of the Cabinet. As a matter of course, the clerks to the secretary of the Admiralty also listed monthly the station of every naval vessel in Disposition Books.[49] This information was conflated in regular statements of dispositions of both French and British forces around the coast of Europe.[50]

By the time of the Napoleonic War, statements of dispositions were global in extent, standardised in format and printed for distribution within the government. In May 1804, when he became First Lord of the Admiralty, Henry Dundas, first Viscount Melville, received an 'Abstract of the Enemy's force from the most exaggerated accounts received from prisoners, neutrals, or others', relating to ports in France, Spain and Holland, in North America, the West Indies, the East Indies and at the Cape of Good Hope. Within this statement, enemy ships were categorised by size from the largest to the smallest – from ships of the line, frigates and corvettes down to gun-boats and invasion craft. The statement also detailed the strength of the opposing British squadrons, with their commanders, broken down by size of vessel, and including ships 'stationary but ready for service', 'in port and fitting' and those 'with convoys or employed on other services'. Melville also had to hand quarterly accounts of seamen raised, and monthly lists of 'ships which will be ready to commission'.[51]

The threat of invasion called forth detailed data on the preparations both of the enemy and of British local forces along every creek and

---

[48] Henry Dundas to Lord Grenville, 12 October 1793. *Historical Manuscripts Commission 14th Report. Appendix Part V: The Manuscripts of J. B. Fortescue, Esq., Preserved at Dropmore* (London, 1894), II, 444.

[49] TNA, ADM. 8/1–134, 1673–1813, 1821–53.

[50] *British Naval Documents*, 397–8, transcribed from the Newcastle Papers, BL, Add. MS. 33048, fos.230–2.

[51] DRO, 152M/c1804/ON, account 18, dated 13 May 1804.

coastline. In mid 1804 statements relating to the enemy included numbers of gun-boats, gun-brigs, armed sloops and troop transports in the
Texel, at Flushing, Ostend and Boulogne.[52] The vessels mobilised to
oppose them were stated in even greater detail. They included a large
proportion in private ownership, even 'country craft' armed with carronades and long guns lining the coasts of England, Wales and Ireland,
and 'sea fencibles' ready to man them.[53]

Melville was concerned with the long term as well as the short. Early
in July 1804, 'for the information of His Majesty and His confidential
servants', he issued statements 'taken from the precis of intelligence
received at the Admiralty' and deliberately 'prepared in the form they
now appear' that they might reveal both 'the real state of the British
navy as it now stands' and 'its progressive improvement to meet future
contingencies'. Melville emphasised that 'exertions for the improvement
and increase of our naval strength . . . must not be on a scale to answer the
exigency of the moment: they must be systematically begun and adhered
to without intermission for a long track of years'. He dealt with not only
the supply of ships but also the supply of naval stores, and enclosed
copies of his correspondence with the Comptroller or chairman of the
Navy Board. At that time there was just one commodity which threatened
to fall short: 'the great article of oak timber'.[54]

This information was used at Cabinet level to inform strategic policy.
Melville aimed to integrate defence with offence, to form 'a complete
system of arrangement':

in order that the whole of the King's naval forces may be so distributed and
appropriated as not only to place the British Empire in a state of security against
the menace of the enemy but also to prevent the movements of his squadrons,
and to put a stop to all his commerce and intercourse both with foreign nations
and such colonies as may yet remain subject to France.[55]

Naturally, different members of the Cabinet had different interests, and
debate over policy tended to spread to the House of Commons.[56] Justly
so, for these matters lay at the foundation of both the survival of the state
and its global power.

---

[52] WLC, Melville Papers, 'General Statement of the Enemy's naval force', 24 June 1804.
[53] DRO, 152M/c1804/ON, accounts 2, 6, 7, 8, 9, 11, 13, 16, 18.
[54] WLC, Melville Papers, 3 July 1804.   [55] *Ibid.*, 12 Sept. 1804.
[56] See Lord Sandwich's warnings in 1777 and defence of his administration in 1782
   and also Lord St Vincent's policies from 1802 and their consequences in 1804, *British
   Naval Documents*, 336–8; R. Morriss, 'St Vincent and reform, 1801–04', *MM* 69(1983),
   269–90.

The defence of the British Isles

Melville's 'complete system' built upon strategic lessons learned during the Seven Years' War and French Revolutionary War. The Channel fleet was used to command the western approaches to the English Channel (see table 2.1) and to blockade the port of Brest.[57] The strength of the blockade depended upon the relative size of the French force sheltering in the port of Brest, the weather conditions, the ships available and navigational knowledge of waters that were dangerously littered with outcrops of rock.

The blockade strategy

In 1756–7, the blockade under admirals Hawke, Boscawen and Saunders was intermittent; in 1757 a change of ministry removed Anson from the Admiralty; but in 1758 he returned to sea, sharpening dispositions; in 1759 the squadron was divided – for inshore and distant stations – and the former was subdivided to watch the different channels that permitted egress from Brest. Ships returned to refit and replenish by turns but from August 1759 supplies of fresh food were delivered to ships off Brest by victuallers.[58] At this time the French were preparing an invasion force along the south Brittany coast, and a gale in November 1759 drove Hawke off station but he returned in time to catch and defeat the French fleet as it entered Quiberon Bay.

Events at that time demonstrated the difficulty with which ships kept station off Ushant at the tip of the Brittany peninsula. Twice during the autumn of 1759, Sir Edward Hawke was driven by westerly gales back to the English coast to take shelter first in Plymouth Sound and then in Torbay. The westerly winds, which kept the British fleet on the English coast, also kept the French in Brest. However, as soon as the wind came round to the east, the French were able to leave Brest and the British western squadron was hard put to get back on station. In November 1759, Hawke only caught the French fleet on account of information

---

[57] R. Middleton, 'British naval strategy 1755–62: The Western Squadron', *MM* 75(1989), 349–67; A. N. Ryan, 'The Royal Navy and the blockade of Brest, 1689–1805: theory and practice', in *Les Marines de guerre européennes XVIIe–XVIIIe siècles*, ed. M. Acerra, J. Merino and J. Meyer (Paris, 1985), 175–93; R. Saxby, 'The blockade of Brest in the French Revolutionary War', *MM* 78(1992), 25–35; D. M. Steer, 'The blockade of Brest and the Royal Navy 1793–1805', unpub. University of Liverpool MA thesis, 1971.

[58] Middleton, 'British naval strategy 1755–62'. For Admiralty instructions to Hawke, 18 May 1758, see *British Naval Documents*, 390–1.

derived from a sighting by an empty victualler making its way back from Ushant.[59]

The British ability to catch the French when they emerged from Brest depended on a number of logistical provisions. The most fundamental was the maintenance of a sufficient number of seaworthy ships. The existence of dockyards at Portsmouth and Plymouth was essential for the refitting and repair of ships.[60] Portsmouth yard was rather too far east in the English Channel for ships quickly and easily to get into its western approaches. Plymouth dockyard had been established in 1689–93 partly as a response to the French creation of a naval port at Brest and, during the eighteenth century, would prove vital to ships operating in the Atlantic and western approaches.[61]

Numbers of ships in commission mattered. During the American War of Independence Britain had insufficient ships to provide an effective western squadron. Partly as a result, the French and Spanish were able to unite their forces and penetrate the English Channel with fleets of forty-five ships of the line in 1779 and of forty-nine in 1781.[62] By the beginning of the French Revolutionary War, Britain again had sufficient.[63] The British fleet had been repaired and rebuilt and was rapidly mobilised.[64] At this time Sir Charles Middleton, as a sea lord at the Admiralty, was sent accounts of the 'supposed force of the enemy' in their different ports and 'the British force required to oppose them effectually'. The forces within and those outside were weighed in terms of gun power, ships of the line being distinguished from frigates, with the ideal blockading force set against the actual number of ships available,

[59] Ryan, 'The Royal Navy and the blockade of Brest, 1689–1805', 183; Corbett, *England in the Seven Years' War*, II, 48–53.

[60] M. Duffy, 'The establishment of the Western Squadron as the linchpin of British naval strategy', in *Parameters of British Naval Power 1650–1850*, ed. Duffy, 60–81.

[61] M. Duffy, 'The creation of Plymouth dockyard and its impact on naval strategy' in *Guerres Maritimes 1688–1713*, 245–74; also his 'Devon and the naval strategy of the French wars 1689–1815' in *The New Maritime History of Devon*, ed. M. Duffy *et al.* (2 vols., London, 1992, 1994), I, 182–91; and J. Coad, 'The development and organisation of Plymouth dockyard, 1689–1815' in *The New Maritime History of Devon*, ed Duffy *et al.*, I, 192–200.

[62] Although there was said to be '63 sail' in 1779 and between forty-four and forty-seven ships of the line in 1781: D. Syrett, *The Royal Navy in European Waters during the American Revolutionary War* (Columbia, SC, 1988), 74–6, 146.

[63] While Britain possessed 158 ships of the line, France had 82. Indeed, in 1793, the British fleet was exactly equal to the combined battle strength of France and Spain, which then possessed 76 ships of the line: W. James, *The Naval History of Great Britain from the declaration of war by France in 1793 to the accession of George IV* (6 vols., London, 1859), I, 56–7.

[64] P. Webb, 'The rebuilding and repair of the fleet 1783–93', *BIHR* 50(1977), 194–209.

the number always to be kept outside and that which could be away refitting.[65]

Yet at this time the blockade strategy was not applied rigorously, for the French navy was reduced to virtual impotence by the political conflicts on shore. The French lacked officers, men and naval stores. The defence of British convoys and interception of French trade consequently assumed priority. In 1793–4 Lord Howe, who commanded the Channel fleet, was required only to keep the port of Brest under occasional observation. Indeed, in May 1794, having established that the French fleet was still in Brest, he focussed his activities upon a search for a transatlantic convoy of grain ships bound for France from the Chesapeake. The French fleet was thus able to leave Brest unobserved and its detection and defeat by the British at the Battle of the Glorious First of June 1794 was to some extent fortuitous.[66] Only in July 1795, with the necessity to prevent French naval interference in British operations on the Brittany coast, was a blockade of Brest and L'Orient instituted under Lord Bridport. And only after December 1796, when forty-four ships, including seventeen ships of the line with 18,000 troops, escaped to mount an invasion of Ireland, did the Admiralty itself attempt to organise the blockade of Brest.[67]

On account of French losses[68] – including captures added to the strength of the British navy[69] – the British fleet enjoyed a margin of numerical superiority which continued to grow with Spanish, Dutch and French defeats of Cape St Vincent, Camperdown and the Nile.[70] The good condition of the British fleet in 1793 also served it well until 1798, for this diminished the number of ships that were absent from sea being repaired. After 1798, however, between a fifth and a quarter of all ships of the line theoretically in a condition for sea service were at dockyards having defects made good and being refitted.[71]

---

[65] NMM, Middleton papers, MID/10/3/20–4.

[66] C. Ware, 'The Glorious First of June. The British strategic perspective' in *The Glorious First of June 1794: A Naval Battle and its Aftermath*, ed. M. Duffy and R. Morriss (Exeter, 2001), 25–45.

[67] Saxby, 'The Blockade of Brest in the French Revolutionary War'.

[68] French losses of ships of the line amounted to 15 in 1793, 8 in 1794, and 10 in 1795: W. L. Clowes, *The Royal Navy: A History from the Earliest Times to 1900* (7 vols., London, 1897–1903, repr. 1997), IV, 552–61.

[69] Of 112 ships of the line (64 or more guns) and 99 frigates (20–60 guns) that served in the Channel fleet in 1793–1801, 8 of the battleships (7 per cent) and 22 of the frigates (22 per cent) were captured from the French during the French Revolutionary War.

[70] In 1798 the French lost 14 ships of the line. By the end of 1798, France had lost 49 ships of the line, which amounted to 60 per cent of her battle strength at the beginning of the war. In total, between 1793 and 1801, the French lost 55 ships of the line, the Spanish 10 and the Dutch 18: Clowes, *The Royal Navy*, IV, 552–61.

[71] The proportion of the British sea-going battle fleet out of commission diminished from 87 per cent in 1793 to 8 per cent in 1797, then increased to 16 per cent in 1798,

Moroever, the port of Brest remained difficult to close. In September 1798 nine French ships with 3,000 troops sailed once more for Ireland. Then, in April 1799, eighteen ships of the line under Admiral Bruix escaped, were reinforced with twenty-five of the line from L'Orient and, after cruising in the Mediterranean, gathered a further reinforcement of fifteen Spanish ships of the line from Carthagena and Cadiz. This combined fleet returned to Brest in August 1799.[72] The presence of at least forty-five enemy ships of the line in Brest demanded blockade in great strength. With France preparing another invasion of Britain, in April 1800 Lord St Vincent replaced Lord Bridport and instituted a close blockade of Brest. As in 1759, St Vincent organised the blockade into inshore and distant squadrons with frigates keeping the entrance to Brest under close observation. At this time the Black Rocks off Ushant became a station. To permit reliefs, by June 1800 the fleet under St Vincent was enlarged to fifty-two ships of the line and seventy-one other vessels.[73]

The maintenance of this huge naval force at the mouth of the English Channel was made possible by three factors. The first was the organisation of a rota for refitting and replenishing ships. Ships returned in small numbers to Portsmouth and Plymouth dockyards as scheduled by St Vincent. A rigorous disciplinarian, he tolerated no delays at the yards, caused either by the ships' officers or by the dockyard officers. Returning ships also, when possible, brought out water and stores for other ships.[74]

The second factor was the establishment of an effective method of feeding and watering the fleet off Brest and in Torbay. By 1800, convoys of victuallers – small vessels, usually 100–150 tons – were escorted to and from Ushant by sloops and cutters. In December 1800 an order required fresh vegetables to reach the inshore squadron off the Black Rocks at least once a month. Although the visits of vegetable sellers were not approved by senior officers on account of the spirits they sold, private trading vessels performed this service, supplementing the cargoes of the victuallers. Otherwise ships re-provisioned by returning into Plymouth Sound, or Torbay, to which provisions were ferried from Plymouth Sound. In Torbay, piped water was made available at Brixham.[75]

---

20 per cent in 1799, 24 per cent in 1800, and 27 per cent in 1801: Morriss, *Royal Dockyards*, 16.

[72] *The Channel Fleet and the Blockade of Brest, 1793–1801*, ed. R. Morriss and R. C. Saxby (NRS, 2001), 163–7, 238–43, 299–304, 346–50, 403–6.

[73] *Ibid.*, 488–93.

[74] TNA, ADM. 1/117, St Vincent to Admiralty, 21 Oct. 1800.

[75] M. Steer, 'The blockade of Brest and the victualling of the Western Squadron 1793–1805', *MM* 76(1990), 307–16.

The third factor was the inauguration in the Channel fleet of the policy of giving seamen lemon juice as a preventative against scurvy. The former naval surgeon, James Lind, had listed lemon juice among various means by which scurvy could be ameliorated. But it was experience in the American and the French Revolutionary Wars that convinced sea officers of its efficacy as a cure. Its availability to the Mediterranean fleet after 1793 was especially important. Its use as a preventative, rather than as a cure, began in the Channel fleet in 1800.[76]

The signature of the preliminaries of the Peace of Amiens in October 1801 permitted a withdrawal of the blockade. But in 1803, when war resumed, the Channel fleet took its former place off Brest and the organisation necessary to its support was reconstructed.[77] After 1803 the French ports of L'Orient and Rochefort were also closed; after December 1804, so too the Spanish ports of Ferrol, Cadiz and Carthagena. With responsibilities stretching as far south as Cadiz, Lord Melville in 1804 referred not to a Channel fleet or a western squadron but to the British Atlantic fleet.[78]

The response to a threat

By September 1804 Britain had 546 naval ships and vessels in commission. In addition to the Atlantic fleet, there was a Mediterranean fleet and squadrons in the West Indies, on the North American coast and in the Indian Ocean. But on account of Napoleon's preparations to invade Britain, three-quarters of the ships in commission in September 1804 were deployed in home waters.[79] When Spain entered the Napoleonic War in December 1804, the number of ships in commission was further enlarged. The coordination and deployment of forces on this scale was a major logistical operation. Their great test came in 1805 when the Toulon fleet under Villeneuve broke out of the Mediterranean and, taking a Spanish squadron from Cadiz, sailed for the West Indies. The interception of Villeneuve's fleet, its re-confinement and subsequent defeat

---

[76] *The Channel Fleet and the Blockade of Brest*, 521–3.
[77] *Papers relating to the Blockade of Brest, 1803–1805*, ed. J. Leyland (2 vols., NRS, 1899, 1902).
[78] WLC, Melville Papers, 12 Sept. 1804.
[79] Melville recorded the disposition in home waters as follows: '34 ships of the line in the Channel fleet, 71 vessels off Ireland, and 9 in the Straights of Gibraltar, all under Lord Cornwallis; 19 off Ireland under Lord Gardner; 26 off Guernsey and Jersey under Sir J. Saumarez; 10 ships of the line to watch the Texel, 41 vessels in the Downes, and 149 smaller vessels all under Lord Keith.' In addition he listed as reserves: '16 at Portsmouth, 15 at Plymouth, 12 for or with convoy, 2 off Weymouth, 12 disposable, 8 stationed': WLC, Melville Papers, memorandum for King and Cabinet, 13 Sept. 1804.

at Trafalgar is well known. The British fleet re-deployments before the battle are little known. Yet they were the foundation of the subsequent British victory.

The Admiralty was not unprepared for the joint operations of the French and Spanish fleets. In 1800 Admiral Bruix had returned to Brest with a combined fleet of forty-five ships of the line. Should they have broken out again, instructions existed that provided for the engagement of the fleet or for its observation wherever it went. These instructions, in successive editions, dated back to at least 1762.[80] In March 1800 the Admiralty up-dated these instructions, requiring a portion of the Channel fleet to pursue any squadron of the enemy that might escape from Brest 'to any part of the world to which it may go', even to the West Indies. The pursuing squadron was to report regularly the course and supposed destination of the enemy to permit the preparation of measures to meet the threat.[81]

In 1805 when the French fleet escaped from Toulon, taking with it a Spanish squadron from Cadiz, and sailed for the West Indies, in conformity with instructions to fleet commanders, Nelson, with the British Mediterranean fleet, followed. As required, he kept the Admiralty informed of his pursuit and, on the return of the Franco-Spanish fleet towards Europe, despatched a sloop with the information, which was received at the Admiralty on 7 July 1805. The sloop had actually sighted the combined fleet and was able to report that it was heading for Ferrol.[82] The information was received at the Admiralty by Sir Charles Middleton, now Lord Barham. Probably on 8 July he made a note of his thinking:

There are now off Brest or sailed for that station	22
Off Rochefort	5
Off Ferrol, including the 2 sent from off Rochefort	12
In port refitting – Goliath	1
If the Defence is continued with the Channel fleet	1

All manned and stored

There are now coming forward:
The Zealous, Orion, Captain, Audacious, Bellona.
These five ought to be able to replace the ships coming in for refitment.

---

[80] *British Naval Dcouments*, 397–9.

[81] Admiralty to Lord Bridport, 17 Mar. 1800, in *The Channel Fleet and the Blockade of Brest*, 435–9.

[82] R. Knight, *The Pursuit of Victory: The Life and Achievement of Horatio Nelson* (London, 2005), 494–9.

By this means there will be from 38 to 40 sail of the line applicable to the Brest, Rochefort and Ferrol squadrons. On the present occasion let 10 sail be detached from off Brest to reinforce the squadron off Cadiz because I take it for granted that only 4 remain there at present.

[That is, because] Admiral Collingwood sailed with 9. Sir R. Bickerton with the Queen = 10. Five were ordered to remain off Cadiz and 5 to go up the Mediterranean, but as 2 were dispatched to W. Indies, I allow 4 for each service only.

That makes a strong force in the very spot where they may be expected.[83]

In anticipation of the arrival of the combined fleet off Europe, Barham deployed two forces for its interception. On 9 July he noted for the Admiralty secretary or board commissioners:

My idea is to send the intelligence immediately to Admiral Cornwallis [commanding the Channel or Atlantic fleet],[84] who may be directed to strengthen Sir Robert Calder's squadron with the Rochefort squadron and as many ships of his own as will make him up to 15, and [to] cruise off Cape Finisterre from 10 to 50 leagues to the west.

[Cornwallis] to stand to the southward and westward with his own ships, at the same distance, for 10 days.

Cadiz to be left to Lord Nelson.[85]

On 22 July Calder intercepted the combined fleet which, with the loss of two ships, withdrew into Vigo, then into Ferrol. Nearly a month later it re-emerged, tried to sail north but turned south and entered Cadiz on 21 August.

Meanwhile, under instructions from the Admiralty, Cornwallis had selected the officers and ships that were deployed south to reinforce the fleet that was taking shape off Cadiz. Collingwood's deployment in May resulted from an Admiralty order of 2 March which directed Cornwallis to complete five ships with stores and provisions for foreign service 'and to hold them in readiness for sailing under the command of a flag officer at the shortest notice'.[86] Cornwallis wanted to retain a fleet equal to that in Brest.[87] But, on 17 August, receiving news of the departure of the

---

[83] *Letters of Lord Barham*, III, 255–6.

[84] Which Barham did, Cornwallis forwarding information as he received it to admirals farther south: NMM, COR/42, 3, 15, 28 July 1805.

[85] Letters of Lord Barham, III, 257

[86] NMM, COR/42, 4, 6, 7 Mar. 1805. Owing to leave, Cornwallis's copybook of letters to the Admiralty lacks correspondence relating to detachments during 18 March – 19 May 1805.

[87] *Ibid.*, 4, 12, 13, 16 Aug. 1805.

Spanish–French fleet from Ferrol, Nothesk and Louis were detached with nineteen ships.[88] The transfer of these ships to Cadiz was duly recorded in the Admiralty's disposition book. On 20 September Barham's private papers also recorded their relocation. By then the squadron off Cadiz had become a fleet of thirty-six ships of the line, while Cornwallis off Brest retained twenty-six. Another eleven ships of the line were recorded as 'in port, fitting and refitting, destined to strengthen the fleets of Admiral Cornwallis and Lord Nelson'.[89]

The battle of Trafalgar relieved some of the pressure upon the ships defending Britain, but less than usually assumed. Napoleon still had in mind the invasion of Britain. In consequence he kept building ships in the territories under his control. From thirty-four ships in French and Dutch ports in 1807, he actually increased the French fleet to over eighty in 1813, with another thirty-five under construction. The Treaty of Tilsit between France and Russia in 1807 agreed to unite the navies of Europe – French, Spanish, Dutch, Danish, Swedish and Russian – into one confederate fleet. The British blockade was thus extended from the Baltic to the Adriatic with strong squadrons off all the enemy's principal naval arsenals.[90] Merchant ports came under increased observation too in response to Napoleon's Berlin decree forbidding the entry of British trade into the continent; the Order in Council of 1807 declared all continental harbours excluding British trade would also come under blockade.

These demands forced ships in the British fleet to remain at sea for longer periods. In 1759 Sir Edward Hawke was expected to return into Torbay only two weeks after sailing for his station off Brest.[91] By the end of the century ships remained at sea until reliefs arrived. In the French Revolutionary War, to permit refitting and resupply, allowance was made for the absence of one ship in four from off Brest and one in three from off Toulon.[92] In 1812, with France controlling much of the coastline of Europe from the Baltic to the Adriatic, the second Lord Melville stretched the blockading force further and thinner. He allowed one British ship in six more than the enemy on 'home stations', with one in four on the Mediterranean station.[93] Yet by then each of the royal dockyards, as it was said at the time, operated 'as a part only of one great machine', so that ships were refitted and returned to sea at a rate that permitted the maintenance in commission of an unprecedented number.

---

[88] *Ibid.*, 16, 19 Aug. 1805.   [89] *Letters of Lord Barham*, III, 119.
[90] Glover, *Britain at Bay*, 17–19, 28–9; Hall, *British Strategy in the Napoleonic War*, 81.
[91] *British Naval Documents*, 390–1.   [92] NMM, MID./10/3/7–8.
[93] *Letters of Lord Barham*, II, 395; *Quarterly Review* 8(1812), 57.

The offensive overseas

Only with effective defence in home waters, and an adequate supply of sea and land forces, could offensive operations be mounted beyond Europe. This combination occurred in the Seven Years', French Revolutionary and Napoleonic Wars. Commitments of land forces on a large scale to Europe during the second half of the latter war, however, reduced overseas expeditions to those of necessity. By then, experience had taught major lessons in large-scale amphibious attack. It also reinforced lessons in long-distance logistical support to the land forces that were garrisoned around the world to hold territories secured against use by the enemy.

The problems of early expeditions overseas

The lessons in logistics began with the earliest expeditions to the Caribbean early in the War of Jenkins's Ear. In 1739 Admiral Vernon, with only six ships of the line and 200 soldiers, captured Porto Bello on the coast of Spanish America – to public acclaim. But in 1740–1, conveyed, protected and purportedly supported by Vernon's fleet, General Thomas Wentworth's army failed take Carthagena, Santiago de Cuba or Panama. At the height of the conflict, Britain had sixty ships, with 20,000 seamen and 8,000 troops, in the Caribbean. Including an American regiment, a total 14,195 soldiers were committed to the operation, but of these 10,126 died by the end of 1742.[94]

The scale of this mortality is a measure of the ignorance that existed in the British military, administrative and political hierarchy regarding the dangers and difficulties that had to be faced in mounting expeditions of this nature. There were problems in the command structure, in the maintenance of ships and in the supply of naval stores, of food and, on account of the mortality, of men. Supplementary were questions about how the supplies of materials were to be managed – either by contractor or by the boards concerned – and these raised further issues about how the materials were to be paid for and shipped to the Caribbean.

Richard Harding has revealed that the death of General Cathcart soon after arriving in the Caribbean placed General Wentworth in command of the troops and that, on account of inexperience, he was prey to the manipulation of Admiral Vernon. The latter in his despatches turned the failures at Carthagena and Santiago de Cuba against Wentworth and maintained his own reputation by the publication of pamphlets. Yet

---

[94] Harding, *Amphibious Warfare*, 86–149; D. Crewe, *Yellow Jack and the Worm: British Naval Administration in the West Indies, 1739–1748* (Liverpool, 1993), 4.

Vernon was the source of some of the difficulties. For, commanding the means of transportation, he had effective control of the expedition until the autumn of 1741 and made a priority of preserving his ships' crews even when they were needed by the army. On occasions, he refused to land seamen or even to permit them to assist in supplying the military camp with water.[95]

Vernon also had to preserve his ships, and the yards at Jamaica and Antigua had only small establishments of artificers and limited facilities. For the ships, the main problem was that their hull timbers were vulnerable to the tropical boring worm *Terredo navalis*, and initially the Admiralty despatched unsheathed vessels to the Caribbean. Small vessels could be careened, but then supplies were needed to repair sheathing and plank. In addition, masts, sails and rigging deteriorated rapidly on account of the heat, humidity, wood ants and sudden hurricanes.

Naval stores in large quantities had to be sent from Britain or North America. But Jamaica, Britain's main Caribbean base, was 4,000 miles from England and a demand for stores took three months to reach London. After orders for loading and delays for convoy, stores took nine months to reach Jamaica and sometimes longer, for the flow of supplies was interrupted between August and October by the hurricane season. In 1740 Vernon himself arranged to collect a supply of stores from Boston. In 1741 the Navy Board also arranged for a contractor to ship stores from New England. Despite these distant sources of supply, shortages still occurred in the Caribbean, where the practice developed of breaking-up defective ships to make others seaworthy.[96]

In the tropical climate, foodstuffs decayed more rapidly than anything else. Long-standing contractors existed but, with the army as well as seamen to feed, they had to enlarge their scale of supplies. However, then their ships were caught in embargoes and convoys, and when they did arrive their cargoes usually had to be stored before distribution. Food was usually still fit for consumption after shipment from Europe or North America but, if stored for long, was not always fit for redistribution within the Caribbean. Food was of central importance, for deficiencies in quantity or quality undermined the strength of seamen and soldiers and reduced their resistance to disease.[97]

This was all the more important because troops, as well as seamen, were deprived of fresh food on long voyages. Then those unaccustomed to the tropics easily succumbed to disease. The 8,000 troops that sailed for the Caribbean in October 1740 were reduced to 5,900 by the time they

[95] Harding, *Amphibious Warfare*, 170–1, 205.
[96] Crewe, *Yellow Jack and the Worm*, 213–84.   [97] *Ibid.*, 4–9, 168–212.

reached Jamaica in December 1740, with 600 dead and 1,500 suffering from scurvy. Following the attack on Carthagena, when over 600 men died, that force had been reduced further to 3,500 effectives, with the majority of losses from yellow fever.[98] The crews of ships were weakened too. By May 1741 the ships in Admiral Vernon's fleet had 1,738 sick, which was 11.5 per cent of his seamen.[99]

The trouble was that, even before the arrival of 28,000 seamen and soldiers, the region produced insufficient food to feed the local inhabitants. The settlers in Jamaica and the Leeward Islands imported flour, biscuit, corn, beef, pork, butter, salt, fish and rice from North America; beef, pork and butter from Ireland; and wine from Madeira. Some cattle were available, for example in Jamaica, but with the demands of the British armed forces their prices rose to 'exorbitant' heights. In consequence, virtually all the provisions for the seamen and troops had to be imported.

The supply of provisions by contract was already sophisticated. The contract for the supply of the navy at Jamaica was held by the partnership of Mason and Simpson, which had operated effectively since 1730. However, in late 1739 they demanded a £10,000 advance in order to carry on their contract. With a further advance of £20,000, in April 1740 they undertook to victual 6,000 soldiers on the same basis as they supplied seamen, the cost of feeding the soldiers being met by the War Office. Between August 1740 and January 1741 the contractors drew food from seven sources: England, Ireland, New York, Philadelphia, the Carolinas, Barbados and Antigua. Nine ships were sent from London and Plymouth, fifteen were hired in Bristol and Cork to bring food from Ireland, twenty-two were despatched from New York and Philadelphia, four from the Carolinas, while rum was sent from Barbados and Antigua.

Yet, even with large advances, the costs of providing food supplies on a war scale at war prices were beyond the contractors' resources. Damage occurred to some cargoes, and the danger of capture or miscarriage required the contractors to over-order. Deficiencies in one market necessitated greater orders in others; and there were complications – the army, for example, would not accept substitutions. Meanwhile, the Victualling Board had entered the same markets as the contractors and on such a grand scale that prices were driven up so far that budgeting, for the contractor, became impossible, 'for there is no withstanding the

[98] J. de Zulueta, 'Health and military factors in Vernon's failure at Cartagena', *MM* 78(1992), 127–41. The suffering was described in the novel *Roderick Random* by Tobias Smollett, then a surgeon's mate on board the *Chichester*.
[99] Crewe, *Yellow Jack and the Worm*, 15.

government's power'.[100] The naval boards competed for shipping too. The food contractors were permitted to increase their contract price per man. But in September 1740, the Admiralty and Victualling Boards acknowledged the business was too great for private management and decided to take the supply of food into their own hands.[101]

The principal consideration in this decision was a financial one. The whole cost of victualling the expedition was estimated at £300,000. Successive advances of £10,000 sterling were required to finance a contractor. This meant that in 1740 the navy needed to raise cash without delay. However, by undertaking food supply itself, the navy could pay with victualling bills, in essence purchasing on credit and spreading the cost into the future. The Victualling Board had anyway already embarked upon large-scale purchases, and to replace a contractor was a matter of procedure. The board's premises and staff in Jamaica, dating from the War of Spanish Succession, were enlarged. It bought storehouses from Mason and Simpson, erected ten more on land at Port Royal acquired for a hospital, and constructed a wharf.[102]

By 1743 the board was in a routine of chartering ships and ordering victuals to be sent to Jamaica about three times a year – in October, February and August – in the expectation that orders would arrive about five months later. This allowed for the loading of victuallers in the Thames, their escort from the Downes to Spithead, the gathering of an outward convoy and frequent adverse winds. Getting out of the Thames and down the Channel ran against the prevailing winds; so too did the voyage from Spithead into the north Atlantic. To begin with, the main convoy went via Cork where it collected victuallers laden with Irish pork, beef and butter. But in 1741 the Admiralty directed the Irish victuallers to join the main convoys at Spithead.[103] In 1741–2 condemnations of provisions at Jamaica were still large, amounting to about one half of all the bread in 1742, four-fifths of all the butter and flour, and one-third of all the beef. But the local agents in Jamaica did much to smooth trans-shipment and make substitutions – for example, of rice for bread and pease. An effective system of state supply to the Caribbean was thus eventually established and it was to continue until 1748 when supply again reverted to contract.[104]

---

[100] In September 1740, for example, the Victualling Board placed orders in Ireland for 15,000 hogs, 10,000 oxen and 30,000 firkins of butter: *ibid.*, 166.

[101] Harding, *Amphibious Warfare*, 52–60; Crewe, *Yellow Jack and the Worm*, 145–60.

[102] Crewe, *Yellow Jack and the Worm*, 160–5.

[103] *Ibid.*, 170–9; D. Syrett, 'The Victualling Board charters shipping, 1739–1748', *IJMH* 9(1997), 57–67.

[104] Crewe, *Yellow Jack and the Worm*, 145–75, 179–88.

Yet it really came too late. Operations in the region before 1742 were a disappointment. The events of 1739–42 revealed that the possession of a fleet and its despatch overseas, even if accompanied by a force of soldiers, was not enough to achieve an objective. Intelligence of what to expect, planning, timing and logistical support were necessary.[105] An experienced state bureaucracy was needed as well as military forces, which knew in advance what was necessary and could make its preparations. That bureaucracy had to think about supplies of every nature, not just ships and men, but naval stores, food, money and facilities in the war zone. Above all, the provision of those supplies had to be guided by policy which guaranteed the maintenance of that bureaucracy and its capability in logistics.

The challenges met

The Caribbean operations of 1739–42 were the first transatlantic operations of any scale undertaken by Britain. Christian Buchet has shown that, during the Seven Years' War, distinct improvements were made on initial practice in the previous war. These included the choice of army and naval commanders with an eye to cooperation and for whom respective authorities and relationships were defined. Ships were despatched that could be careened for repair and that were sheathed for protection against the worm. Expeditions were timed to avoid the hurricane or monsoon season, and periods of rest were included to permit troops to recover from long sea voyages. These were shortened by the supply of men and rations from North America to the West Indies, an easier voyage than from Europe. If possible, local contracts for provisions were arranged, to include fresh food, with the supply of a surplus to cover decay and delays in delivery. Finally, convoys were scheduled and local facilities were prepared to meet the needs of both ships and men.[106]

Experience from the previous war determined these improvements. Some were simply finishing touches to practice that had proved necessary during the preceding war, less than a decade earlier. They became important when placed within the framework of a campaign aimed at seizing the colonies of Britain's enemies. During the Seven Years' War, such a campaign became possible when British strategy succeeded in limiting the operational capabilities of the French navy and in creating

[105] P. Woodfine, 'Ideas of naval power and the conflict with Spain, 1737–1742' in *The British Navy and the Use of Naval Power in the Eighteenth Century*, ed. Black and Woodfine, 71–90.

[106] Buchet, 'The Royal Navy and the Caribbean'.

the security necessary to convey land forces to North America, the West Indies, Africa and the Philippines. The campaign benefited from the experience of a cadre of capable officers who knew what to expect and were selected for that knowledge. Likewise, the logistical arrangements benefited from a bureaucracy that was developing the capability which henceforth would characterise British overseas operations.

During the Seven Years' War, the main campaigns against the French outside Europe took place around the Atlantic basin. There the colonies of France in North America and in the West Indies were the initial target. Those of Spain were added after entering the war late in 1761. The dominant objective was to weaken Britain's enemies economically. So valuable were the perceived benefits that, even at the end of the century, Henry Dundas, Secretary for War and Colonies, advocated a similar campaign, believing that for Britain the maintenance of 'an extensive and complicated war' was dependent on 'destroying the colonial resources of our enemies and adding proportionately to our own commercial resources, which are, and must ever be, the sole basis of our maritime strength'.[107]

In 1756 France's Canadian possessions appeared vulnerable, having low population and sheltering behind the isolated fortress of Louisburg on Cape Breton Island near the mouth of the St Lawrence River. During the previous war in 1745, 4,000 New Englanders had taken the fortress with the assistance of just three frigates under Sir Peter Warren based in the Leeward Islands. Louisburg was exchanged for Madras at the peace in 1748, but from 1749 a naval vessel was based at Halifax to support a 'sea militia' of armed vessels equipped by settlers. A naval squadron paid visits from 1755 and a base was begun in 1757.[108] Halifax was to prove highly useful in the campaign between 1758 and 1760 which saw the British again take Louisburg and then occupy the main centres of French settlement in the St Lawrence valley.

Between 1758 and 1762 campaigns against French and Spanish colonies become more ambitious as the war proceeded. In North America, Louisburg fell in July 1758, Quebec in September 1759, Montreal a year later. In West Africa, the French colonies in Senegal and Goree also fell in 1758 and 1759. In the West Indies, the French islands fell like nine-pins: Guadaloupe in 1759; Dominica in 1761; Martinique, Grenada, St Lucia and St Vincent in 1762. That year, Spanish Havana was also taken. In India, French Pondicherry fell in January 1761; and in

---

[107] Dundas to Richmond, 8 July 1793, British Library, Bathurst Papers, Loan 57/107; Dundas to York, 28 July 1795, BL, Add. MSS. 40102, quoted in Duffy, *Soldiers, Sugar and Sea Power*, 25, 161, 371.

[108] G. S. Graham, *Empire of the North Atlantic: The Maritime Struggle for North America* (London, 1958), 116–42; Gwyn, *Ashore and Afloat*, x, 3.

the Pacific, after a voyage from Madras to the Philippines of seven weeks, an expedition took Spanish Manila in October 1762. The achievements fulfilled the apparent promise of British sea power. But they resulted from decisions based on experience.

Much was due to Anson at the Admiralty. As leader of the circumnavigation of 1740–4, Anson appreciated what was possible on a global scale. Of the Manila expedition, he observed with remarkable sanguinity: 'as the voyage from the coast of Coramandel thither is chiefly thro' Straits, no difficulty can occur with regard either to the victualling or transport of such a number of troops on board the squadron in a run of no more than six weeks'.[109] Anson ensured that the men appointed to lead expeditions were tried, tested and known to him. Hence, at Quebec, we find that Saunders had been on the 1740–4 voyage; so too had Augustus Keppel, who took Goree and Senegal, and was second-in-command at Havana; so too had Hyde Parker who went to Manila. This last expedition also benefited from Richard Kempenfelt and Samuel Cornish who were at Carthagena in 1740.[110]

A feature of all the large-scale expeditions – to Quebec, Havana, Manila – was that the officers of the army and the navy strived to cooperate, and afterwards mutually praised their harmony or unanimity. Not only did the officers have to work together, their men did too. Seamen were indispensable in landing and hauling guns, powder, equipment or provisions on shore to wherever batteries were formed and the army encamped. It has been calculated that seventy men were needed to haul a 32-pounder along European roads, fifty-six for a 24-pounder, and forty to move a 12-pounder. Over rough terrain in a tropical environment, the manpower demands were far greater. Just supplying the powder demanded a constant flow of porters. To permit a 32-pounder gun to remain active for the course of a single day needed thirty-two to forty-eight men to make one journey each from the beachhead to the battery, carrying powder. The construction of the battery, supply of water and so on took more men.[111]

The British transatlantic campaign gained impetus from the fact that the French were unable to concentrate their sea power following the defeats of their Toulon fleet at Lagos Bay near Gibraltar in August 1759 and of their Brest fleet at Quiberon Bay on the coast of Brittany in November 1759. The blockade of their ports not only checked their

[109] N. Tracy, *Manila Ransomed: The British Assault on Manila in the Seven Years War* (Exeter, 1995), 25.
[110] S. Keppel, *Three Brothers at Havana 1762* (Wilton, 1981), 6; Tracy, *Manila Ransomed*, 22.
[111] Harding, *Amphibious Warfare*, 166–74; Harding, *Seapower and Naval Warfare*, 211–12.

intention to mount an invasion of Britain but handicapped their efforts at new construction and repair through shortages of raw materials. These efforts were further undermined by inadequacies in the system of French finance: seamen and dockyard artificers deserted their posts; payments to French colonies were suspended; commercial companies failed; funding for overseas expeditions fell short; and, despite desperate measures like melting plate owned by the French Crown and East India Company, the French navy was reduced to inactivity.[112]

By contrast, the British economy thrived. British shipping expanded despite the hostilities. This enlarged the pool of seamen available for naval service and helped the navy to enlarge its manpower by nearly 33,000 over the previous war.[113] The Elder Pitt, Earl of Chatham, inspired a confidence in the British war effort that maintained investment in government stocks, loans to the government, subsidies to allies and new ship construction. Over forty ships of the line were built during the war, adding to Britain's ability to mount the blockade of French naval bases.[114] Meanwhile the boards providing logistical support to the British navy and military operations overseas became accustomed to demands on a global scale. The logistical arrangements and innovations of the Navy, Ordnance, Victualling and Treasury boards, and of their subordinate bureaucracies, are examined in succeeding chapters. Common to them all was a new global perspective and a confidence in long-distance operations that had not existed before.

The growth in scale and distance

The new expertise

The new attitude was evinced in successive voyages of exploration from 1763.[115] In 1770 the Admiralty listed five ships 'at the Falkland Islands and discovering': ships involved in the first voyage of James Cook and the crisis in 1770-1 between Britain and Spain over their respective claims to islands which could support ships entering and leaving the Pacific. During the crisis Britain was prepared to commission sixty-four ships, including twenty-two ships of the line. From intelligence, it was known that Spain could mobilise forty-four ships, and France thirty-five, but

---

[112] J. Pritchard, *Louis XV's Navy 1748–1762: A Study of Organisation and Administration* (Kingston and Montreal, 1987), 184–205; see also J. C. Riley, *The Seven Years' War and the Old Regime in France: The Economic and Financial Toll* (Princeton, NJ, 1986).

[113] Davis, *Rise of the English Shipping Industry*, 27, 43, 321, 394.

[114] Kennedy, *The Rise and Fall of British Naval Mastery*, 105–6.

[115] Williams, '"To make discoveries of countries hitherto unknown"'.

that neither could prepare their ships for sea as quickly as Britain. It was a factor that gave the British government confidence to declare war if necessary, and discouraged the French from backing Spain.[116]

Although in 1771 both backed down, in 1778 and 1779 France and Spain joined the War of American Independence and diverted British naval resources from the effort to suppress the American colonists. The latter gained their independence but Britain honed a growing expertise in despatching and maintaining soldiers overseas. According to Charles Jenkinson, by December 1779 there were 38,203 British troops in America (excluding provincial soldiers), 7,059 in Canada, 10,510 in or on their way to the West Indies, 369 in West Africa, 4,930 in Gibraltar, and 2,134 in Minorca.[117] In 1776 the Treasury signed contracts to feed about 60,000 soldiers. By 1781–2 the number had risen to 92,000. In addition, owing to a need to feed loyalist civilians and Indians in North America, in 1782 the Treasury was funding rations for 28,000 more than the army's strength in North America.[118]

The ability to maintain 120,000 people overseas reflected a new logistical capability. These rations had to be shipped to America with every other requirement. Initially the troops were expected to 'live off the land'. However, the evacuation of the army at New York from New Jersey in January 1777 critically reduced the land available for it to live from. As a result, thereafter, 'every bullet and every biscuit' had to be shipped from Britain. As early as 1776, the shipping available to transport troops and foodstuffs had proved insufficient. That summer the Navy Board alone hired 416 vessels, amounting to 128,427 tons of shipping, more than had ever been hired at any one time during the Seven Years' War.[119]

The scale of the supplies and the amount of shipping that was employed raised awareness of logistics to a new height: 'There was during the American War a simple interrelation and interaction between strategy, logistics and shipping which geography imposed upon the British effort.' Initially, according to David Syrett, 'this was not seen in Whitehall and only dimly perceived at the Navy Office'.[120] Senior administrators had at their right hand, however, those who knew what was involved in long-distance supply.

One of these experts was George Teer who, from 1779 until 1789, was the Navy Board's agent for fitting transports in the River Thames. A

---

[116] N. Tracy, *Navies, Deterrence, and American Independence: Britain and Sea Power in the 1760s and 1780s* (Vancouver, BC, 1988), 94–8.

[117] BL, Add. MSS. 38212, fos. 309–10, cited in Syrett, *Shipping and the American War*, 139.

[118] N. Baker, *Government and Contractors*, 4.

[119] Syrett, *Shipping and the American War*, 90–3.   [120] *Ibid.*, 243.

lieutenant in 1767, he had commanded a storeship that supplied provisions and stores to the garrison at Port Egmont in the Falkland islands until 1772.[121] Otherwise undistinguished, Teer had become a captain in 1783, mainly through his practical contribution to the American war effort in the River Thames. By 1789 he believed he had supervised the fitting of 'upwards of six hundred hired transports . . . the greater part in loading with all kinds of naval stores, military equipment, provisions, troops'.

Teer's expertise was called upon in 1789 after British colonisation in Australia demonstrated the possible necessity of sending troops to the antipodes. In 1787, as agent for transports at Deptford, he had fitted the first fleet to Australia. In April 1789 Teer addressed a memorandum to Middleton at the Navy Board on the supply of stores to Botany Bay. Teer looked upon Middleton as his protector, and the Navy Board commissioners Samuel Wallis and Edward Le Cras as his professional masters. Wallis was the circumnavigator of 1766 who, though subject to criticism for lack of initiative,[122] gave the Navy Board actual experience of global voyages.

In 1789 Teer argued that troops were best carried in 44-gun ships.[123] They were 'the best constructed for carrying troops, provisions, naval or military stores, than any other class of his majesty's ships; . . . the troops will be carried in health and fit for service to any climate, which never was the case in hired ships'. He claimed a single ship would carry 'to Gibraltar a thousand men or eight hundred tons of heavy provisions, and return in six weeks, in the presence of all the frigates that Spain can equip at sea at any time of the year'. Such 44-gun ships were later used for troop deployments in the French Revolutionary War.

He went on: assuming twenty-four ships, all of 44 guns, were employed, they would measure about 21,000 tons and demand the services (at the rate of 10 officers and men to 100 tons) of 2,100 seamen. They would 'carry 12,000 troops to the East Indies with their baggage, camp equipage, and return back in a year; to the West Indies 18,000 troops on a pressing occasion, and return back in four months; . . . to Halifax or Canada in the summer, the same number' of months. He believed ships

---

[121] For reference to Teer's private letter book and logs of his voyages to the Falkland Islands, see R. Knight, 'The first fleet – its state and preparation 1786–1787' in *Studies from Terra Australis to Australia* (Canberra, 1989), 121–36, 256–62, note 45.

[122] G. Williams, *The Great South Sea: English Voyages and Encounters 1570–1750* (Newhaven and London, 1997), 272.

[123] On 29 September 1787, Teer had already provided Middleton with an appraisal of the accommodation a 44-gun ship could provide for troops: NMM, MID.7/1. See also MID.1/184.

could reach Botany Bay in six months, and make the round trip 'much within a year':

On the whole a copper bottomed ship may be expected to arrive at the Cape, allowing for a weeks stop at Tenerif to refresh and water,... in about 9 or 10 weeks; suppose at the Cape or False Bay according to the season a week or ten days more, a single ship may from thence to Botany Bay in or about six weeks. Six weeks to discharge and ballast, from Botany Bay to Cape Horn about six weeks.

There a ship could stop at Staten Island or, two days further, in the Falkland Islands. After perhaps two weeks in the Falklands, it was then eight weeks home. 'Should a ship miss Falklands also, by steering well to the eastward, it is almost certain at any season of the year to arrive at St Helena in or about three weeks; and from that place to England is a well known passage of from 6 to 9 weeks...'

In Teer's experience, scurvy was no longer a problem: over a period of five years between 1767 and 1772, the storeship *Florida* had not lost a man to the disease on three trips to the Falklands. On Staten Island at Cape Horn or in the Falklands, penguins permitted crews to refresh 'in such a manner as to be in a better state of health at sailing home than on going out from England'.[124] Hence, through experience, and well-staged stops, he argued that troops could be carried to, and supplied in, the most remote corners of the earth.

There is every reason to believe Teer's observations were heeded. The first fleet had been accompanied by small vessels, the *Sirius* and *Supply*, a 22-gun frigate and a brig-rigged sloop. Following Teer's memorandum, in July 1789 the second fleet to Australia was accompanied by two 44-gun warships converted to storeships, the *Guardian* and *Gorgon*. After their return, the same ships sailed independently for Australia in 1789 and 1791, respectively.[125]

The great expeditions

Experiences during the American War of Independence of shipping food and soldiers overseas gave rise to new responsibilities at the beginning of the French Revolutionary War. In 1793 responsibility for despatching food to the army overseas was placed with the naval Victualling Board. In 1794 a Transport Board was formed, which would hire all transports and make them available to whichever department of government required their use. Meanwhile, supplies to the army in the field overseas were

---

124 NMM, MID.7/1, G. Teer to Middleton, 25 Apr. 1789.
125 *Select Documents in Australian History 1788–1880*, ed. G. M. H. Clark (Sydney, 1950), 42; C. Bateson, *The Convict Ships 1787–1868* (Sydney, 1983, repr. 1985), 124, 131.

managed by the army Commissariat, responsible to the Treasury. The difficulties preceding the creation of these new responsibilities and the operation of these departments will be discussed in later chapters. Suffice to say here that the new responsibilities contributed to the readiness to mount expeditions overseas of growing scale after 1793.

In 1794–5, over 22,000 soldiers were sent to the West Indies. General Grey's expedition against the French islands in 1794 was intended to employ 8,590 officers and men. By April 1794, after the capture of Martinique and St Lucia, sickness and garrison duty reduced Grey's force to 6,810 effective rank and file. With never less than 2,000 men sick and a rapidly rising death toll, he had only 4,761 fit for duty in June 1794 and only 2,065 in September. In consequence, in December 1794, an intended assault on Guadaloupe was abandoned. Between October 1794 and August 1795, 13,500 more men were sent out to the West Indies, but they arrived too late and in contingents too small to aid Grey's main force.[126]

The necessity for an even greater number of soldiers in the initial assault force made the 1795 expedition to the same destination 'the largest single long-distance overseas expedition to depart from British shores'. An army of more than 30,000 men was envisaged. However, in June 1795 there were only 31,154 men in the 79 line regiments in the British Isles and, after home defence and minor deployments, fewer than 20,000 were available for the West Indies. Still, 16,000 were planned for the Windward and Leeward Islands, and 15,114 sailed from Southampton with General Abercromby in the autumn of 1795. Another 13,500 men were intended for an attack on St Domingue, and 7,500 sailed from Cork in early December 1795, to be followed shortly by three times as many. Between 9 December 1795 and 20 March 1796, 30,818 rank and file departed for the West Indies.

The size of these expeditions demanded extraordinary logistical efforts. The 1795 expedition demanded about 100,000 tons of shipping – 'something like a seventh of all British merchant ships capable of long-distance voyages'. Just the pork, beef, butter and flour shipped in August 1795 amounted to 5,824 tons. The results too were extraordinary. Between 1793 and 1801 British forces took French Tobago, Martinique, St Lucia, St Vincent, briefly held Guadaloupe in 1794, and maintained a foothold on St Domingue until 1798. The ability of France to turn European neutrals into allies added further conquests. In 1797 Spanish Trinidad was taken, and between 1796 and 1801 Dutch Demerara, Surinam, Curaçao

---

[126] Duffy, *Soldiers, Sugar and Sea Power*, 70–1, 100, 130, 153; Barnett, *Britain and her Army*, 234–5.

and St Martins. In 1800 Sweden's St Bartholomew was taken, and in 1801 Denmark's St Thomas, St Johns and St Croix. Prevented from trading with their mother countries, the plantation owners and merchants of these islands were obliged to sell to British shippers and the value of imports from the region to Britain rose sharply.[127]

The campaign in the West Indies was complemented by campaigns too in the east. As both President of the board of Control for India and Secretary for War and Colonies, Dundas was able to employ the resources of the East India Company against the French in the Indian Ocean and the sub-continent. The subordination of Holland to France in 1795 resulted in a regular army expedition from Britain to secure the Cape of Good Hope. But it was a company force, despached from Madras with the help of the British navy, which took control of the Dutch bases in Ceylon: Colombo and Trincomalee. In 1798, when Napoleon invaded Egypt – and on the premise that the French army might reach India – Dundas permitted Richard Wellesley, Earl of Mornington and Governor-General of India, to launch an assault on Mysore, the sultan of which, Tipu Sahib, had always favoured the French. Over the next few years, Wellesley brought the entire Carnatic region of India under company control, settling the balance of power in the south and west of India. The expansion of British control in India and the Indian Ocean provided for the security of both company shipping and company revenues.[128]

The success of these operations cannot be separated from other British naval achievements. The Mediterranean was abandoned in 1796 when Spain took sides with France, only to be reoccupied in 1798 after the battle of the Nile. The great battles of the war helped to reduce the naval forces of France, Spain, Holland and Denmark, though less than usually assumed. The four states lost 558 vessels, including 85 ships of the line, during the war. But most were lost from wreck or foundering. Only 9 per cent of their total losses resulted from battle, which accounted for only 36 per cent of the ships of the line. By contrast, Britain lost 200 vessels, only 54 in action. Britain's total loss included 20 ships of the line, with none lost in battle.[129] Numerous captured ships were added to the British fleet. With further additions from new building, the British fleet steadily developed a greater margin of strength over her enemies.

The deployment of the greater part of this strength in home waters made possible the success of overseas operations. The navy served as their shield. The navy also protected expeditions nearer home: that to the

[127] Duffy, *Soldiers, Sugar and Seapower*, 221–315.
[128] Fry, *The Dundas Despotism*, 219–22.   [129] Clowes, *The Royal Navy*, IV, 548–61.

Helder in north Holland in 1799 with 35,000 British and Russian troops to relieve French pressure on Austria; and Abercromby's expedition to Egypt in 1800–1 with an intended 15,000 men to destroy Napoleon's army of invasion. Supply in European waters was easier than that overseas as transports could make numerous trips. However, the more distant Egyptian campaign posed serious supply problems, which are examined in the final chapter.

The global deployment of British forces

Necessary imperialism

Between 1793 and 1801, nearly 89,000 white officers and men served in regular military units in British pay in the Caribbean. However, 62,250 of them died: that was 70 per cent of all the British soldiers sent to the region.[130] Yet in 1802 at the Peace of Amiens, all the territories seized overseas, except Spanish Trinidad, were returned. The sacrifice of so many men for little long-term advantage contributed to a change of policy towards overseas expeditions in the Napoleonic War.

After the invasion threat was lifted in 1805, Britain's naval forces deployed around the world – in the Mediterranean, the West Indies, on the coast of North America and in the Indian Ocean – were all enlarged. After the Treaty of Tilsit between France and Russia in 1806, a significant British naval force was located in the Baltic where the supply of naval stores was threatened by the French occupation of the southern Baltic shore. From 1808 another squadron was located off Portugal to defend and support the military campaign in the Iberian Peninsula.[131] This global distribution of naval forces facilitated local military expeditions. But large-scale amphibious operations were sanctioned with caution, especially after the debacle at Buenos Aires.

To protect British trade to the Indian Ocean, in 1805 6,500 men were sent to the Cape of Good Hope, where the Dutch colony on the southern tip of Africa was again seized. But from here in 1806, on his own initiative, Commodore Sir Home Popham deployed 1,600 soldiers to take Buenos Aires. He thought it would open a window into South America for British trade. But the British occupying garrison was soon forced out by local popular revolt and the government had to send 2,000 troops in reinforcement, who took Montevideo. Buenos Aires was again

---

[130] Duffy, *Soldiers, Sugar and Seapower*, 162, 170, 179, 181, 191, 196, 180–4, 191, 196, 254, 328–33.
[131] TNA, ADM. 8/99, 100.

assaulted but, with 3,000 casualties, the whole of the River Plate region was evacuated.[132]

This reminder of the great losses overseas in the 1790s contributed to the new policy of economy in assaults upon enemy colonies. Napoleon helped in being relatively inactive overseas. At the same time, British troops were needed in Europe. Thus, when enemy colonies were attacked, local troops were generally used, and used sparingly, while the occasional expedition from Britain was highly organised, professionally conducted and usually rapid.

Early in the Napoleonic War, off Canada, the small French islands of Miquelon and St Pierre were taken to eliminate bases of privateers. So too was St Lucia which could be used for a British base to blockade French vessels in Martinique. Tobago was taken under British control, having been British prior to 1783 and again in 1793–1802, so containing British settlers and investments. British financial interest and the value of their sugar, coffee and cotton prompted the seizure of the Dutch Guiana settlements of Demerara, Essequibo and Berbice, which had been taken in 1796 and received some £5 million from merchant houses in Liverpool, London, Glasgow and Bristol. In 1807 the Danish islands of St Croix, St Thomas and St Johns were taken to prevent them becoming havens of privateers.[133]

Only 166 soldiers, with seamen and marines, were used to take the French fort of St Louis in the Senegal River in July 1809. This removed France's only remaining African territory. There were complaints of privateers operating from other French colonies, in particular the large French West Indian islands. But the government resisted pressure to secure their products for the British economy at a time when the European market was virtually closed on account of Napoleon's continental system. Eventually, when the government did succumb to pressure to seize the largest islands, this was achieved with local troops and others from Bermuda and Nova Scotia. After Napoleon reinforced his garrison, Martinique was taken with 10,000 men in February 1809. Guadeloupe was also taken in February 1810, with the small islands of St Eustatius and St Martins.

By then, the only remaining French privateer bases outside Europe were in Mauritius, Bourbon and Rodriguez in the Indian Ocean, from where trade along the Coromandel coast of India was virtually eliminated in 1807. After four large French frigates arrived in 1809, the French

---

[132] *The Royal Navy in the River Plate, 1806–1807*, ed. J. D. Grainger (NRS, 1996), 7–14; Hall, *British Strategy in the Napoleonic War*, 144–8.
[133] Hall, *British Strategy in the Napoleonic War*, 95–6, 104, 108–9.

islands were taken in operations from the Cape of Good Hope in 1810. The threatened arrival of another French squadron prompted the seizure of the Dutch Moluccan islands and of Java in 1810–11.[134]

The largest expeditions of the Napoleonic War were confined to the waters of Europe and aimed more directly at confining and weakening Napoleon. In the Mediterranean, Malta had been retained since the Peace of Amiens. Sicily was also garrisoned with a large force in early 1806. The island produced large quantities of grain, was the refuge of the independent Neapolitan court, and served as a base to attack the French in Italy and the Ionian islands. From there too, Alexandria might be occupied if Napoleon decided to repeat his 1798 invasion of Egypt.[135]

The largest expeditions were short-distance affairs. In 1807, 19,000 men were used to attack Copenhagen to prevent Demark joining the Franco-Russian alliance against Britain and to prevent the closing of The Sound by the Danish fleet. In 1809 44,000 men were sent to the island of Walcheren in the Scheldt estuary to divert French pressure from the Austrian campaign in the Danube valley, to terminate the French naval building at Antwerp and to prevent the river being used for an attack on the coast of England. At the same time, about half that number of soldiers was being used to support Britain's Iberian allies in Portugal.[136] The subsequent enlargement of the forces under Wellington in the Iberian Peninsula to about 54,000 constrained deployments elsewhere. Understandably so, for by 1811, aside from Wellington's Peninsular army, garrisons on Mediterranean islands and in Canada, 'there were 76,000 regular troops scattered in outposts extending from Africa through the Orient to Australia'.[137]

The war with America in 1812 forced the British government to think about a large-scale deployment of British troops outside Europe. And, indeed, between 1811 and 1813 the troops in North America grew from 7,625 to more than 20,000. Yet, on the whole, these reinforcements did not come from Europe. They came from the West Indies where, after Napoleon's disastrous Russian losses, no threat was expected. Few more troops were released for the American war until Napoleon abdicated and, even in 1814, they failed to arrive in the numbers expected. The British naval Commander-in-Chief, Sir Alexander Cochrane, was led to believe 30,000 men would be released for his campaign against American coastal cities. In the event, in July 1814 he received only 3,700 men for

---

[134] *Ibid.*, 96, 184–9.   [135] *Ibid.*, 86–7, 142–3, 156, 194.

[136] *Ibid.*, 157–63, 175–9.

[137] D. Gates, 'The transformation of the Army 1783–1815' in *The Oxford History of the British Army*, ed. D. G. Chandler and I. Beckett (Oxford, 1994), 138; Hall, *British Strategy in the Napoleonic War*, 184–98, 212.

the attack on Washington, and in December he had only 7,500 for that on New Orleans.[138]

For soldiers were still much needed in Europe, if only for diplomatic purposes. In March 1814, the European allies were kept together by the Treaty of Chaumont which agreed all signatories would maintain 150,000 men in the field until the war was won.[139] The achievement of peace in Europe thus took priority over the conduct of British affairs overseas. The same priority shaped the Treaty of Vienna in June 1815. Captured colonies were used to purchase the return of the French monarchy and of France to her 1792 frontiers, leaving Britain with the strategically important islands of Tobago, St Lucia and Mauritius. Another settlement with the Dutch agreed the attachment of Belgium to Holland and return of the Dutch East Indies, but Britain retained Ceylon, the Cape of Good Hope and trading centres in Guiana. In addition, the Treaty of Vienna permitted Britain to keep Malta and the Ionian islands.

The system of global supply

These wartime acquisitions, with Canada, Australia, New Zealand and India, established Britain as the world's principal maritime power in Europe. That situation reflected the exhaustion of the other European states but also Britain's ability to mount overseas expeditions, establish garrisons and station squadrons on a global scale. This ability to project the state overseas came from developing practice throughout government. Collaboration between departments was necessary. Efficiency at their individual tasks was important. So too was the supply of munitions, ships and food by the private sector.

By July 1813, 135,889 seamen and marines were recorded as serving on 624 British ships and vessels which were distributed between 15 stations, allotted to particular functions or remained 'unappropriated' (see table 2.2). Of the 624 vessels, 27 per cent (containing 28 per cent of the men) were committed to European stations, including the Baltic, while 41 per cent of the ships (containing 48 per cent of the men) were located in the Mediterranean and on stations as far distant as the East Indies and South America. The weighting of ships and men to stations indicates British naval priorities.[140]

---

[138] R. Gardiner, ed., *The Naval War of 1812* (London, 1998), 148, 174; R. Morriss, *Cockburn and the British Navy in Transition: Admiral Sir George Cockburn, 1772–1853* (Exeter, 1997), 100–3; Hall, *British Strategy in the Napoleonic War*, 197–8.
[139] Christie, *Wars and Revolutions*, 322–6.   [140] TNA, ADM. 8/100.

Table 2.2 *The disposition of the British navy and its seamen on 1 July 1813*[a]

Station	Rate	Number of ships	Number of men
European waters			
Channel Fleet	1	3	2511
	3	13	7820
	5	8	2427
	Sloops	6	610
	Gun brigs	3	150
Ireland	5	2	599
	Sloops	10	1149
	Gun brigs	3	150
Downes	Stationary ship	1	280
	Sloops	10	868
	Gun brigs	7	405
Jersey and Guernsey	6	1	140
	Sloops	2	221
	Gun brigs	5	255
Texel and Scheldt	2	1	738
	3	11	6760
	5	2	499
	Sloops	9	715
	Gun brigs	7	350
Portugal	3	1	491
	5	2	589
	6	4	646
	Sloops	8	756
	Gun brigs	5	252
	Receiving ship	1	380
Baltic	3	8	4664
	4	1	345
	5	2	579
	6	2	294
	Sloops	14	1285
	Bombs	3	212
	Gun brigs	15	772
	Total	**170**	**37,912**
Mediterranean	1	4	3424
	2	4	2952
	3	21	12750
	5	19	5483
	6	6	834
	Sloops	28	3067
	Bombs	3	223
	Gun brigs	4	160
	Receiving ship	1	121
	Hospital ship	1	121
	Total	**91**	**29,135**

(*cont.*)

Table 2.2 *(cont.)*

Station	Rate	Number of ships	Number of men
West Indies			
Jamaica	3	2	1180
	5	1	340
	6	5	767
	Sloops	6	726
	Gun brigs	2	100
	Receiving ship	1	70
Leeward Islands	3	2	1180
	4	1	343
	5	7	1998
	6	3	363
	Sloops	20	2228
	Gun brigs	6	290
	Total	**56**	**9585**
North America			
American Coast	3	11	6491
	5	16	4675
	6	2	296
	Sloops	25	2525
	Schooners	3	90
	Receiving ships	2	126
	Prison ship	1	97
Newfoundland	3	1	590
	5	4	1223
	6	2	242
	Sloops	4	423
	Cutter	1	42
	Prison ship	1	11
	Total	**73**	**16,831**
East Indies	3	2	1180
	5	11	3105
	Sloops	5	605
	Hos. & Rec. ship	1	62
	Total	**19**	**4952**
Cape of Good Hope	3	1	491
	5	3	852
	Sloops	2	242
	Total	**6**	**1585**
South America	3	1	590
	5	5	1479
	6	3	431
	Sloops	3	363
	Cutters, etc.	2	95
	Total	**14**	**2958**

Table 2.2 *(cont.)*

Station	Rate	Number of ships	Number of men
British Isles			
Convoys and Particular Services	3	4	2460
	5	10	2950
	6	3	417
	Sloops	10	1165
	Cutters	4	190
At home ports: Plymouth, Portsmouth, Sheerness, Yarmouth and Leith	Receiving ships	6	1202
	3	2	1180
	5	5	1329
	6	1	135
	Sloops	31	2749
	Gun brigs	16	767
Troop ships	3	3	600
	4	2	350
	5	10	1270
	6	3	315
Stationary ships	Yachts	6	117
	Prison ships	12	1152
	Prison Hosp. ship	1	58
	Hospital ships	4	230
	Convalescent ship	1	70
	Slop & con. ships	3	367
	Station. & rec. ships	8	580
	Tenders	3	117
Unappropriated	2	1	738
	3	9	5310
	4	2	930
	5	14	4196
	6	4	484
	Sloops	12	1391
	Bomb	1	67
	Schooners	4	155
	Total	**195**	**32,941**
	Grand total	**624**	**135,899**

[a] TNA, ADM. 8/100.

At almost the same time, in August 1813, a total of 156,448 soldiers – rank and file and non-commissioned officers – were stationed outside Britain and Ireland in twenty-five different locations around the world. In January 1813, there were even more (see table 2.3): 68,240 (29%) were committed to Europe. But 31,533 (13%) were in the Mediterranean; 21,236 (9%) in the West Indies; 13,552 (6%) in North America; 29,534 (13%) in the Indian sub-continent and Ocean; 6,071 (3%) in South and West Africa; and 1,250 (0.5%) in Australia.[141]

These troops were supplied in part by the army's Commissariat organisation which spawned new branches wherever troops were deployed. Some food was sent to the army garrisons by the Victualling Board, everything sent from Britain being shipped by the Transport Board. By the time of the Napoleonic War, army and navy victuallers were routinely being despatched to the southern hemisphere. We know because, from May 1806, the Admiralty demanded from the Victualling Board on the first day of each month a list of the victuallers likely to require convoy during the course of that month. From these lists may be gleaned knowledge of the developing geography of British military power and the capabilities of bureaucracy responsible for supplying Britain's armed forces.

In October 1809 the Transport and Victualling Boards despatched forty-three victuallers. Twenty-eight served the navy and fifteen the army. The destination of twenty army and navy supply ships was the Walcheren. Otherwise, naval victuallers were despatched to the Baltic (three) and Gibraltar (four) while army victuallers went to Gibraltar (three), Newfoundland, Bermuda, the Bahamas (two), Sierra Leone, Goree and Honduras.

In February 1811 the Iberian Peninsula had become the principal focus of supply. Thirty victuallers went there that month, including fourteen destined for Gibraltar and eight for Portugal. Otherwise, one sailed for Sicily and Malta, one to Barbados, two to the Cape of Good Hope, and four to the Indian Ocean.

By September 1813 the focus had shifted once more, to North America. That month, twenty-five victuallers were despatched: eight for the army in Canada and three for the navy at Newfoundland; the navy needed six others at Gibraltar and Cadiz, four in the Baltic, one in the Leeward Islands and one at Rio de Janeiro. Meanwhile the army received one at Madeira and one in Honduras.[142]

---

[141] Hall, *British Strategy in the Napoleonic War*, 212. Here the locations of troops are listed by region in table 2.3.

[142] NMM, ADM. DP/29, 2 Oct. 1809; ADM. DP/31A, 1 Feb. 1811; ADM. DP/33B, 1 Sept. 1813.

Table 2.3 *The disposition of the British army overseas in 1813 and 1815[a]*

	1 Jan. 1813	1 Jan. 1815
Europe		
Anholt, Denmark	322	–
Cadiz	2,918	–
Gibraltar	3,783	2,891
Heligoland	382	210
Spain and Portugal	60,835	–
Flanders	–	18,868
Total	**68,240**	**21,969**
Mediterranean		
Malta	4,574	2,411
Sicily and Ionian Islands	26,979	–
Total	**31,553**	**2,411**
West Indies		
Honduras	305	292
Jamaica	3,915	–
Leeward and Windward Islands	16,380	–
Curaçao	636	664
Total	**21,236**	**956**
North America		
Bahamas	950	815
Bermuda	838	–
Canada	7,887	27,505
Newfoundland	–	–
Nova Scotia	3,877	6,794
New Orleans	–	8,376
Total	**13,552**	**43,490**
East Indies		
Bengal	5,444	6,926
Bombay	3,032	2,975
Ceylon	4,629	5,240
Madras	9,977	9,690
Mauritius	4,101	3,161
Java	2,351	1,406
Total	**29,534**	**29,398**
Africa		
Cape of Good Hope	4,517	4,980
Sierra Leone, Senegal, Goree	903	–
Madeira	651	–
Total	**6,071**	**4,980**
Australia		
New South Wales	1,250	628
Total	**1,250**	**628**

(*cont.*)

Table 2.3 *(cont.)*

	1 Jan. 1813	1 Jan. 1815
British Isles		
South Britain	42,117	36,925
Scotland	9,127	3,354
Ireland	12,492	26,806
Total	**63,736**	**67,085**
Grand total	**235,172**	**170,917**
Army overseas	171,436	103,832
Percentage overseas	72.9%	60.75%

[a] TNA, WO.17/2814, Monthly summaries and annual abstracts of effectives and casualties of British and Foreign Corps at home and abroad.

Voyages of this nature added to the experience of ship masters and accrued to the benefit of ship owners. Michael Henley and Son, who enlarged their fleet from nine ships in 1790 to twenty-two in 1810, chartered between 60 and 70 per cent of their vessels to the Transport Board. Their charters took Henley ships to the West Indies in the middle of the 1790s; to the Mediterranean during the Egyptian campaign of 1800–1; to the Baltic in 1805–6; to South America, the Cape of Good Hope and India in 1807–8; and to the Mediterranean once more in 1809–11. The experience gained by their masters paid off. Subsequently, the Henleys extended their voyages under private charter from the North Sea to the Baltic, to Quebec, New Brunswick, Honduras, the West Indies, Surinam and Demerara.[143]

Some of these voyages were under charter to commercial contractors serving the Victualling, Ordnance and Treasury Boards. Merchants of all nationalities tried for supply contracts. In 1810 Mr Hippolyto Joseph de Costa of 'the Brazils' – a property owner in Rio Grande province – offered to supply the British navy on the West India and Brazil stations with salt beef, portable soup and other provisions. He also wanted to supply fresh provisions at Rio de Janeiro. The Victualling Board demurred, claiming beef and pork was insufficiently cured in that country to survive long storage at sea and that anyway the Portuguese government already supplied fresh beef at Rio de Janeiro free of charge.[144]

[143] Ville, *English Shipowning during the Industrial Revolution*, 6, 56–8.
[144] The board also wanted 'to promote a fair and open competition for contracts', a general practice from which departure would be warranted only if 'very peculiar circumstances pointed out its expediency': NMM, ADM. DP/30A, 30 Aug. 1810.

Despatching between 25 and 43 victuallers a month during the last half of the Napoleonic War, some to the other side of the world, required the Transport Board to have under charter sometimes as many as 400 vessels each year.[145] This was between 8 and 10 per cent of the British ocean-going merchant fleet. Resources were stretched but never exhausted, for the number of ships registered in Great Britain rose by about 50 per cent between 1793 and 1815, their total tonnage by 70 per cent.[146] Naturally enough, freight rates went up with the onset of war and there was growth of about 50 per cent over the whole length of the French Revolutionary and Napoleonic Wars, but there was no period of marked instability in those rates.[147]

This supply of shipping in the private sector, drawn upon as required, permitted over 45 per cent of the men in Britain's armed forces in 1813 to be dispersed outside Europe. This distribution denied Britain's enemies the use of their colonial territories, protected those of Britain, and permitted offensive operations against powers like the United States of America who challenged British policies. Although the Napoleonic War saw few large-scale long-distance expeditions outside Europe, that capability existed, developed during the wars of American Independence and the French Revolution. That capability in turn rested on the expertise of men who, since the mid eighteenth century, had become practised in the business of despatching transports into the southern hemisphere, and the Indian and Pacific oceans. Although operating in different environments, the capability was shared by bureaucrats, naval officers, seamen, ship masters and owners. In wartime, the secure despatch of expeditions overseas rested on effective defence in home waters, which in turn demanded an appropriate strategy, armed services capable of implementing that strategy, and a governmental organisation that gave expression to those with the vision, experience and administrative talent. Yet none of this, of course, would have been possible without finance, the supply of which is examined in the following chapter.

---

[145] Ville, *English Shipowning during the Industrial Revolution*, 12; see TNA, ADM. 108/148–67, Ships and Freight Ledgers.

[146] Mitchell and Deane, *Abstract of British Historical Statistics*, 220; see also C. N. Parkinson, ed., *The Trade Winds: A Study of British Overseas Trade during the French Wars 1793–1815* (London, 1948), 83.

[147] Ville, *English Shipowning during the Industrial Revolution*, 171, Appendix J, 'Freight rates in the transport service, 1775 to 1822'.

3 Economy and finance

The geographical expansion of the British state in the eighteenth century benefited from a pre-existing maritime economy of global dimensions, within which was constructed a structure of trade and navigation laws intended to promote trade between mother country and colonies. Trade within and beyond the empire, built upon networks of credit, helped to support the state's system of credit, debt and taxation, of which some of the leading beneficiaries were contractors operating in the colonies. The maritime economy and financial system proved resilient and adaptive to wars of growing scale. Indeed, the combination gave Britain her special power. Moreover, reform and rising trust enhanced confidence in Britain's economic capacity to sustain war. But the process of reform was a long one. Private profiteers had to be checked and control of public expenditure shifted slowly from Parliament to the Treasury. By their spending, the military departments during the second half of the eighteenth century were able to exert considerable influence on the growth of the British economy, for they added to domestic demand, had some flexibility in the management of their money and audited their own accounts. Yet from the late 1790s public insistence on individual responsibility and a growth of Treasury control gradually checked the power of the bureaucracy and counteracted fears of the uncontrolled growth of debt.

The British maritime economy

The British maritime economy had three important advantages over a land-based economy. Firstly, it relied for much of its long-distance transport on rivers and the sea which cut the costs of conveyance of bulk cargoes dramatically. One ship carried what twenty to thirty wagons did on land and, per ton, one seaman did the work of six to twelve drivers and labourers on shore. Moreover, as most other costs, like port fees, were sustained at the beginning and end of a voyage, the cost per ton mile declined as voyages became longer.[1] The economy of water transport made savings for manufacturers, merchants and other industrialists that were passed on to consumers, making for greater purchasing power and consumption. The savings on a commodity like coal in the second half of the eighteenth century benefited both industrial and domestic purchasers.[2]

Secondly, the sea supplied a variety of markets, which was vital in wartime and permitted flexibility in the development and use of commercial connections. Not surprisingly, overseas trade grew more rapidly than the internal economy. Especially important were colonial branches like the American and West Indian trades from the late seventeenth century,[3] but also non-colonial trades like those in naval stores from the Baltic, and those into central and South America in the early nineteenth century. Overseas demand never exceeded domestic demand but it was vital to the development of manufacturing and commerce.[4]

Thirdly, Britain's economy possessed colonies which, mercantilists believed, added to the labour and materials of the mother country for which they also served as a market. This imperial economy was encouraged – in the view of many mercantilist advocates until at least 1815 – by the Navigation Laws.[5] The most important of these laws were those of 1651 and 1660 which had been aimed at removing the Dutch from the English carrying trade and required goods to be carried to and from England and her colonies in English ships.[6] The 1660 Act also enumerated certain colonial products that could only be carried to England or another English colony, and was followed in 1663 by the Staple Act which stipulated that specified goods entering the colonies had to come from an English port. Colonial trade thus sheltered shipping from foreign competition and reinforced domestic demand for British manufactures, encouraging innovations that overcame bottlenecks in production.[7]

---

[1] T. S Willan, *The English Coasting Trade, 1600–1750* (Manchester, 1938), xiii–xvi, 189–93; P. Colquhoun, *A Treatise on the Commerce and Police of the Metropolis* (London, 1800), 11.

[2] Ville, 'Total factor productivity in the English shipping industry'.

[3] N. Zahedieh, 'London and the colonial consumer in the late seventeenth century', *EconHR* 2nd ser. 47(1994), 239–61; and 'Economy' in *The British Atlantic World, 1500–1800*, ed. D. Armitage and M. J. Braddick (Basingstoke, 2002), 51–68.

[4] R. Davis, *The Industrial Revolution and British Overseas Trade* (Leicester, 1979), 9–10.

[5] K. Morgan, 'Mercantilism and the British empire, 1688–1815' in *The Political Economy of British Historical Experience, 1688–1914*, ed. D. Winch and P. K. O'Brien (Oxford, 2002), 165–91.

[6] For the origins of the 1651 law, see J. E. Farnell, 'The Navigation Act of 1651: the First Dutch War and the London merchant community', *EconHR* 2nd ser. 16(1963–4), 439.

[7] Price, 'The imperial economy, 1700–1776'.

The growth of global trade

The Navigation Laws were loosely enforced where British and colonial interests benefited from trade into French, Spanish and Dutch territories. Indeed, attempts after 1763 to tighten trade regulations to enhance revenue were regarded by American colonists as anachronistic, misconceived and counter-productive.[8] The British government was out of touch with American attitudes.[9] It was slow too in catching on to a growing desire for less regulation among merchants and ship owners excluded from trades monopolised by chartered companies. In 1776, Adam Smith, the economist, criticised the Navigation Laws as protective and thus inhibitive of free trade and profit-making.

But merchants and ship owners could not avoid the influence of international trade where improvements in ship size and manning ratios were necessary to make their vessels economical and competitive.[10] These improvements enhanced the potential profitability of long-distance trades and the shipping engaged in them tended to grow. By contrast the relative amounts of tonnage engaged in trade over shorter distances declined. Early in the eighteenth century, nearby Europe, Spain and Portugal were the destination of over half of all the ships that cleared the port of London. However, 100 years later only a quarter of the voyages from London, 14 per cent of London's ship tonnage, and about 10 per cent of the value of London's trade was with nearby Europe and the Iberian peninsula (see table 3.1). This decline was matched in other ports, and Europe's share of Britain's total overseas trade fell from 74 per cent in 1713–17 to 33 per cent in 1803–7.[11]

Instead, by 1755, nearly half of all the shipping sailing from English ports was engaged in long-distance oceanic trades. Domestic demand fostered these trades and was the initial engine of their growth. But, especially after 1783, foreign demand enlarged re-exports and exports, in particular manufactured cotton goods based on cotton imports. These supplemented Britain's traditional export, woollen cloth.[12] So also did other imports. From North America came naval stores, sugar, cotton, rice and tobacco, four-fifths of the latter being re-exported into Europe. From the West Indies came mahogany and other hard woods, but above all sugar, imports of which doubled in the two decades after 1756. From

---

[8] Baugh, 'Maritime strength and Atlantic commerce'.
[9] I. R. Christie, *Crisis of Empire: Great Britain and the American Colonies, 1754–1783* (London, 1966), 8–14.
[10] R. Unger, 'Warships, cargo ships and Adam Smith'; Ville, 'The growth of specialization in English shipowning, 1750–1850'.
[11] Cain and Hopkins, *British Imperialism*.   [12] *Ibid.*, 88–9.

Table 3.1 *The regional distribution of London's shipping and trade values in 1797*[a]

Region of trade	% of ship voyages	% of total tonnage	% of import value	% of export & re-export value
Northern Europe	49	44	21	36
Nearby Europe	15	6	3	9.5
Spain & Portugal	10	8	5.2	2.3
Mediterranean	2	3	1.7	0.5
East India	2	8	28	15
West India & Africa	12	20	31	16
North America	7	9	7.8	20
Whaling trade	2	2	1.4	0

[a] Based on figures in 'A general view of the whole commerce and shipping of the River Thames' in Colquhoun, *A treatise on the commerce and police of the metropolis.*

India and China, among a great variety of exotic goods, came tea, silks and porcelain in particular. The hold space required for these goods doubled the tonnage of the East India Company's ships between 1756 and 1776.[13]

With colonies stretching along the eastern seaboard of North America, by the mid eighteenth century Britain had built up an oceanic trading empire stronger than that of either of her principal mercantile rivals, France and Holland.[14] Chains of credit linked provincial producers to colonial consumers; colonial merchants became partners in banks founded in the ports of Bristol, Liverpool and Glasgow; transatlantic relations increased and coordinated markets, products and capital, contributing to a growth in size of British merchant houses.[15] The Peace of Paris in 1763 consolidated Britain's American territories, established British control of their eastern sea border and removed the French from North America. It sowed the seeds of both rebellion and revenge but it also reinforced the transatlantic trade and stimulated the shipping industry.

[13] Davis, *Rise of the English Shipping Industry*, 40–2.
[14] K. Morgan, *Slavery, Atlantic Trade and the British Economy, 1660–1800* (Cambridge, 2000), 13–17.
[15] K. Morgan, 'Atlantic trade and British economic growth in the eighteenth century' in *International Trade and British Economic Growth in the Eighteenth Century*, ed. P. Mathias and J. A. Davis (Oxford, 1996), 8–27.

The shipping industry

Between 1748 and 1788 the English shipping industry doubled in size, and in the subsequent thirty years it doubled again.[16] (See table 3.2.) The growth was most marked in the three most distant trades – those to America, the West Indies and the Far East – and in that to the Baltic and northern Europe. The latter was important because it was stimulated by the growth in the shipping industry and in the British navy. Both required raw materials from the Baltic to build and maintain ships. Ralph Davis, quoting William Sutherland in 1711, observed that 'a ship built entirely of foreign materials – foreign timber, iron, pitch and tar, hemp – would call for the transport services of as many as two or three ships of its own size to carry the materials'. At an average growth rate of 2 or 3 per cent per annum, a merchant fleet 'would require to import (without allowing for naval demands) nearly a hundred tons of "naval stores" for every two hundred tons of goods which it brought in for other purposes'.[17] In this way, the shipping industry was self-stimulating.

The growth in the quantity of shipping produced competition and held down freight rates, reducing the cost of shipping cargoes. This was vital to industries consuming materials like timber, iron and coal in bulk. Cheap coal was a major factor in the profitability of a wide range of industries at the heart of the industrial revolution, and the coal trade employed a greater volume of shipping than any other trade (see table 3.3). Frequently the tonnage of coal shipped from the ports of north-east England exceeded the total volume of English imports.[18] The shipping involved in this coastal trade was thus as important as that in the transatlantic trades, especially as colliers were also used in trade to the Baltic for naval stores and timber for use in canal and mill building.[19]

The growth in importance of the trade to the Baltic and northern Europe is evident in the voyages of ships using the port of London. In 1686 only 15 per cent of ships cleared that port for the north of Europe.[20] But by 1797 49 per cent of all ship voyages were made to that region (see table 3.1). Return bulk cargoes accounted for only 21 per cent of the value of London's imports. The trade nevertheless employed 44 per cent of London's shipping tonnage excluding coasters, and by far the greatest proportion of English shipping – 30 per cent of English tonnage in 1788 – was owned in the port of London.[21]

---

[16] Davis, *Rise of the English Shipping Industry*, 40, 43.   [17] *Ibid.*, 19–20.
[18] Ville, 'Total factor productivity in the English shipping industry'.
[19] C. E. Fayle, 'The employment of British shipping' in *The Trade Winds*, ed. Parkinson, 72–86.
[20] Davis, *Rise of the English Shipping Industry*, 201.   [21] *Ibid.*, 33.

Table 3.2 *The English merchant fleet, 1751–1820*

	Ships	Total tons	Tonnage: London	Tonnage: outports
The tonnage of English-owned shipping, 1751–1788[a]				
1751		421,000	119,000	302,000
1755		473,000	131,000	342,000
1763		496,000	139,000	357,000
1765		543,000	134,000	409,000
1770		594,000	150,000	444,000
1775		608,000	143,000	465,000
1786		752,000	186,000	566,000
1788		1,055,000	315,000	740,000
The number and tonnage of ships registered in the United Kingdom, 1790–1820[b]				
1790	13,557	1,383,000		
1793	14,440	1,453,000		
1794	14,590	1,456,000		
1795	14,317	1,426,000		
1796	14,458	1,361,000		
1797	14,405	1,454,000		
1798	14,631	1,494,000		
1799	14,883	1,551,000		
1800	15,734	1,699,000		
1801	16,552	1,797,000		
1802	17,207	1,901,000		
1803	18,068	1,986,000		
1804	18,870	2,077,000		
1805	19,027	2,093,000		
1806	19,315	2,080,000		
1807	19,373	2,097,000		
1808	19,580	2,130,000		
1809	19,882	2,167,000		
1810	20,253	2,211,000		
1811	20,478	2,247,000		
1812	20,637	2,263,000		
1813	20,951	2,349,000		
1814	21,550	2,414,000		
1815	21,869	2,478,000		
1816	22,026	2,504,000		
1820	21,969	2,439,000		

[a] Davis, *Rise of the English Shipping Industry*, 27.
[b] Mitchell and Deane, *Abstract of British Historical Statistics*, 217. Registration of merchant shipping began in 1787.

Table 3.3 *London's import and export trade in 1797*[a]

Foreign Trade	No. of ships including repeat voyages	Aggregate tonnage (tons)	Value of imports (£ s d)	Value of exports and re–exports (£ s d)
Northern Europe				
Germany	235	37,647	2,658,011 8 2	8,014,260 3 0
Prussia	608	56,955	220,827 14 0	211,662 12 0
Poland	69	17,210	207,477 0 0	35,468 18 3
Sweden	109	14,252	152,707 6 10	169,293 18 4
Denmark and Norway	202	48,469	94,821 3 6	711,082 10 8
Russia	230	56,131	1,565,118 7 6	452,106 16 7
Nearby Europe				
France	56	5,573	15,951 17 8	859,974 16 0
Austrian Flanders	66	5,104	21,027 3 2	118,064 2 2
Holland	329	19,166	673,241 17 4	1,538,120 3 6
Spain and Portugal				
Spain and Canaries	121	16,509	776,686 13 2	171,073 4 6
Portugal and Madeira	180	27,670	414,359 7 2	438,877 16 2
Mediterranean				
Mediterranean and Turkey, etc.	72	14,757	390,794 19 10	118,914 3 7
East India				
East Indies	53	41,456	6,544,402 10 2	3,957,905 5 1
West Indies and Africa				
West Indies	346	101,484	7,118,623 12 8	3,895,313 18 7
Africa and the Cape	17	4,336	82,370 15 0	419,075 19 3
North America				
British North America	68	13,986	290,894 4 10	1,347,250 0 0
States of America	140	32,213	1,517,386 2 8	3,898,864 12 9
Whaling trade				
Southern fishery	29	7,461	250,689 3 2	54 16 4
Greenland fishery	16	4,769	64,142 0 8	0 0 0
Total foreign Trade	**2,946**	**525,148**	**23,059,533 7 6**	**26,387,363 18 4**
Coasting Trade				
Foreign coasting				
Channel Islands	46	5,344	218,916 12 8	83,281 12 1
Ireland	276	32,824	1,878,971 7 2	659,922 14 1
British coasting				
Coal trade	3,676	656,000	1,700,000 0 0	10,000 0 0
England & Wales	5,816	500,000	3,900,000 0 0	2,200,000 0 0
Scotland	684	60,000	200,000 0 0	300,000 0 0[b]
Total Coasting Trade	**10,498**	**1,254,168**	**7,897,887 19 10**	**3,253,204 6 2**
Total Foreign & Coasting Trade	**13,444**	**1,779,316**	**30,957,421 7 4**	**29,640,568 4 6**

[a] From a 'General View of the whole commerce and shipping of the River Thames, taken from authorities and documents applicable to the year ending the 5th January, 1798; with the true valuation of the merchandise imported and exported from and to parts beyond seas, ascertained on the new principle established by the convoy duties; exhibiting also the number of vessels and the aggregate tonnage employed in each particular branch of the foreign and coasting trade', printed in Colquhoun, *A treatise on the Commerce and Police of the Metropolis.*

[b] 'The value of imports and exports in the coasting trade cannot be ascertained by the public accounts; what is here stated is merely the supposed value on the best data that could be found' (*ibid.*).

Table 3.4 *Merchant ships built and first registered in Britain and the British empire, 1787–1818[a]*

	In Britain		In empire		Britain and empire	
	Ships	Tons	Ships	Tons	Ships	Tons
1787[b]	943	91,700	484	26,600	1,427	118,300
1788	848	73,500	479	29,800	1,327	103,300
1790	577	57,100	148	11,600	725	68,700
1795	540	63,200	179	9,000	719	72,200
1800	845	115,300	196	18,900	1,041	134,200
1805	714	71,400	287	18,200	1,001	89,600
1810	–	–	–	–	685	84,900[c]
1815	877	101,000	–	–	864	97,900[d]
1818	704	84,700	378	19,700	1,082	104,400

[a] Mitchell and Deane, *Abstract of British Historical Statistics*, 220.
[b] The official registration of British shipping began in 1787.
[c] Figures for 1810 are incomplete.
[d] Figures for 1815 are also evidently incomplete, those for Britain alone exceeding those for Britain and the empire.

The great import of 'naval stores' fed the ship-building industry. At the end of the seventeenth century this prospered along the River Thames up to London, with the demand for large ocean-going ships for the Levant, East India and West India trades. The colonies – especially Newfoundland, Massachusetts and the Chesapeake region – also supplied a large quantity of shipping. In 1730, about one English ship in every six was built in America; by 1760, about one ship in four. In 1774, a contemporary estimated that nearly one-third of British-owned ships were American-built.[22] The loss of control of the thirteen American colonies in 1776 thus reduced the ship-building capacity of the empire severely. Figures for ships built and registered after the American War of Independence indicate production elsewhere in the empire also declined (see table 3.4).

Yet, with the demand for naval stores from the Baltic and for coal from domestic consumers and industrial furnaces, ship-building in the northeast of England boomed. By 1790–1 the London region was building only 10 per cent of the ships and 16 per cent of the tonnage in England. By contrast, the north-east, along the Rivers Tyne and Wear near Newcastle and Sunderland, was building 21 per cent of the ships and 40 per cent of the tonnage, which was twice the production of the Thames region. The

[22] *Ibid.*, 57, 68, 70.

north-west, mainly along the River Clyde near Glasgow, also built more ships than the River Thames and almost as much tonnage.[23]

The ship-building capacity of the merchant yards was fully exploited by the state for naval construction in wartime. The great number of ships owned in the port of London also served as the main source for hired transports. War indeed increased the demand for ships to be built and made available to the state. Some ship owners suffered severely. Enemy privateers took 1,855 merchant ships in the Seven Years' War, at least 3,386 in the American War of Independence, and as many as 11,000 between 1793 and 1814. But the annual loss was never more than 2.5 per cent of all British-registered ships.[24]

Shipping generally adapted to the dangers and opportunities of war. The intelligent use of information, the flexibility of shipping, insurance, the employment of port agents and sensible ship masters diminished risk in the industry. Income could be assured from the lease of some ships to government while others were leased into trades that reaped profit. The experience of one coal merchant – Michael Henley and Sons – indicates that owners were capable of making profits from their ships in most years in wartime and that their profits permitted them to enlarge their fleet. Henley had an eye for captured prizes and took advantage of their low prices when hostilities ceased.[25]

War and state expenditure

Outside the shipping industry, war also brought more benefits than damage to the British economy. Stephen Conway points out that, early in the mid-century wars, exports fell and trade suffered in the regions affected by hostilities when the supply of merchant ships and seamen were also seriously depleted. Then the state as well as the private sector suffered from money shortages for, while the state efficiently raised loans, there were great losses of money in interest payments to foreign investors and in subsidies to allies. Yet conway admits, at least for the Seven Years' War, that victories boosted confidence and exports, and new markets and sources of raw materials took the place of those inaccessible, while foreign carriers and seamen took the place of those recruited into the navy.[26] A similar re-adjustment seems to have taken place in later wars too. At the beginning of each period of hostilities exports tended to slump, but

[23] *Ibid.*, 70.
[24] P. Crowhurst, *The French War on Trade: Privateering 1793–1815* (Aldershot, 1989), 31.
[25] Ville, *English Shipowning during the Industrial Revolution*, 29, 148–54.
[26] Conway, *War, State and Society*, 100–14.

they generally recovered and by the end exceeded pre-war levels. Exports reached new heights in both 1759–60 and in 1796–1800.[27]

Before 1763, despite increased taxation during hostilities, capital for investment in industry, agriculture and transport was not lacking. Industries supplying war munitions, especially the iron and ship-building industries, enjoyed expanding demand which prompted innovation and increased production. Agricultural improvement, including enclosures, continued, and there were no interruptions to the improvement of turnpikes or canal building.[28] After 1763, despite the loss of the American colonies, merchants were able to find new sources of raw materials and new markets.[29]

Between 1755 and 1793, the value of imports and exports for England and Wales roughly doubled, while re-exports increased by 60 per cent. However, during the French Revolutionary War, while imports to Great Britain increased overall by 70 per cent, exports doubled in value, and re-exports tripled. By region of trade, imports from the East and West Indies enlarged most notably, while exports to those regions increased by 173 per cent between 1790 and 1798. Significantly, the greatest growth in re-exports was to north-west Europe and 'the north', paying for the import of naval stores.[30]

During the Napoleonic War, trade suffered from the closure of markets on the continent, particularly after Napoleon's Berlin and Milan decrees of 1806 and 1807. Conflict with America also brought an embargo on British trade and recession in the cotton and tobacco industries. Yet, almost at the same time, major new markets opened in Spanish America and in captured French and Dutch West Indian islands. Thus, while the value of trade with the British West Indies more than doubled between 1793 and 1814, that with the foreign West Indian islands and with Latin America rose from virtually nothing to £10.5 million in the same period.[31]

On the whole, therefore, the British economy prospered between 1756 and 1816 despite wartime disruptions, and probably to a great extent because state demand compensated in some sectors for the disruptions. Average annual state expenditure stepped up in each war between 1755 and 1815, from £15.75 million in the Seven Years' War to £103.5 million

[27] J. Hoppit, *Risk and Failure in English Business 1700–1800* (Cambridge, 1987), 127.
[28] A. H. John, 'War and the English economy, 1700–1763', *EconHR* 2nd ser. 7(1954–5), 329–44.
[29] P. Deane, 'War and industrialisation' in *War and Economic Development*, ed. J. M. Winter (Cambridge, 1975), 91–102.
[30] Mitchell and Deane, *Abstract of British Historical Statistics*, 280–1, 289, 312, 388.
[31] Kennedy, *The Rise and Fall of British Naval Mastery*, 144.

in the Napoleonic War. This compares to £10 million in the 1764–75 peace, and £18 million in the 1784–92 peace.

Among historians, the importance attached to the state's impact on the economy has gradually been growing. During the 1930s rising real income among domestic consumers and the growth of overseas markets were thought to be far more important.[32] However, the Second World War altered opinion.[33] In 1955, A. H. John acknowledged that the state took only the capital surplus to the requirements of the private economy.[34] In 1975, P. Deane recognised that state expenditure was 'at least as likely to have had a growth-promoting as a growth-retarding effect'.[35] By 1990, P. K. O'Brien was able to argue that public expenditure on security, trade and empire, in conjunction with that of merchants and industrialists, 'formed preconditions for the market economy and night-watchman state of Victorian England, as well as the liberal world order which flourished under British hegemony'.[36]

Breaks in data and differences between indices make for difficulty in differentiating state from domestic and overseas demand and in measuring the relative importance of state expenditure in the period 1755–1815.[37] Overseas demand has been reckoned to account for about one-fifth of industrial output in 1700, rising to one-third around 1800.[38] Relatively speaking, the importance of domestic demand thus declined, from about four-fifths to about two-thirds of industrial production. Nevertheless, during hostilities between 1712 and 1815, the state took between 15 and 20 per cent of gross national product in revenues, a large proportion of which were spent on war supplies.[39] As a share of national income, state military expenditure rose from 15.5 per cent in

---

[32] E. W. Gilboy, 'Demand as a factor in the Industrial Revolution', originally printed in *Facts and Factors in Economic History*, ed. A. H. Cole *et al.* (Newhaven, CT, 1932), repr. in *The Causes of the Industrial Revolution in England*, ed. R. M. Hartwell (Bungay, Suffolk, 1967), 121–38.

[33] J. L. Anderson, 'Aspects of the effect on the British economy of the wars against France, 1793–1815', *Australian Economic History Review* 12(1972), 1–20; Bowen, *War and British Society*, 63–72.

[34] John, 'War and the English economy'.   [35] Deane, 'War and industrialisation'.

[36] O'Brien, 'Central government and the economy, 1688–1815'.

[37] R.V. Jackson, 'Government expenditure and British economic growth in the eighteenth century: some problems of measurement', *EconHR* 2nd ser. 43(1990), 217–35.

[38] K. Morgan, 'Atlantic trades and British economic growth in the eighteenth century', and S. L. Engerman, 'Mercantilism and overseas trade, 1700–1800' in *International Trade and British Economic Growth from the Eighteenth Century to the Present Day*, ed. P. Mathias and J. A. Davis (Oxford, 1996), 182–204.

[39] P. K. O'Brien and P. A. Hunt, 'The rise of a fiscal state in England, 1485–1815', *HR* 66(1993), 129–76.

1760 to 17.6 per cent in 1780, declined to 10.4 per cent in 1800, but rose again to 14.1 per cent in 1810.[40]

Credit finance and the National Debt

While state spending provided stimulus in wartime, the means by which the state raised capital and paid for its supplies also appears to have benefited the economy, for the state relied on loans for much of its capital and purchased a great deal on short-term credit. Both had benefits. These features of state finance were, moreover, entirely apiece with the British mercantile economy. The state's financial system is usually viewed in isolation. But viewed as part of the wider maritime economy, it becomes just part of a greater culture and practice by which it was supported and sanctioned.

Britain's maritime economy was built upon credit and debt. The growth of the transatlantic trade, for example, was naturally accompanied by an expansion in the amount of credit allowed to merchants. The Atlantic trade, the North America – West Indies – West Africa complex, may have contained credit to the amount of £9 million in 1774. That allowed to American tobacco planters in 1774 has been estimated at £4 million.[41] Well-established American importers were allowed up to twelve months' credit. 'So general' was the allowance of credit that debt became a problem on account of the competition among vendors in America. Some in 1776, were unable to sell their imports and pay off their debts to English exporters who in turn remained in debt to the manufacturers who supplied them. In 1776, planters and merchants in the tobacco colonies owed their British creditors about £2 million.[42]

Warehousemen in England, stocking exports and imports, were at the centre of the vast web of trade credit relations extending through middle men to producers and retailers both at home and abroad. During the second half of the eighteenth century, many warehousemen turned to banking.[43] They occupied the pivotal point in trade relations where balances were kept and debts settled, and were trusted by merchants otherwise reluctant to take out loans. They contributed to the growth

---

[40] O'Brien, 'Political preconditions for the Industrial Revolution'.
[41] L. Neal, 'The finance of business during the industrial revolution' in *The Economic History of Britain since 1700*, ed. R. Floud and D. McCloskey (2nd edn., Cambridge, 1994), 151–91.
[42] K. Morgan, *Bristol and the Atlantic Trade in the Eighteenth Century* (Cambridge, 1993), 110–14.
[43] For example, Barclays in London and Lloyds in Birmingham: Neal, 'The finance of business during the industrial revolution', 160–2.

in the number of private banks during the second half of the eighteenth century: from eighteen to fifty-two in London between 1754 and 1774. By 1797 there were sixty-nine, almost all of which were linked by 'correspondents' with provincial banks and printing their own promissory notes, postponing the date at which they had to pay cash. Moreover, they were increasing their range of financial transactions and assisting the accumulation of capital for investment either in trade or in government stocks.[44]

Credit and banking arrangements facilitated business without hard cash or paper money, for accounts could be settled with bills of exchange which could circulate, usually for sixty or ninety days, and increase the money supply. The growth of the British economy accustomed the British public to purchases on credit, payments with bills, short-term debt, loans, banking and debt management.[45] The state used some of the same financial processes. For example, in 1796 the Victualling Board recorded bills of exchange to the value of £573,181 19s 1d drawn upon it by agents, captains and pursers around the world.[46] (See table 3.5.) Similarity in proceedings also accustomed the public to the means by which the state funded war expenditure.

Short-term debt in bills

Most payments were made with bills. There were 'ready money' bills which, when the Treasurer of the Navy had funds to hand, could be cashed immediately. However, the inability of the Treasurer to cash bills resulted in the growth of a backlog of bills, numbered and paid in the order or 'course' in which they were issued.[47] The first had been issued by the Treasury and the navy in the time of the English Commonwealth and Protectorate.[48] By the end of the seventeenth century, those issued by the Exchequer were superseding tallies as records of debt to be set against future tax revenues. The Bank of England, founded in 1694, undertook the circulation and encashment of Treasury bills. The Navy,

---

[44] D. Hancock, *Citizens of the World: London Merchants and the Integration of the British Atlantic Community, 1735–1785* (Cambridge, 1995), 247–58.

[45] P. James, *Population Malthus: His Life and Times* (London, 1979), 101–2; Neal, 'The finance of business during the industrial revolution', 157–63.

[46] NMM, ADM. DP/17, 30 Mar. 1797.

[47] P. G. M. Dickson, 'War finance, 1689–1714', in *New Cambridge Modern History*, vol. VI: *1688–1725*, ed. J. S. Bromley (Cambridge, 1970) 284–93; Baugh, *British Naval Administration in the Age of Walpole*, 470–5; C. Wilkinson, *The British Navy and the State in the Eighteenth Century* (Woodbridge, Suffolk, 2004), 220.

[48] Rodger, *The Command of the Ocean*, 41.

Table 3.5 *Bills of exchange drawn, accepted and paid for victualling services in all parts of the world, 1793–1796ᵃ*

Station	1793 £	1794 £	1795 £	1796 £
East Indies	16,473	12,987	34,918	139,737
West Indies	3,385	37,032	98,336	125,317
America	21	14,957	65,748	53,511
Newfoundland	4,532	9,488	5,249	6,085
Mediterranean, Corsica	40	30,051	46,007	27,238
'other places'	23,948	177,481	149,037	169,734
Spain and Portugal	2,185	3,474	2,422	14,003
Coast of Africa	806	659	1,872	1,414
North Seas (Continent)	3,032	3,088	7,236	20,411
Jersey, Guernsey, I of Man	2,955	3,567	10,218	15,731
New South Wales	–	988	–	–
Totals	**57,378**	**293,772**	**421,042**	**573,182**

ᵃ An account prepared by Admiralty order of 29 March 1797, on Treasury direction, to be laid before the Secret Committee of the House of Lords: NMM, ADM.DP/17, 30 Mar. 1797. Sums rounded to nearest pound. Final totals include shillings and pence omitted in preceding figures.

Victualling and Ordnance departments also then began issuing their own bills. Navy Bills were the most common.[49] Registers of the bills paid by the naval boards still survive. Those for the Navy Board in the period 1755–1815 record payments in bills for everything from travelling expenses to stores.[50]

During the eighteenth century, Navy and Victualling bills formed the greater part of the unfunded debt of the navy. Their course of payment sometimes extended nearly three years. Nevertheless, because the bills were transferable, they circulated as public currency and a regular market for bills developed, with published quotations of *pro rata* prices which fluctuated according to the length of time to encashment. The longer that time, the greater was the discount price. But as the discount increased, so contractors had to raise their prices to cover their losses on discounted bills. In 1778 the discount stood at 22 per cent; and the

[49] For the manner in which these bills were issued in course, see instructions to the Commissioners for victualling the navy, 4 Jan. 1700/1, in NMM, CAD.B/10.
[50] TNA, ADM. 18/104–26, 1756–1815.

interest on bills that accrued for payment in 1779 amounted to nearly £130,000.[51]

To maintain the confidence of contractors, the 'course' of the navy could not become too long. It thus became necessary for Parliament at regular intervals to fund the debt in bills from revenue or loans. Between 1722 and 1747 the debt was reduced by large-scale payments of more than £1 million every five or six years. During subsequent wars, the debt was funded almost annually; hence £1 million was paid each year between 1776 and 1778; £1½ million in 1780 and 1782; and more than £3 million in 1781. Nevertheless in 1783 the Navy, Victualling, Transport and Ordnance bills circulating at a discount of 21 per cent were still thought to amount to £14 million. In 1784 Pitt the Younger funded nearly £4 million of naval and transport bills, then £2 million of victualling debt, paying off the bills with 5 per cent consolidated stock, known as Bank Annuities.[52] This permitted a considerable reduction in the interest payable. He was able the following year to fund the rest of the bills – then totalling £10 million – also at 5 per cent.[53] By 1789 the debt was reduced to less than £2 millions.[54]

During the French Revolutionary War, the navy's unfunded debt again rose and was regularly funded at times when the length of the course demanded the support of contractor confidence. However, by 1795, the total unfunded naval debt, arising from bills issued by all departments on the course of the navy, stood at £8,848,293 carrying £283,377 interest. From December 1796 the course was cut by an Act of Parliament stipulating payment no later than three months from the date of the bill.[55] Thereafter, most 'in course' bills were paid uniformly ninety days after

[51] J. E. D. Binney, *British Public Finance and Administration, 1774–92* (Oxford, 1958), 141–2. For the concern over the rate of discount, see NMM, MID.6/4, 'Navy, Victualling and Transport Bills', *c.* 1785; also 'Nature and Progress of Navy Bills', *c.* 1782. The latter is accompanied by a copy of a Navy Bill, No. 2523, made out to Mure Son and Atkinson for freight of 5,244 tons of provisions and stores in consequence of a contract made 23 Nov. 1780.

[52] Binney, *British Public Finance*, 98, fn. 1; NMM, ADM. BP/5, 3 Dec. 1784; ADM. BP/6A, 3 Jan. 1785. The problem of debt reduction was not new. See M.W. Flinn, 'Sir Ambrose Crowley and the South Sea Scheme of 1711', *Journal of Economic History* 20(1960), 51–66; P. K. Watson, 'The Commission for victualling the navy, the Commission for sick and wounded seamen and prisoners of war, the Commission for transports, 1702–14', unpub. University of London Ph.D. thesis, 1965, 348–64; *Naval Administration, 1715–1750*, ed. D. A. Baugh (NRS, 1977), 456.

[53] NMM, ADM. BP/6A, 19 Oct. 1785; NMM, MID.1/176, T. Steele to C. Middleton, 28 July 1785; J. Ehrman, *The Younger Pitt: The Years of Acclaim* (London, 1969), 259–60.

[54] Binney, *British Public Finance*, 142.

[55] J. Sinclair, *The History of the Public Revenue of the British Empire* (3 vols., 3rd edn, London, 1804), III, 48–9.

issue so that the interest charges paid by the navy after the expiry of that period diminished.[56]

In 1797 sharpening price rises, exacerbated by the fluctuating cost of foodstuffs, and the demand of seamen for higher basic rates of pay, demonstrated the inadequacy of the rates at which the navy was granted money on an annual basis. In the same year, the system of naval finance was thought to be on the point of collapse and the rate per man, by which Parliament granted naval funds, was sharply increased from £4 a man to £7.[57] This increase in rate per man had a salutary effect. In 1798 financial supplies actually exceeded expenditure, and the naval debt declined.[58] That year over 52 per cent of the value of bills passed at the Navy office was paid in ready money.[59] Similarly, in other departments, ready money was paid to meet a proportion of immediate costs. In 1812 the Victualling office paid 85 per cent of its costs 'in course' but the remainder in 'ready money'.[60]

Long-term debt in loans

A large part of the money needed to pay these bills came from loans. The Bank of England performed the vital functions of raising and paying the interest on loans. From its beginning, the Bank was closely associated with money raised for military services. The Bank was incorporated in May 1694 by Act of Parliament for the purpose of lending £1,200,000 to the government at 8 per cent interest. More than half the sum was allotted to the navy, and almost alone financed it until the end of 1694.[61] Part of the loan was in cash, but part was in 'Bank bills' or notes which could be spent immediately and at no discount. For the Bank had the support of some of the largest and most important institutions of the

---

[56] NMM, ADM. BP/39B, Navy Office, dated 18 Aug. 1819, 'An account of the total amount of the unfunded debt of Great Britain in Navy Bills... from the year 1785 to 1818 inclusive'; 24th Report of the Select Committee on Finance, 1798, *Reports of Committees of the House of Commons*, 1797–1803, 1st ser., XII–XIII, Appendix E.4, p. 57; Binney, *British Public Finance*, 143.

[57] The system of funding is examined below, p. 114. See also P. K. Watson, 'The Commission for victualling', 332; Baugh, *Naval Administration 1715–1750*, 453; C. M. Bruce, 'The Department of the Accountant-General of the Navy', *MM* 10(1924), 256.

[58] NMM, ADM. BP/19A, 11 Apr.1799.

[59] TNA, ADM. 49/40. 'Ready money' bills amounted to £1,417, 951 while 'in course' bills amounted to £2,679,491 and carried £34,393 interest.

[60] TNA, ADM. 112/198.

[61] Ehrman, *The Navy in the War of William III*, 540–1.

City of London, in particular great merchant companies such as the East India, Hudson's Bay and Royal Africa Companies.[62]

The foundation of the Bank signalled a new era in public finance. It developed first its relationship with Parliament, which commanded revenues that included customs duties (revised in 1643), excise duties (reformed 1683), the land tax (first raised 1693) and stamp duties (first required 1694). From 1694 Parliament continued to grant revenues to pay interest on loans raised by the Bank of England, the strength of which lay in the confidence of depositors, investors and merchant bankers that Parliament would respect the state's debt to the Bank and vote the revenues assigned to pay the interest. In effect these revenues were used by the Bank as security to raise new loans. Subscribers to these loans included the general public, chartered companies and government contractors. The state's collective debt at this time became known as the National Debt and, owing to its funding from tax revenues, came to be regarded as the responsibility of Parliament.[63]

The Bank of England not only made money available to the government in London, but remitted money abroad to selected companies who acted as the Bank's agents and made available their own credit facilities to those who needed to purchase supplies for the navy abroad. From 1695, companies in Cadiz, Alicante, Madrid and Leghorn facilitated the provision of supplies to the Mediterranean fleet which was being maintained from Cadiz. In so doing, the Bank's agents relieved the navy temporarily of the necessity to issue Navy, Victualling or other bills, and so relieved it of 'unfunded debt'.[64]

The security created by the integration of Bank and state functions permitted the British government to raise very large sums at relatively low rates of interest. Much of the money was raised by subscription to irredeemable interest-bearing bonds or stock on London's capital market. The investments formed a long-term loan to government. Their annual scale grew from £8.5 million to more than £20 million between 1756 and 1815. As a proportion of total expenditure, they rose from 37.4 per cent to 39.9 per cent at the end of the War of American Independence, then declined during the Napoleonic War to 26.6 per cent. In relation to revenues, British debt in the international money market

---

[62] Sainty, *Admiralty Officials*, 21, 132; Ehrman, *The Navy in the War of William III*, 512, 542, 570.

[63] The success of the arrangement was marked by the foundation in 1695 of the Bank of Scotland which performed some of the same functions as the Bank of England, including the circulation of notes.

[64] Ehrman, *The Navy in the War of William III*, 543–4.

peaked in 1797 before direct taxes were increased by the introduction of the income tax.[65]

Central to the state's success in raising loans was a group of City business men known as 'the moneyed interest'. In 1757 they represented Dutch and Jewish interests; and the directors of the Bank of England, the South Sea Company, the East India Company and the insurance companies. Less central, but important, was a range of bankers, government contractors and other business men.[66] A large part of the money borrowed by the Bank of England came from abroad. On its first foundation, agents of the Bank had been established in Amsterdam and Antwerp, and numerous aspects of Dutch financial practice were copied in England. Early in the eighteenth century only short-term loans had been raised abroad. However, by mid-century, foreign investors, predominately Dutch, were responsible for significant portions of long-term debt – perhaps 14 per cent in 1747 and 25 per cent in 1762. Conflict with Holland in 1779–83 stifled this source of investment, but it only petered out during the Napoleonic War when the spread of French power through Europe so reduced foreign loans that they became 'marginal'.[67]

But by then Britain's domestic economy was generating wealth. Surplus business capital provided the core of investment in government debt. A drawback for the British economy was that, in the first years of war, investment in government funds removed capital from private companies causing a scarcity of capital and spates of bankruptcies. Debts arising from interruptions to trade by hostilities were more often dishonoured in wartime.[68] But subscribers to loans had always come from a variety of backgrounds. When country banks came into existence, they remitted surplus funds to London which were made available to the state through the purchase of consols and securities.[69] As the economy grew, money for loans to government always seemed available: in December 1796 a 'loyalty loan' of £18,000,000 was subscribed in four days.[70]

In wartime the armed forces were the greatest public expense. Indeed, owing to the frequency of war in the second half of the eighteenth century, government spending made the military sector of the British economy

---

[65] Kennedy, *Rise and Fall of the Great Powers*, 105; Mitchell and Deane, *Abstract of British Historical Statistics*, 405; J. C. Riley, *International Government Finance and the Amsterdam Capital Market, 1740–1815* (Cambridge, 1980), 116–17.

[66] Marshall, *The Making and Unmaking of Empires*, 71.

[67] P. G. M. Dickson, 'War Finance', 288, 293; J. C. Riley, *International Government Finance and the Amsterdam Capital Market, 1740–1815* (Cambridge, 1980), 119–26.

[68] Hoppit, *Risk and Failure in English Business*, 122–6.

[69] L. S. Pressnell, *Country Banking in the Industrial Revolution* (Oxford, 1956), 416.

[70] J. Ehrman, *The Younger Pitt: The Reluctant Transition* (London, 1983), 639.

grow faster than either agriculture or manufacturing.[71] In wartime the army tended to cost more than the navy; by 1815 the cost of the army was more than double that of the navy. (See table 3.6.) But in peacetime the navy tended to cost more owing to maintenance and rebuilding costs:[72] between 1688 and 1815 about 60 per cent of the revenues available for military purposes were allotted to the navy.[73] Separate from both army and navy, but complementary to both, was the cost of ordnance. These three expenses made up military expenditure and as a whole always exceeded state expenditure for civil purposes. Moreover, there were always interest payments to meet, which in peacetime grew in scale relative to the other costs of government.

In peace, when military expenditure dropped away, interest payments became the principal charge on the state (see table 3.7). Peace was used to clear debt and reduce interest payments. For the honouring of that debt was vital to the financial credit of the British state and to its ability to raise funds in subsequent wars. Expenditure on debt in peacetime was thus as important to the funding of war as expenditure during the hostilities themselves. Indeed, as the size of the National Debt was a public obsession, its control and reduction became a principal task of the Treasury.

Taxation revenue and debt management

To pay the interest on these debts, the British government primarily used the revenues from three forms of tax: customs duty on articles imported for trade, excise duty on goods and services of domestic production, and the land or property tax assessed on visible and immovable 'manifestations' of wealth and income. Other taxes like stamp duties on newspapers and documents of insurance produced further but smaller quantities of revenue.[74]

Tolerance of these taxes is attributed by P. K. O'Brien to their 'flexible administration, complemented by an expedient tolerance of evasion and a prudent selection of the commodities and social groups "picked upon" to bear the mounting exactions of the state'. He points out that Britain's

---

[71] Cain and Hopkins, *British Imperialism*, 71.
[72] P. L. C. Webb, 'The rebuilding and repair of the fleet 1783–1793'.
[73] O'Brien, 'Political pre-conditions for the Industrial Revolution'.
[74] Sinclair, *The History of the Public Revenue of the British Empire*, II, appendix 1, states the revenue for the year ending 5 Jan. 1803, dividing it into customs, excise, stamps, and land and assessed taxes.

Table 3.6 *British military expenditure, 1756–1820*

	Navy	Army	Ordnance	Civil govt	Debt charges	Total expend.
1756–1801[a] (in £000 sterling)						
Seven Years' War						
1756	2,714	2,396	426	1,292	2,761	9,589
1757	3,595	3,210	520	1,083	2,805	11,214
1758	3,893	4,586	547	1,279	2,895	13,200
1759	4,971	5,744	729	991	2,947	15,382
1760	4,539	8,249	682	1,152	3,372	17,993
1761	5,256	9,923	853	1,256	3,823	21,112
1762	4,892	8,781	746	1,218	4,404	20,040
1763	7,464	4,067	470	1,056	4,666	17,723
Inter-war peace						
1764	2,150	2,234	279	1,137	4,887	10,686
1765	3,154	2,702	282	1,050	4,828	12,017
1766	2,467	1,815	276	1,069	4,686	10,314
1767	1,657	1,696	243	1,022	5,020	9,638
1768	1,431	1,472	296	1,036	4,911	9,146
1769	1,527	1,438	303	1,498	4,803	9,569
1770	2,082	1,545	236	1,223	4,836	10,524
1771	2,061	1,514	361	1,057	4,611	10,106
1772	2,738	1,497	334	1,017	4,686	10,725
1773	1,787	1,581	327	1,032	4,649	9,977
1774	2,030	1,532	298	1,095	4,612	9,566
1775	1,765	1,765	349	1,211	4,674	10,365
American War of Independence						
1776	2,745	4,248	549	1,271	4,632	14,045
1777	3,531	4,677	573	1,769	4,709	15,259
1778	4,563	5,464	957	1,425	5,030	17,940
1779	4,271	7,112	1,074	1,158	5,618	19,714
1780	6,329	7,210	1,330	1,251	5,995	22,605
1781	6,589	8,928	1,546	1,530	6,917	25,810
1782	10,807	7,755	1,546	1,263	7,364	29,234
1783	6,994	5,332	1,341	1,383	8,054	23,510
Inter-war peace						
1784	9,447	3,301	1,014	1,324	8,678	24,245
1785	11,851	2,390	551	1,451	9,229	25,832
1786	3,127	1,984	372	1,513	9,481	16,978
1787	1,991	1,803	384	1,513	9,292	15,484
1788	2,262	2,099	547	1,522	9,407	16,338
1789	2,073	1,899	475	1,664	9,425	16,018
1790	2,482	2,197	545	1,703	9,370	16,798
1791	3,400	2,009	769	1,886	9,430	17,996
1792	3,331	1,829	417	1,565	9,310	16,953
French Revolutionary War						
1793	2,464	4,829	844	1,835	9,149	19,623
1794	6,127	9,209	1,501	1,572	9,797	28,706

(cont.)

Table 3.6 *(cont.)*

	Navy	Army	Ordnance	Civil govt	Debt charges	Total expend.
1795	9,626	14,651	1,996	1,751	10,470	38,996
1796	11,518	14,236	2,500	2,014	11,602	43,372
1797	11,984[b]	15,327	2,122	2,527	13,594	46,053
1798	12,793	14,142	1,780	2,178	16,029	47,422
1799	11,614	14,289	1,980	2,180	16,856	47,419
1800	3,843	4,151	465	537	3,387	12,383
1801	14,707	15,297	1,663	2,072	16,749	50,991

	Navy	Army and Ordnance	Civil govt	Debt charges	Total expend.
1802–1816[c] (in £000,000 sterling)					
Inter-war peace					
1802	17.3	20.1	5.6	19.9	65.5
Napoleonic War					
1803	12.0	13.3	6.7	20.4	54.8
1804	8.1	15.5	5.1	20.7	53.0
1805	11.9	22.2	5.2	20.7	62.8
1806	14.3	25.8	5.2	22.3	71.4
1807	16.3	24.8	4.7	23.2	72.9
1808	16.9	24.0	5.3	23.8	73.3
1809	17.6	27.2	4.7	23.1	78.0
1810	19.4	28.9	5.2	24.2	81.5
1811	20.0	28.0	5.1	24.4	81.6
1812	19.6	33.8	5.2	24.6	87.3
1813	20.8	36.5	5.4	26.4	94.8
1814	22.5	49.6	5.3	27.3	111.1
1815	22.8	49.6	5.8	30.0	112.9
Post-war peace					
1816	16.8	39.6	6.1	32.2	99.5
1817	10.2	18.0	5.5	32.9	71.3
1818	6.6	11.1	5.0	31.5	58.7
1819	6.6	9.1	6.0	31.3	57.6
1820	6.4	10.3	5.4	31.1	57.5

[a] Figures 1756–1801 from 'Public Finance 2. Net Public expenditure – Great Britain 1688–1801' in Mitchell and Deane, *Abstract of British Historical Statistics* (Cambridge, 1962), 389–91. Their figure for total net expenditure is sometimes rounded up or down – as in 1757. Occasionally the difference of the total from the sum of its constituent parts is somewhat bigger – as in 1770–3 and 1775 – and is not explained. However, they do add an explanation in 1784–5 and 1797, when they note the total contains an element of debt funded in these years but contracted previously. For 1800 and 1801 they observe that the financial quarter and year, respectively, ended on 5 January.

[b] Excludes £11,596,000 for debt contracted earlier.

[c] Figures 1802–20 from 'Public Finance 4. Gross Public Expenditure – United Kingdom 1801–1939' in Mitchell and Deane, *Abstract of British Historical Statistics*, 396–8. They excluded payments for capital investments, and expenditure on debt redemption other than through payment of terminable annuities. Costs of collection, though given by Mitchell and Deane, are also excluded here.

Table 3.7 *State expenditure for military services, civil purposes and interest payments on debt 1756–1815*[a]

	Military expenditure	Civil expenditure	Interest payments
	%	%	%
1756–63 (war)	70	8	22
1764–75 (peace)	37	20	43
1776–83 (war)	62	8	30
1784–92 (peace)	31	13	56
1793–1815 (war)	61	9	30[76]

[a] O'Brien, 'The political economy of British taxation, 1660–1815', esp. p. 2.

tax receipts increased more than sixteen times between 1660 and 1815[75] but argues the British people were able to bear this increase because indirect taxation helped to spread the burden, because Britain's national income rose by a factor of about three between 1670 and 1810,[76] and because, with economic and population growth, the tax base was slowly expanding. Higher income per capita in Britain made tax payment there easier than in some other European states where wealth was distributed with greater inequality. Moreover, taxes were collected efficiently, the excise service being held up as a model organisation, that tax consistently being regarded as 'the best and easiest tax'.

Nevertheless, between 1755 and 1815, the relative importance of excise and stamp duties declined as the importance of customs duties and direct taxes gradually increased, until each was contributing about one-third of the state's revenues (see table 3.8). In 1799 the Younger Pitt introduced income tax, aiming to divert nearly one-tenth of private income into government revenue. This was not achieved but it pushed up direct tax yields, and permitted the existing tax structure to be preserved until the end of the Napoleonic War.

O'Brien suggests that the political management of taxation was critical to its acceptance by the British population which, in consequence, shouldered a greater burden than the citizens of France.[77] By his estimate, the British burden of taxation multiplied five times between 1660

[75] Tax receipts rose fivefold in 1690–1780, from £2.05m to £11.75m. By 1815 they had risen again to £62.67m: Bowen, *War and British Society*, 24.

[76] Taxation took 6.7 per cent of national income in the early 1690s; 11.7 per cent in 1778–82; and 18.2 per cent in 1812–15: *ibid.*, 24.

[77] By the 1780s, per capita taxation was two and half times greater than in France; during the Napoleonic War it had risen to three times as much: Bowen, *War and British Society*, 25. See also E. A. Wrigley, 'Society and economy in the eighteenth century' in *An Imperial State at War*, ed. Stone, 72–95.

Table 3.8 *Direct taxes, excise, stamp and customs duties as proportions of total state revenues, 1755–1815*

	Excise/stamp duties on domestic products/services		Customs duties on retained imports		Direct taxes on manifested wealth and income	
	£m	%	£m	%	£m	%
1755	3.8	54	1.7	24	1.5	21
1760	4.1	49	1.9	23	2.3	28
1765	5.5	55	2.2	22	2.3	23
1770	6.0	57	2.6	25	1.9	18
1775	6.3	58	2.6	24	1.9	18
1780	6.6	56	2.6	22	2.6	22
1785	7.8	57	3.3	24	2.7	20
1790	7.5	43	6.3	36	3.6	21
1795	8.9	44	7.2	36	4.0	20
1800	11.5	36	11.5	36	8.8	28
1805	19.4	41	16.4	35	11.2	24
1810	22.9	36	18.8	30	21.2	34[a]

[a] O'Brien, 'Political economy of British taxation'. For the sources and the manipulation of the data, see O'Brien's notes, p. 9.

and 1789 without provoking political upheaval 'except among those fiscally privileged colonies of North America'. Between 1780 and 1810 the tax burden doubled again.

Political management appears all the more important when the decisions made in France are compared to those made in Britain. France, during the Seven Years' War, adopted the same financial method as Britain and Holland in paying for hostilities predominantly from loans: government borrowing met 59 per cent of French war expenses, taxes only 29 per cent. However, the French monarchy managed its debt less wisely, opting to fund it with short-term annuities with high-interest rates rather than using long or perpetual loans with lower rates. It therefore increased the cost in interest and the amount of the peacetime debt to be serviced. Like the British, the French attracted foreign investment, especially Dutch. But after the American War of Independence, despite similar amounts of revenue and expenditure, the British were able to service a much larger debt (£240m in 1787) than the French (£201m in 1788) on lower interest rates (3.7 per cent) than the French (6–6.5 per cent).[78]

[78] J. C. Riley, *International Government Finance and the Amsterdam Capital Market*, 101–12; see also Riley's *The Seven Years War and the Old Regime in France*.

The anachronistic organisation of French finances also told against the French state. The reputation of its government suffered from the perpetuation of large-scale tax farming, venal tax collectors, the regional management of money, and the arbitrary character of monarchical decisions regarding payment.[79] Britain, on the other hand, developed professional tax collectors, a respected system of local property tax assessment, the centralised management of money, and decision-making in matters of finance by a political figure-head answerable to the public in the House of Commons.

The First Lord of the Treasury and the Chancellor of the Exchequer were the ministers of the Crown principally involved in decisions affecting revenues, expenditure and debt. The Exchequer was the repository for funds raised from the revenues, and the Chancellor was responsible for monitoring their levels. Meanwhile the First Lord of the Treasury matched revenues to expenses, including interest payments on debt, and ensured all the services of the state were paid for.

The Treasury's authority over state revenue and expenditure dated to Orders in Council of November 1667 and February 1668, which directed that any grant, order or warrant signed by the Crown should 'first take rise and begin' with the Treasury, and conferred on the Treasury autonomy from the Privy Council.[80] By 1685 the Treasury had gained effective control over the collection of revenue, abolished tax-farming and centralised receipts at the Exchequer. By 1714 it was exercising loose control over expenditure, especially that which controlled the public debt, and the First Lord of the Treasury was the most important man in government.[81]

Effective Treasury management was especially necessary because of the public dread of excessive debt.[82] Much ingenuity was needed to reduce it between the wars and it is a tribute to Treasury that after 1783, although sometimes a concern, the National Debt was no longer dreaded and was eventually acknowledged as a state asset.[83] By 1815 Patrick Colquhoun was able to observe that 'the present domestic debt, enormous as it appears to be, yields an increase of wealth to the country in proportion to its magnitude'. He concluded that 'the interest of the domestic public debt, although in some respects a pressure upon the country, is the main spring by which its general industry is stimulated and promoted. It is the

---

[79] Pritchard, *Louis XV's Navy 1748–1762*, 184–205.
[80] H. Roseveare, *The Treasury 1660–1870. The Foundations of Control* (London, 1973), 26–7, 113–15.
[81] P. G. M. Dickson, 'War finance, 1689–1714', 284–5.
[82] Wilkinson, *British Navy and the State in the Eighteenth Century*, 211.
[83] Cain and Hopkins, *British Imperialism*, 68–75.

seed sown to produce a bountiful harvest of newly created property every year.'[84]

State expenditure and naval contracts

In 1797 state expenditure amounted to £46 million, of which over £29 million (64 per cent) was spent on the armed forces. Of that £29 million, £15 million (52 per cent) was spent on the army, nearly £12 million (41 per cent) on the navy, and £2 million (7 per cent) on the ordnance.[85] Naval expenditure, examined here, was divided between the four boards managing supply of the navy, and in 1797, according to the Select Committee on Finance, they spent another £2 million, possibly using bills which, for funding, became the responsibility of the Treasury. Thus, altogether, in 1797 the four subordinate boards actually spent £14,065,979. The Navy Board spent the most, £7,433,685; the Victualling Board the next greatest, £4,578,788; the Transport Board £1,613,336; and the Sick and Wounded Board, £440,170.[86]

Included within the expenses of the Navy Board were the costs of the navy at sea. During a year when there were 125 ships of the line and 411 frigates and other vessels commissioned for sea and harbour services, with 122,192 men on the books of those ships,[87] £1,352,910 was spent on wages to officers and men and another £462,538 on the marine corps. But that was only 13 per cent of the costs of the navy.

The shore establishment consumed far more. In 1797 £757,599 was paid in wages within the dockyards, and £1,951,623 on naval stores. A large proportion of the navy's stores came from abroad, in particular the Baltic. These imports included timber, masts, hemp, iron, tar and tallow, the whole in 1797 costing £980,686. Hemp, which came mainly from Russia, in 1797 cost the most, £651,269, almost precisely two-thirds the total amount. The range of materials supplied to the navy from within Britain cost about the same as those imported from abroad: £970,938. Domestic oak and other timber cost the most, £191,463, and after 1800 became increasingly scarce and expensive so that much was also imported.[88] Other domestic supplies comprised raw materials

---

[84] Colquhoun, *Treatise on the Wealth, Power and Resources of the British Empire*, 284–5.

[85] Mitchell and Deane, *Abstract of British Historical Statistics*, 391.

[86] 24th Report of the Select Committee on Finance, 1798, Appendix E.1, pp. 49–53. For a detailed account of naval expenditure in 1798, see TNA, ADM. 49/40; for 1804, see NMM, ADM. BP/25B, 16 July 1804.

[87] These figures are for January 1798. They therefore represent what ships and men were maintained in commission during the course of the previous year.

[88] See Morriss, *Royal Dockyards*, 80–3; also Albion, *Forests and Sea Power*, which tends to exaggerate the shortage of domestic oak.

like copper, lead and stone, and processed materials or manufactured equipment like canvas, cordage, hammocks, pumps, flags, blocks, paint, ironmongery, sails, fire hearths, tanners' wares, ballast, anchors, pitch, glass, bricks and varnish.

Wages in the Victualling yards were unspecified but the provisions purchased, processed and packaged within those yards cost £1,758,529; those supplied by contractors abroad cost £627,898, and those paid for by pursers, agent victuallers and consuls abroad cost £688,308. Altogether naval food supplies cost £3,074,735. From 1793 the Victualling Board also purchased and freighted provisions for the army; in 1797 they cost £1,069,296, a third of the cost of the navy's foodstuffs. Victuals of all types, secured by every means, cost a total of £4,144,031.

Large quantities were also spent by the Transport Board – on hiring transports, £883,999, and on caring for prisoners of war, £402,513. The Sick and Wounded Board spent £440,170, of which £341,371 went on caring for sick seamen.

But the two principal costs were for naval stores and victualling provisions which consumed £6,095,654. That was 43 per cent of total naval expenditure, 13 per cent of total state expenditure. Virtually all this expenditure went into the hands of naval contractors. These were immense sums passing into their hands. Hitherto, little has been known about the business conduct of these contractors.[89] The following reveals how some made their money, how they used it and how they were managed.

The manoeuvres of naval contractors

Before the American War of Independence, the naval boards were still relatively inexperienced in managing contractors for distant stations. As a result, the Victualling Board was systematically imposed upon by their contractors. How they did so is revealed in rare and extraordinary surviving correspondence between one supplier and his agent who engaged to victual ships in the Royal navy on the coast of North America in 1773: that of Alexander Brymer at Boston to his partner or agent, Christopher Champlin in Rhode Island. Champlin was new to the business and Brymer was anxious to advise him how he should operate.

First and foremost, in his management of ships' officers, Champlin was told on 29 March 1773 to adopt an attitude of complaisant indifference to anything they might want or say: 'never be uneasy about their making

[89] Hancock, *Citizens of the World*, 28, 41, 115, 144, 221–39; see also N. Baker, *Government and Contractors*.

demands; if they once find that they can plague you in this or any other way they will make a point of doing it'; 'never betray the least uneasiness before these gentlemen, on the contrary, if you would wish to have weight with them, you must seem quite unconcerned at every thing they can do or say'.

Second, Champlin was to get the better of the purser:

when the purser finds you, go to the Fountain Head [the captain]; he will in future act more cautious and make a seeming mint of obliging when 'tis not intended as such . . . Between us there is not a purser in twenty has the least influence with the captain and this you may easily observe in the course of a little conversation with the captain by sounding at a distance.

Third, both captain and purser were to be induced into accepting what was convenient and cheaper to Champlin:

I am certain if you only mention calavances to Capt. Inglis or the commanding officer and tell either of them that the ships here [at Boston] take them that they will not hesitate ordering them to be received on board . . . Real pease and oatmeal! Do you make yourself uneasy at this? A finesse of the purser! Laugh him out of it and tell him that I have the admiral's leave to substitute indian [meal] for oatmeal, and likewise let him know that I may send half of the demand of oatmeal in rice.

Champlin was obviously more scrupulous about supplying what was wanted. But on 1 April 1773 Brymer retorted: 'It gives me nor ought it you any concern whether they like indian meal and calavances or not, as both are countenanced here by the admiral; if they should complain 'twill be to no purpose; therefore act indifferently on this head. If they had rather prefer rice in lieu of oatmeal let them have it.'

Brymer was careful to ensure he profited from any substitution: 'By supplying 4 bushels pease for 112 lbs bread is getting 2 of [*sic*] or at that rate for it, because there is 4/- difference between the pease and bread price; altho' in fact I only receive 16/- yet 4/- ought to be taken from that [as extra profit] as the pease is only worth 12/-.'[90]

Of advantage to the contractor, the capacity or weight that constituted a barrel or bushel was open to interpretation:

There is no obligation to measure the oatmeal, 'tis only to charge a barrel pre-sumptuously so many bushels and 'tis the purser's business to do himself justice if he thinks he is overcharged: and in measuring 'tis to be observed that the meal is to be sifted thro' the hand in the lightest manner . . . : heaped measure is a

---

[90] Rhode Island Historical Society Library, A. Brymer to C. Champlin, 29 Mar., 1 Apr. 1773.

mistake as is that of my having allowed him 42 [lbs] to the bushel at Halifax. I have a certificate under the commissioners' hands that settles 36 [lbs] Irish or English oatmeal to the bushel, and from the trial made here I find this country oatmeal much the same.

Ignorance of the difference between English and local measures made for complacency in the purser and advantage to the contractor: 'As to the method of measuring or weighing in the colonies, the pursers have nothing to do; the contract is made in England and the supplies are to be agreeable to the standard weight and measure of that country; if any advantage can be made this way the contractors reap it, not the pursers.'

Should questions arise, Champlin was advised to mollify his customers:

I would not wish nor do I ever desire to dispute but when I am morally certain of being right, and even then instead of aggravating I sooth or endeavour to do it before I become positive or give a definitive answer. As I act on this principle so I would not pretend to say any thing positive for you to go by in critical points, as I am sure your good sense and experience will direct you on such occasions much better than I can pretend to do.[91]

More generally, prior knowledge of the personalities and proceedings of officers and pursers aided the contractors. 'The Mercury[92] will leave you in a very short time', Brymer informed Champlin on 2 August 1773, 'and I firmly believe that would have been the case ere now if Captain Keeler had not shewed so great an impatience'. But such communications were to remain secret: 'This however sub-rosa, and everything relative to the navy department that I may communicate must never be mentioned as from me, as it would deprive me of the confidential intelligence I stand in need of to regulate matters.'[93]

And this Brymer did, to a degree of which naval officers could hardly have been aware. He obviously worked in conjunction with contractors as far north as Halifax. When the *Mercury* left Rhode Island for Boston, on 9 August Brymer observed that no vessel was yet appointed to replace her; however, the *Kingfisher* was appointed to relieve the *Swan* which was to be ordered to Halifax to refit. 'As that is the case', he lectured Champlin,

and as bread, pork and butter bear a high price there, it is a great loss to the contractors that the Swan has had such a large supply of them from you; an attention to this is the great thing on which turns the contractors getting a profit;

---

[91] Rhode Island Historical Society Library, Brymer to Champlin, 12 July 1773.
[92] Sixth-rate with 20 guns built in 1756 and wrecked in December 1777 near New York.
[93] Rhode Island Historical Society Library, Brymer to Champlin, 9 Aug. 1773.

Table 3.9 *A comparison of food contract prices for the supply of St Lucia, Barbados and Jamaica, 1776–1783*[a]

	At St Lucia	At Barbados	At Jamaica
Bread, per cent	44s 0d	38s 0d	31s 6d
Rum	5s 0d	3s 6d	4s 6d
Beef, per piece	1s 9d	10d	11d
Pork, per piece	6d	1s 4d	6d
Peas, per bushel	3s 0d	10s 0d	3s 0d
Oatmeal per bushel	10s 0d	3s 0d	3s 0d
Butter, per lb	1s 6d	1s 6d	1s 0d
Vinegar, per gallon	6d	6d	6d
Bags, per bag	5d	5d	2s 0d

[a] 8th Report of the Commissioners appointed to inquire into fees, gratuities, perquisites and emoluments which are, or have been lately, received in the several public offices, 1788, repr. in *House of Commons Sessional Papers of the Eighteenth Century*, ed. S. Lambert (Wilmington, DE, 1975), 15–16.

therefore think of it another time, and supply those articles sparingly when any vessel is bound to Halifax, and ply them with rum, beef, cheese etc.[94]

These manoeuvres on the part of Champlin were not unusual among contractors. The Commissioners on fees, in 1788, noticed that contractors made a regular policy of varying the quantities actually supplied according to the contract prices they obtained for the commodities – moreover, that they arranged their prices when tendering with this in mind. They observed:

If the contractor perceives that he shall be enabled to issue a greater quantity than the due proportion of one article which affords to him a large profit, he will, under that expectation, reduce the price of another which he can withhold and regulate his tender to the board accordingly; but he can neither, in his issues, exceed in one instance, nor diminish in another, without the connivance of pursers, whose duty it is to prevent such practices.

To explain the degree of loss suffered by the public, the Commissioners compared the contract prices for the supply of St Lucia and Barbados, held by the same person, with those for Jamaica during the American War of Independence (see table 3.9).

The Commissioners on fees made two observations. Firstly, they noticed how much cheaper provisions supplied at Jamaica were compared to those at St Lucia. At Jamaica they cost on average only 12.31 pence

[94] *Ibid.*

per man per day while at St Lucia they cost 15.8 pence. They pointed out that if 8,000 men were victualled at St Lucia for one year, the difference in cost from victualling them at Jamaica was an extra £42,500. Secondly, they compared prices of the same individual commodities in the contracts for St Lucia and Barbados, two islands which were situated close together and which were supplied by the same contractor. They noticed that it was in the interest of the contractor to deliver the larger proportion of beef at St Lucia, and of pork at Barbados, which was exactly what he did![95]

Of course, contractors could not be expected to hold stocks of provisions sufficient to victual a large fleet, such as that which rendezvoused at Jamaica towards the end of the American War of Independence. As a result the contractors had to make extra purchases 'in a very scanty market' in which prices were much enhanced by numerous persons who appeared as competing purchasers. The extra supplies made available to the British fleet consequently cost £15,543, when the same, had they been supplied under contract, would have cost £2,829. As the fees commissioners observed, however,

it was evidently for the interest of the contractor to refuse all supply [by contract], or at least to furnish the same as sparingly as possible, because all such provisions or stores as the contractor could reserve or keep back might be supplied through the medium of the pursers at nearly six times the price for each article which he would have received under his contract.[96]

The profits of army contractors

The private manoeuvres of army contractors are less well documented. But, from the work of Norman Baker and David Hancock, something is known of their public conduct during the Seven Years' War and American War of Independence.[97] Until 1793 the British Treasury was responsible for the supply of food to the British army overseas and it allowed two means by which contractors could enhance their profits. It provided an advance of £3 for every man to be supplied. It also permitted prompt payment, before contractors' accounts were closely examined, on receipt of satisfactory reports of compliance from its Comptroller of Army Accounts. Payments were reduced by the advances made to the contractors, and by the value of provisions remaining undelivered

[95] 8th Report of the Commissioners on Fees, 15–16.

[96] 9th Report of the Commissioners on Fees, 1788, *House of Commons Sessional Papers*, 9.

[97] See also 'A narrative of the Sir James Cockburn, Bart., Services in Germany', his memorial of 1782 in Wentworth Woodhouse Muniments, Sheffield City Library, R108–137–1.

or condemned and not replaced. Nevertheless, contractors delivering within the British Isles or Europe could receive payment within months of undertaking their contracts. It was a system that encouraged the rapid performance of contracts and further undertakings.[98]

But at the beginning of the American War of Independence some contractors were paid far in excess of what they spent. In 1777 excess payments were totalled at £75,643, nearly 17 per cent more than was properly due; a later recalculation put the total at £136,123 or 30 per cent more than was due. This was because deliveries of provisions were computed in terms of man-rations and excesses were valued by the contractors themselves. Like naval contractors, some contractors also delivered more of one commodity than another to take advantage of market price fluctuations. They also substituted one type of provision for another. A group of contractors delivering to Canada received excesses of at least £22,859 on account of excess poundages that ran to five or six figures. From 1777 the Treasury attempted to recoup over-payments, and revised its contracts and procedure.[99]

Nevertheless, the profits of contractors were substantial. Norman Baker believed they made between 15 and 20 per cent profit before 1780 and around 10 per cent profit after that. He found that investments in government stocks coincided with Treasury payments for contracting, and that over the course of a war successive investments laid the foundation of significant fortunes. Between 1776 and 1782, from virtually nothing, William James accumulated stock worth over £116,000, while over the same period John Blackburn accumulated £123,000. In 1778, after three years of provisioning troops in Canada, William Baynes purchased stock worth £13,000. At his death, Richard Atkinson's fortune was estimated at £300,000.[100]

These amounts may be compared to the amount Richard Oswald made from supplying bread to the British army in Germany in 1758–62. Oswald produced each of the 6-pound loaves he supplied to the army in Germany for 3.5 pence and sold them for 8.5 pence, making a profit of 143 per cent on each loaf. This earned him £112,000 over four and a half years.[101] Thus contractors could and did accumulate small fortunes which permitted them to rise up the social scale.

---

[98] N. Baker, *Government and Contractors*, 56–7. The distance in time between payment for the performance of a contract and the finalisation and declaration of accounts was a problem that was common to naval administration. See Morriss, *Naval Power and British Culture*, 96–100.

[99] N. Baker, *Government and Contractors*, 58–62.   [100] *Ibid.*, 236, 242–54.

[101] This amount has been calculated to have been approximately equal to £8,589,452 in 1994: Hancock, *Citizens of the World*, 237.

Contractors and their management

What sort of men were they? The majority of army contractors had no earlier experience in supplying food to government forces. In 1776 only one had held a Navy Board contract, and only in 1781 did four more with similar experience get employed. Most were substantial London merchants, a significant proportion already working in finance. Of forty-six army supply contractors during the American War, ten were partners in London banking houses, while another three had banking interests outside London. Four were directors of insurance companies. The bankers were prominent in subscribing to government loans, but thirteen who were not bankers subscribed a total of £348,000 between 1777 and 1783. Ten contractors were, or had been, directors of the East India Company, or had acquired wealth in the company's service as a ship's captain, or husband who equipped vessels for their voyages. Indeed, between 1771 and 1782 there were on average four contractors on the company's Court of Directors. Twelve contractors had West India interests, either owning plantations or conducting business in the trade. Other interests included the trades to Africa, Portugal, the Levant and the American colonies, and the trades in linen and timber. Sixteen contractors had shares in the ownership of ships, some of which were hired into government service during the war.[102]

In 1793 the naval Victualling Board took over the supply of the army overseas. That board benefited from a fund of experience dating back to the seventeenth century and, unlike the Treasury, managed food supplies in peace as well as war. It was wholly dedicated to the task. Between the American War of Independence and the end of the Napoleonic War its policies show how, on the North America and West Indies stations, a body of reliable food suppliers was gradually developed. While closing loopholes for abuse, the board took into account contractors' risks in wartime and assisted those whom it trusted.[103] It tried to ensure the public received value for its money but that the terms of contracts were equitable, ensuring seamen and soldiers were well fed while rewarding good service.

In these endeavours, the Victualling Board was always handicapped by distance from the scene of operations and dependent on the return of accounts. Thus there was always a time lag before it could terminate irregularity, and always a chance it would emerge elsewhere. Regulation

[102] Baker, *Government and Contractors*, 225–37.
[103] See, for example, the board's discussion in 1794 of the uncertainties and risks involved in the arrangement of contract provisions for ships at Malta: NMM, ADM. DP/14, 31 Jan. 1794.

was an on-going battle. However, individual commissioners accumulated experience in spotting abuse and some seem to have had a nose for it. In the case of the St Lucia contract examined by the Commissioners on fees, mentioned above, one board commissioner, Montague Burgoyne, recorded that he suspected the contractor had deliberately issued larger quantities of high-priced provisions and that this suspicion resulted in the termination of the St Lucia contract.[104]

Consistently, the board attempted to deter deviants from accepted practice, including those who wished to escape their obligations simply owing to unforeseen circumstances.[105] At the same time, the Victualling Board was generous to contractors who had developed a reputation for honesty, fair dealing and faithful service. Thus in July 1779 the Victualling Board informed the Admiralty of the 'misfortune' of Bobert Biggins, contractor for the supply of HM ships at Jamaica, 'whose family have been contractors with this Board for near forty years, with great distinction'. The board was pleased to note that his relation, William Ward of London, 'who has also been a contractor with us for many years, and in whom we put great confidence', had undertaken to assist him.[106] The Biggins contract was terminated six months later when it was put up for tender and obtained by Ward. It took him into the big league. Since 1773 he had supplied fresh beef to ships putting into Newcastle, and from July 1783 he provided sea provisions at Leith.[107] Board favour thus permitted Ward to expand his business at a rapid rate.

However, contract conventions were still being beaten out.[108] In June 1784 Ward became the object of a prosecution filed in the Exchequer Court against the company of Blackburn, Shirley and Ward for substitutions in supplying victuals to the British fleet at Jamaica and the Leeward Islands. The contractors had supplied no olive oil or butter during the period of their contract but sugar and money instead – when the sugar was worth only 15 per cent of the value of the butter. It was alleged that substitutions were inconsistent with the terms of their contract, and that the money making up supplies to the value of the contract was paid privately by the contractor's agent to ships' pursers. Ward denied any inconsistency and, although the Crown Prosecutor asserted that he had had adequate reminders of the terms of his contract, the prosecution failed: arbitrators were called in and they decided in Ward's favour.[109] Yet thereafter few problems arose from substitutions.

---

[104] 8th Report of the Commissioners on fees, Appendix 7. See original MS copy, NMM, CAD.D/14.
[105] NMM, ADM. DP/103, 11 Feb. 1771.   [106] NMM, ADM. DP/111, 14 July 1779.
[107] NMM, ADM. DP/2, 22 Apr. 1782.   [108] NMM, ADM. DP/4, 18 Nov. 1784.
[109] NMM, ADM. DP/9, 28 July 1789.

In succession to Ward on the Jamaica station, Alexander Donaldson won the confidence of the board, which, in 1797, stood him in good stead. That year the unexpected detention in England of his victuallers for the supply of ships at Jamaica and St Domingue obliged Donaldson to make purchases of new provisions in North America for 9,374 men. By purchasing supplies twice over, Donaldson laid out over £40,000 more than he anticipated. The Victualling Board was thus persuaded, with Admiralty sanction, to make Donaldson an advance of £35,000 on account.[110]

Two years later the board went to Donaldson's help again. The quantity and quality of his supplies was the subject of complaint from Sir Hyde Parker, the Commander-in-Chief on the Jamaica station. Parker made the mistake of alleging the contractor had 'a superior interest' at the Victualling Board. Insulted by the 'style and language' of Parker's complaint, the board went on the offensive, claiming that

during the period of fourteen years in which he [Donaldson] has held the said contract we never had a complaint of his not having been prepared to comply with the demands made upon him or his furnishing any article of an improper or inferior quality until the command of H.M. ships on the Jamaican station devolved upon Vice Admiral Sir Hyde Parker.[111]

Soon after this, Donaldson once more got into financial difficulties. The board permitted him to transfer his commitments at Jamaica temporarily to another pair of London contractors, Messrs Jordan and Shaw, from whom, in October 1803, when Donaldson's embarrassment had passed, he was permitted to retrieve his contract.[112]

During the Napoleonic War, contractors suffered from an unusually difficult international situation which had the effect of reducing supplies of foodstuffs from both Europe and North America.[113] Prices rose, and with them the financial difficulties of contractors. This placed the Victualling Board in a dilemma. To exact penalties was often to ruin the contractors; not to do so was to abandon their sanction. It was a dilemma appreciated by the contractors. At least one proposed a compromise: that the partial fulfilment of his contract should be rewarded by the exaction of penalties in proportion to the degree of his failure.

At a time of scarcity, the Victualling Board had to be careful to prevent contractors reneging on engagements which might be difficult to place elsewhere. In October 1807, only a few weeks after signing their contract, the Flowers partnership backed out of an agreement to supply butter and

[110] NMM, ADM. DP/17, 25 Nov. 1797.   [111] NMM, ADM. DP/19, 30 Jan. 1799.
[112] NMM, ADM. DP/23, 26 Nov. 1803.
[113] J. Steven Watson, *The Reign of George III, 1760–1815* (London, 1960), 464–6.

cheese even though they had by then bought up all the available supplies on the English and Irish markets, by which they 'possessed upwards of two hundred tons of each'. The board commissioner handling the contract, Mr N. Budge, believed all the butter and cheese should be delivered to victualling yards and paid for at prime cost with merely an allowance to reduce penalties and cover the cost of carriage. In the event, the illness of Budge permitted Flowers to escape their contract and keep all the butter and cheese, which they resold to their advantage. But by July 1809 Budge had recovered enough to urge upon the Admiralty a standing order to prohibit any person or persons whatever from becoming a contractor or contractors with the Victualling Board after they had relinquished a contract under circumstances similar to those of which the Flowers had taken advantage.

Budge maintained that 'contractors might not in a Court of Law be considered as bound to conform to the principle I have urged; yet, I am firmly persuaded, that in a Court of Equity they would be declared to be so'.[114]

Conditions were difficult for contractors during the War of 1812 with America too. In July 1813 Andrew Belcher, contractor for ships at Nova Scotia, Halifax, Quebec, Norfolk in Virginia, and Bermuda, wanted an increase in his prices. The Victualling Board was sympathetic, observing that, when he made his undertakings in 1812, Belcher must have reckoned upon securing his supplies by indirect means from America. But when war began he had been completely shut out of the United States' markets, then out of Canadian sources. In consequence he was obliged to obtain his flour, pork, biscuit and pease from Britain where prices were unusually high. Potentially, his losses were ruinous because he was also engaged to supply the British army in North America. The board could have re-advertised Belcher's contract. But instead, having 'entered into a most strict and minute inquiry' of Belcher's costs, it persuaded the Admiralty to grant Belcher increased prices. Moreover, these were back-dated to 1 January 1813 for all commodities.[115]

Other instances of assistance occurred.[116] The Victualling Board defended its payments to contractors as the most economical means by which His Majesty's armed forces could be fed: 'It is scarcely necessary for us to remark that upon a *general principle* Government obtain their supplies by contract at a cheaper rate than if they were procured by

[114] NMM, ADM. DP/29, 25 July 1809.
[115] Subsequently, an anonymous letter condemned the board for increasing the prices without opening the contract to new tenders but the board insisted its policy was expedient: NMM, ADM. DP/33B, 30 July 1813.
[116] NMM, ADM. DP/34A, 17 Feb., 24 Mar. 1814.

their own immediate servants.'[117] On this principle, the board pursued a policy that was fair to contractors while achieving value for the public. At the same time, it supported reliable merchants and by this means achieved continuity and stability in its suppliers.

The bureaucratic control of expenditure

Because abuses were rumoured to exist, especially before 1780, Parliament did not trust the management of contractors by the naval bureaucracy. The estimates of military expenditure sanctioned by Parliament were relatively complex and parliamentary grants of funds were often exceeded by the issue of bills for payment in due course. Departmental accounting was scrupulous, but it could not easily be checked because there was no detailed public audit of expenditure. Meanwhile the public boards regarded their accountability for expenditure in detail as a matter secondary to the necessity to maintain the armed forces.

The system of funding

Parliamentary sanction for war expenditure had been a constitutional necessity since the Bill of Rights in 1689. However, the process by which funds were granted for state military operations was complex. Each department of government placed its estimates for the funding of its operations before Parliament. The Navy Board drew up the navy's annual estimates in consultation with the Admiralty. The Treasury was informed, as a matter of courtesy, before they were submitted to the King for his approval and then to Parliament.[118]

The naval estimates had three parts. They were calculated to meet the Ordinary costs of maintaining the navy, its Extraordinary costs and those incurred by Sea Service.[119] The Ordinary costs comprised the fixed costs of the naval establishment: they varied little from year to year, although they crept upwards with the size of the fleet laid up 'in Ordinary', the size of the officer corps and the scale of the bureaucracy that was maintained

---

[117] NMM, ADM. DP/34B, 30 Oct. 1814.

[118] P. K. Crimmin, 'Admiralty relations with the Treasury, 1783–1806: the preparation of naval estimates and the beginnings of Treasury control', *MM* 53(1967), 63–72. For earlier periods, see P. K. Watson, 'The Commission for victualling . . .', 341–2; and the *Calendar of Treasury Books and Papers, 1742–45*, ed. W. A. Shaw (London, 1908), xxvi–xxxix.

[119] For the Extra and Ordinary estimates of the Navy 1750–76, see TNA, ADM. 49/49, 50.

between wars. The Extraordinary costs comprised new building, repairs to ships and facilities, costs that mushroomed in wartime.[120] The Sea Service costs included the pay of officers and seamen, the refitting costs of ships, and the cost of food supplies and of ordnance supplies.

The Sea Service costs were estimated in terms of the number of seamen required to man the navy during the year. According to constitutional theory, the King in Council advised the Admiralty of the number of men that would be required and the Navy Board simply multiplied that number by the sum allowed for each man. From January 1649/50 until the end of the eighteenth century the sum allowed for each man was £4 per man per lunar month for thirteen lunar months, each of twenty-eight days. Of these £4, at the end of the seventeenth century 30 shillings was allotted to Wages, 27 shillings and 6 pence to the 'Wear and Tear' of the ships, 20 shillings to Victualling, and 2 shillings and 6 pence for Ordnance.[121] By the 1740s this allocation was amended to 30s/27s/19s/4s. But the total parliamentary allowance remained at £4 a man until 1797 when it was raised sharply from £4 to £7. From that time until 1832 the vote per man was divided 37s/60s/38s/5s.[122]

Otherwise, the method of calculating total annual costs in the Sea Service budget remained the same. The allowance per man was simply multiplied by the number of men to be fed and the number of lunar months in a year. The estimates in 1756, when 40,000 men were needed and the vote per man was 30s/27s/19s/4s, would have been calculated as follows:

Wages	30s	×	40,000	× 13 =	£780,000
Wear and Tear	27s				£702,000
Victualling	19s				£494,000
Ordnance	4s				£104,000
	80s				£2,080,000[123]

The Sea Service, the Ordinary and Extraordinary estimates for the navy were submitted to the House of Commons together, considered by the Committee of Supply, and voted in annual acts of appropriation. The

[120] *Naval Administration, 1715–1750*, ed. Baugh, 454–7; P. Webb, 'Construction, repair and maintenance in the battle fleet of the Royal Navy, 1793–1815' in *The British Navy and the Use of Naval Power in the Eighteenth Century*, ed. Black and Woodfine, 207–19.
[121] Ehrman, *The Navy in the War of William III*, 159.
[122] There was no distinct allowance for the Sick and Wounded, who were three-fifths funded from Wages and two-fifths from Victualling: P. K. Watson, 'The Commission for victualling', 332; *Naval Administration, 1715–1750*, ed. Baugh, 453; Bruce, 'The Department of the Accountant-General of the Navy'.
[123] Baugh, *British Naval Administration in the Age of Walpole*, 456–7.

money voted was issued, as needed, by the Exchequer and lodged in the appropriate bank accounts of the Treasurer of the Navy whose Paymaster and clerks paid the bills drawn on him by the spending departments. The bills issued by those departments provided the state with credit for as long as the course took to pay. It thus permitted the state to pay for goods and services for which it did not have the cash.

The system of accountancy

Accountancy echoed the system of credit funding and is well illustrated in the organisation of the Victualling Office. This was divided into the control of public money and the control of provisions. Both were managed on an expenditure–income basis. That of money was overseen by the Accountant for Cash; that of provisions was overseen by the Accountant for Stores. Beneath both of these chief Accountants, there was a variety of offices, each headed by a chief clerk.[124]

In each half of the office, debt for goods and services was incurred before bills promising payment were issued as required. The duty of the Accountant for Cash was to keep a register of all bills drawn upon the Treasurer of the Navy. These encompassed the whole expense of the navy so far as victualling was concerned. They included bills in course; bills of exchange drawn upon the board; advances, known as imprests, made for example to pursers; bills for clearing debts (also termed imprests) incurred by sub-accountants like pursers, ship owners and contractors; and all other bills incurred in the name of the board.

Of course, not one of these claims of expense was registered as a valid item of expenditure until vouchers were checked to justify payment. Many of the bills were actually made out in the Accountant's office. Knowing the costs currently being incurred, his office also checked the rates upon which contracts were being undertaken and, before 1794, the tonnage and freight of shipping employed by the board.

Debts incurred had a very high priority in the office – both those deliberately entered by the board through imprests or advances to sub-accountants, and those incurred by the latter through expenditure for which they were responsible. In each case the reasons for it, and the voucher attesting it, had to be passed as valid. Hence, under the Accountant was an office for examining and stating imprest accounts. The sub-accountants involved included agents, consuls, correspondents, pursers of ships, commanders of cutters, masters of transports, storekeepers and

---

[124] This was the logical division according to which the Commissioners on fees in 1788 wrote about the Victualling Office. See their 8th Report, 5–9.

messengers. They incurred debts, and thus made claims for refunds, in payment for provisions, water, stores, hire of boats, necessary money, and short-allowance money.[125] Bills to clear the debts were made out if the explanations and vouchers accompanying claims were approved; if not, the debt became an imprest or charge against the sub-accountant.[126]

Other debts included those for work performed by artificers and labourers processing and packing foodstuffs, for materials delivered into the yards, and for the work on buildings and machinery performed by contractors. Artificers and labourers were paid according to attendance by quarterly accounts of wages prepared by the Muster Master / Clerk of the Cheque. A clerk from his office also attended the receipt of all materials at the Deptford yard, the mills at Rotherhithe, and the brewery in Wapping to examine and certify quantity, quality and condition of deliveries; subsequently he certified the bills for payment of the contractors. Physical work on buildings and machinery was checked by the Office surveyor who had prepared the plans and estimates. He joined with the officer in charge of the department where the works were performed to sign the bill for payment, certifying that rates were correct, workmanship and materials satisfactory.

Bills of payment had to be signed by three board commissioners. They were then numbered and assigned to be paid by an issue of money from the Exchequer into one of the Treasurer of the Navy's accounts. Assignation was done in the Office for keeping a charge on the Treasurer. All other payments were similarly assigned to an account for payment.[127]

The debits and credits monitored by the Accountant for Cash had their parallel in the offices managed by the Accountant for Stores. Their primary task was to monitor the state and remains of stores in every storehouse. Always there had to be enough in store to feed the number of seamen voted by Parliament for that year. For this purpose, returns were made into the Accountant's office weekly, monthly and quarterly – all three from storekeepers at home, the last alone from storekeepers abroad. Each detailed what remained in store. In addition the weekly returns specified deliveries by contractors during the previous week, with remainders due; the monthly specified provisions returned from ships, distinguishing serviceable and decayed; while the quarterly

---

125 A separate office dealt specifically with the checks necessary to the payment of short-allowance money. Here short-allowance lists received from pursers were compared to the muster books at the Navy Office; abstracts and certificates were made out and lists for payment compiled in order to permit the clearance of pursers' short-allowance accounts, their payment if necessary by bill, and the payment of the seamen at the Pay Office in Broad Street.
126 8th Report of the Commissioners on fees, 5.   127 Ibid., 6–7.

comprehended all transactions, treating the responsible officer as a debtor for his receipts and a creditor for issues.

Having established the stock of provisions in store, the Accountant was in a position to order more. He did this when he donned the hat of Secretary to the board. In that role, with his chief clerk taking the minutes, he read to the board letters that had been received, registered the response, prepared replies, signed orders and undertook to have them executed. During the course of this business, when provisions were wanted, he secured the sanction of the board to advertise for tenders. He had the contracts drawn up, attested and sent to the departments where the materials were to be received. On the receipt of provisions, certificates of conformity to contract and of satisfactory standard were returned to the Accountant for Stores who had the Accountant for Cash make out a bill for payment. As a double check, the bill was returned to the receiving officer to certify on its back the receipt and fitness for use, upon which the bill was signed by the board and despatched to the contractor.[128]

Against the income of materials, the Accountant for Stores monitored their expenditure. Clerks to the Clerk of the Issues made out and signed bills of lading for victuallers proceeding from Deptford to the outports – Dover, Portsmouth and Plymouth – and overseas. They also registered the distribution of provisions to yards and ships within the Thames estuary, no further than the Nore. They drew up bills for river lighterage and home freights. They registered orders for victualling ships within the Thames and Medway and, when those ships were supplied, they recorded these issues as credits in their accounts of stock.

The Office for examining and stating yard agents' and storekeepers' accounts ensured the latter were received at the proper times and checked all provisions received against those issued, ensuring they were supported by valid vouchers. It looked for errors or omissions, surpluses and deficiencies. It formed separate statements for beer, bread, beef and pork 'manufactured' and the casks made or 'raised' from staves. It calculated any resulting waste and obtained comments from the yard officers on the causes, for the information of the board. During the late eighteenth century, this office extended its work to the accounts of army commissaries of stores and provisions received from the navy.

Storekeepers, of course, issued provisions to ships, tenders or transports, and their receipt and issue on board these vessels by the pursers was monitored by two offices working closely together. At a micro level, the Office for stating and balancing pursers' accounts examined

<hr />

[128] *Ibid.*, 7.

all transactions in detail, including provisions received and issued, their proportions, further purchases, and loans from one purser to another. Debits were recorded for provisions remaining the responsibility of the purser. Credits were given for the numbers of men victualled, loans, losses, condemnations and returns.

By comparison, the Office for taking cognisance of pursers' accounts worked at macro level and balanced the issues/credits of pursers and masters against their charges/debits. Their books were compared with those of captains and allowances were made for the number of, for example, soldiers who were victualled while on board. At the same time the validity of the captain's books and papers was verified for the purpose of passing their accounts. Purchases, for which pursers had made imprests, were checked and the final reckoning settled for the purser either to receive a balance bill or to make a payment himself to the Treasurer of the Navy. Occasionally, of necessity, certificates had to be prepared to allow wages to pursers whose ships may have been lost or taken.[129]

So receipts were checked against issues, debits against credits, on an income and expenditure basis, the whole supervised by the Accountant for Stores who could, when required, represent to the Victualling Board, for the benefit of the Admiralty, the state of the victualling both at home and abroad. At the same time the Accountant for Cash could report all expenditure, remaining debts, and the state of the Treasurer's accounts for clearing those debts. Accountancy was well organised and sophisticated.

The absence of public audit

However, accountancy primarily served the bureaucracy. There was no detailed public audit. As late as 1810 the parliamentary Select Committee on public expenditure observed that 'the duty of examining and passing the different public accounts is distributed among a great variety of offices, in some of which, namely in the Customs, Post Office, Stamp Office, Ordnance, Navy and War offices, the officers themselves will be found to be the auditors of their expenditure'. An Audit Office existed, a subordinate office of the Treasury, but it had 'no general power . . . to embrace and comprehend' all these offices' accounts.[130]

Hence, accounts of ship-building materials in the dockyards, of provisions issued by the Victualling Office and of money passing through

---

[129] *Ibid.*, 7–9.
[130] 5th Report from the Committee on the Public Expenditure of the United Kingdom, printed in Roseveare, *The Treasury 1660–1870*, 154–6.

the hands of the paymasters were audited in the Navy Office by the Comptrollers of Storekeepers' Accounts, Victualling Accounts and Treasurers' Accounts, respectively. The arrangement tended to conceal inconvenient backlogs of uncleared accounts.[131] But it persisted until 1829 when an Accountant General was instituted and required to state to the House of Commons his proceedings regarding the audit of departmental accounts. Only after 1832 were all the navy's accounts of expenditure subject to the audit of independent commissioners.[132]

Audits of naval accounts were carried out in the Navy Office partly because that office had existed longer than the other departments and partly because the Navy Board traditionally assembled the annual naval estimates. It was logical to locate the control of estimating and expenditure together. The only accounts that were sent farther than the Navy Office were those of the Treasurer of the Navy. Even these, after passing through the Navy Office, were not subsequently examined in detail. Before 1785 they were simply declared to the Exchequer's auditors of the imprests. After 1785 they were declared to auditors under Treasury supervision.[133] Yet they did only an arithmetical check; and no vouchers guaranteeing the propriety of expenditure were required.[134]

Meanwhile the Navy Board considered itself the navy's financial authority and was by no means deferential to requests concerning procedure or data from other branches of government, whether those departments were the Admiralty, the Treasury or the House of Commons.

In 1784, with regard to supplies secured by contract, the Admiralty questioned the grounds upon which the Navy Board had 'adopted the measure of Government allowing the discount on Navy Bills without it being communicated to their Lordships for their approbation'. The Navy Board responded that 'the same authority which vests the power of making purchases in this board must necessarily include the manner of payment', for without this means the board would be unable to take advantage of market circumstances as they changed. The Navy Board went on to observe that the authority it so assumed was not new: it had been 'practiced in contracts of the greatest magnitude, both in the former and the last war'.[135]

[131] For the management of pursers' accounts, see NMM, ADM. DP/2, 15 Feb. 1782; surgeons' accounts, NMM, ADM. BP/6A, 14 Sept. 1785; victualling accounts, NMM, ADM. BP/7, 25 Sept. 1787.

[132] Morriss, *Naval Power and British Culture*, 218–19.

[133] The board of professional auditors was established by 25 Geo. III *c*. 52; see Roseveare, *The Treasury 1660–1870*, 63.

[134] Binney, *British Public Finance*, 146, 149.   [135] NMM, ADM. BP/5, 3 Nov. 1784.

Likewise, in October 1783 the Navy Board stood on its dignity with respect to a Treasury investigation of naval expenditure and the authority for it between 1779 and 1782. The Navy Board provided the necessary information but then added:

We now beg to assure their Lordships that we are animated in the whole business with the same spirit that dictated their Lordships' enquiries, having in the first place the effectual execution of the public service in our view, and being at the same time under full conviction that a rational economy is essentially necessary in the whole conduct of it, both of which points shall as far as is in our power be duly and diligently attended to.[136]

However, it was the House of Commons which was treated with the most distance. In June 1805, when the effort to get a greater number of ships to sea was at its height, the Navy Board responded to a request from the House of Commons for the cost of repairs to ships in commission with evident irritation. It pointed out that accounts already supplied had been prepared 'with very considerable exertion and some inconveniences and delay to the current business of the office'; it then went on to explain:

The expense of repair from year to year given even to an individual ship cannot be ascertained without reference to many official books and papers, and generally to several dock yards . . . if difficulties exist in obtaining a correct account of the expense annually of a single ship, how much time and labour must be required to the detriment of the current business of the office to obtain a similar account for nearly every ship in service.[137]

In effect, the Navy Board suggested the practical work of equipping and maintaining the navy in commission had a higher priority at that time than statements of expense called for by the House of Commons.[138]

The growth of public control

The relative independence of state bureaucracy from public control was facilitated by the increasing pre-occupation of Parliament with other business: inquiries into, and legislation affecting, the structural fabric of finance, trade, and local and central government.[139] Its workload also doubled: the number of statutes passed rose from around 70 a session in the 1780s to over 150 by the mid 1810s.[140] Meanwhile, naval

---

[136] NMM, MID.2/55, Navy Board to Treasury Secretary, 2 Oct. 1783.
[137] PRO, ADM. 106/2237, 17 June 1805.
[138] See also NMM, ADM. BP/7, 27 Nov. 1787; ADM. BP/10, 3 Dec. 1790.
[139] D. Eastwood, '"Amplifying the province of the legislature": the flow of information and the English State in the early nineteenth century', *BIHR* 62(1989), 276–94.
[140] Harling and Mandler, 'From "fiscal-military" state to laissez-faire state, 1760–1850'.

bureaucracy enjoyed a period of expedience in the use of its money and financial instruments. Yet, especially after 1793, public opinion increasingly demanded personal accountability and, for a variety of reasons, expected the Treasury to exercise more detailed oversight over the proceedings of spending departments.

A vacuum in control

Despite the submission of estimates of expenditure, both the Admiralty and the House of Commons accepted they were conjectural. The House of Commons in the mid eighteenth century maintained some sense of control by rounding down the naval Extraordinary estimate. Between 1751 and 1765 this estimate was consistently rounded down to either £100,000 or £200,000. From 1766, however, this estimate was granted in full with no reduction, the House of Commons recognising that improvements in the dockyards were needed and new ships had to be built.[141] Full grants lasted until after the American War of Independence when there was no question that the fleet had to be repaired and rebuilt and the estimates were roughly amended in minor ways or passed without question as votes of confidence in the government of the day.

Even when amended, the parliamentary process suggested no close examination. The estimates were passed unaltered in seven of the nineteen years from 1774 to 1792.[142] Paul Webb observes that between 1783 and 1793 'parliamentary approval for Extra and Ordinary request became almost automatic'. They might be trimmed to a round £10,000 but were never seriously cut. Indeed, so generous were grants that they were not all spent. In the ten peacetime years, Ordinary estimates totalled £6,312,625 and parliamentary grants were £6,247,659. Extraordinary estimates totalled £6,198,130, grants £6,093,760.[143]

Meanwhile, expedience prevailed over discipline in the use of public money. In theory, money issued from the Exchequer for payments under a particular head of expense should not have been used for payments under another head. This was the Treasury's view with regard to both army and navy expenditure.[144] And the principle was employed by successive Treasurers of the Navy to justify holding large sums of money on account of one head while drawing more for a second or third. Had the

[141] Wilkinson, *British Navy and the State in the Eighteenth Century*, 135, 213.
[142] Binney, *British Public Finance*, 248–9.
[143] P. Webb, 'Construction, repair and maintenance in the battle fleet'.
[144] Binney, *British Public Finance*, 140.

money all been combined and applied indiscriminately to any service, the total sum held could have been much smaller.

The principle was criticised in the third report of the parliamentary Commission on accounts in 1781. As a result, an Act of Parliament in 1782 stipulated that the sums held by the Paymaster of the Forces or Treasurer of the Navy should be kept as small as possible and in the Bank of England.[145] However, certainly within the Sea Service budget, the Admiralty continued to transfer sums from one account to another.[146] In 1831 the First Naval Lord claimed 'it had always been the practice to consider that the gross sum voted was applicable to all purposes indiscriminately in detail, provided the total amount of the vote was not exceeded'.[147] In addition, when budgets ran dry, boards could run up debts in bills which the Treasury was responsible for clearing.[148]

The naval departments thus had a degree of flexibility in their expenditure. Members of Parliament sitting on committees of supply noticed anomalies, but were baffled by the information available to them and guided by traditional prejudices.[149] The navy was favoured while the army and ordnance were not.[150] In 1784 the House declined to pass the Ordnance estimates, which were still recalled twenty years later when Sir John Sinclair observed that 'plans of fortification, when they are brought forward by the board of Ordnance . . . have often proved a great and useless source of public extravagance'. He noted that the expenses of the Ordnance were 'in general extremely unpopular'. It was 'natural to suppose that, when once a country is sufficiently provided with artillery

---

[145] Roseveare, *The Treasury 1660–1870*, 63, 151.

[146] Wilkinson, *British Navy and the State in the Eighteenth Century*, 178.

[147] *The Parliamentary Debates . . . from 1803*, ed. T. C. Hansard, 3rd ser., II (1831), 983.

[148] In 1767 the former Treasury commissioner, Charles Jenkinson, was appointed to the Admiralty and attempted to align estimates, appropriation and expenditure so as to control debt. The Navy Board was 'sensible of the difficulty that must attend confining every branch of such [an] extensive concern as the Navy within exact bounds'. It pointed out that many unforeseen items of expenditure never appeared on any estimate. For this reason, Jenkinson acknowledged that his plan of alignment could only work in peacetime when unexpected contingencies did not occur on a large scale. Jenkinson's plan for controlling debt did not last beyond 1771. In the wake of the Falklands crisis, for which Parliament had made large grants, naval funds were in surplus and obviated any likelihood of growth in the naval debt. Thereafter the plan was forgotten: Wilkinson, *British Navy and the State in the Eighteenth Century*, 149–54.

[149] Sir John Sinclair grumbled in 1804 that the Extraordinary estimate 'frequently contains the names of ships and the sums they are to cost respectively, which are never expended for that purpose, whilst no mention is made of other vessels on which part of that very money is laid out': Sinclair, *History of the Public Revenue*, III, 214.

[150] Tomlinson, *Guns and Government*, 188–9, 192–7, 204–5.

and arms, it cannot require any great additional charge to keep up the stock'.[151]

The demand for accountability

Despite popular prejudices, from about 1780 Parliament consistently worked to create a fabric of regulations which preserved public money from both private and illegitimate bureaucratic use. For example, before 1785 Treasurers and paymasters enjoyed the interest from the public sums held in their private accounts, sums that were not inconsiderable. During his first twenty years as paymaster of marines, 1757–77, over £2¼ million was imprested to John Tucker.[152] An Act of Parliament of 1781 required Treasurers of the Navy to surrender balances in existing accounts within three months of their departure from office.[153] Another of 1785 vested the balances of a previous incumbent in his successor at each change of Treasurership. It also required money issued from the Exchequer for naval services to be lodged in the Bank of England.[154] It failed, however, to say where or for how long subsequent withdrawals from the Bank of England might be kept, which permitted Alexander Trotter, Paymaster during the 1780s, to lodge more than £1 million at a time in his private account at Coutts Bank for the convenience of making his payments, but from which, it was later discovered, large amounts were invested for his own personal profit.[155]

The depositing of public property in the hands of office-holders occurred in numerous other parts of the public service. Naval stores entering the royal dockyards became the responsibility of the yard storekeepers.[156] Other storekeepers included ships' warrant officers – the pursers, boatswains, carpenters and gunners who took charge of

[151] A. G. Olson, *The Radical Duke: The Career and Correspondence of Charles Lennox, Third Duke of Richmond* (Oxford, 1961), 81–6; Sinclair, *History of the Public Revenue*, III, 180, 216.

[152] NMM, ADM. BP/2, 12 Sept. 1781.

[153] 21 Geo. III c. 48. Initially the act was not rigidly enforced. By March 1783 only half the outstanding balances had been remitted; £3,000 held by John Tucker, Paymaster of Marines 1757–78, was not remitted until November 1785: NMM, ADM. BP/6A, 7 Nov. 1785. See also Binney, *British Public Finance*, 146, fn. 3.

[154] Certificates of receipts by the Governor and Company of the Bank of England on account of the Treasurer of the Navy survive in the papers of the Secretary to the Admiralty. They reflect the long-standing system of allotting the proceeds of taxes as well as the equally well-established system of naval estimating. Duties on malt, monies from the sinking fund, contributions to the lottery are allotted to headings under Victualling, Wages, Wear and Tear. See particularly the series in NMM, ADM. BP/6A, 4, 6, 20 Feb., 7 Mar., 8 May 1786.

[155] Binney, *British Public Finance*, 147–9.   [156] PRO, ADM. 106/2227, 14 Aug. 1801.

stores on board ships. Private ship builders constructing naval vessels in private yards and engineers carrying on building works at the dockyards were also sub-accountants. Their contracts, running on for several years, were funded by advances and final balance bills. Contractors who received advances for the purchase of naval stores or victuals also became sub-accountants of the Navy and Victualling boards. Accounts with contractors were run up by officers overseas, both yard storekeepers and commanders of ships.[157]

All could became debtors to the state, some for large amounts. Furthermore, indiscipline in the submission of accounts at timely intervals meant that central government did not always know how much money it had spent. In departments with sub-accountants overseas, this was a perpetual problem. With military forces dispersed around the world, it was a problem implicit in the maintenance of global power.[158]

Coincident with, and almost in reaction to, the 1805 peak in British naval power, there was a demonstration of the public demand for financial accountability in its public servants. The impeachment of Henry Dundas, Lord Melville, was an unmistakable signal to every official and employee of the British state that money or materials issued to him were to be appropriately employed. While he was Treasurer of the Navy between 1780 and 1800 Dundas failed to supervise the financial proceedings of his subordinate Paymaster, Alexander Trotter, who had contravened the 1785 Act of Parliament by lodging unspent naval funds in his personal bank account at Coutts.[159] Dundas's culpability appeared all the greater because he had himself sponsored the 1785 Act. His public trial in 1806 was enhanced by party animosity but was part of the movement through British bureaucracy corresponding with the Benthamite call for individual responsibility in state offices.

The furore raised by Dundas influenced the dispute into which the Victualling Board entered in 1806 with the Hon. Basil Cochrane, which lasted until 1818. The Victualling Board stood on the principle that Cochrane remained responsible for deficiencies in his accounts until he was cleared of that responsibility. Eventually, Cochrane appealed by publication to the people, clearly hoping to use the deep-seated distrust of state efficiency and military power to appear a victim.[160] But the

---

[157] 'Sub Accountants with the Navy Board...', the observations of the Navy Board for the information of the Treasury, c. 1784: NMM, MID.6/4.

[158] NMM, ADM. BP/4, 15 Dec. 1783.

[159] Morriss, *Naval Power and British Culture*, 96–101, 181–4.

[160] TNA, T.64/207, 1–9; see also Knight, 'Politics and trust in victualling the navy, 1793–1815'.

Victualling Board had the approval of utilitarian opinion and the dispute was settled by arbitration on an equitable basis.[161]

The growth of Treasury control

The same concern for the security of state assets underlay the gradual growth of Treasury control of government finance. What this meant should not be exaggerated. In 1989 Brewer in *The Sinews of Power* spoke of 'the overweening power of the Treasury'. Certainly in the seventeenth century, to achieve economy for the King, it gained oversight of state revenues and disbursements.[162] But the Treasury secured a 'comparatively weak grip on departmental expenditure' and, even after the Glorious Revolution, it had to work discreetly against branches of government jealous of their autonomy. Moreover, 'the relationship between parliament and the Treasury was not yet one of trust and collaboration'. During the eighteenth century the department was as closely associated with the political deployment of Crown patronage as with the maximisation of revenues, and neither endeared the Treasury to the tax-paying country gentleman, the City merchant or the colonist.[163]

Only with 'economical reform' did the Treasury gain more power. Critical was the Commission for examining, taking and stating the public accounts. Led by an equity lawyer, empowered to take evidence upon oath and recommend reforms, the commission issued fifteen reports between 1780 and 1787. Setting the 'benefit of the state' above all other considerations, it 'developed a comprehensive and quite radical philosophy of the public interest'. Utilitarian in appeal, it aimed to replace anomalies and anachronisms 'with machinery that was uniform, impartial, speedy and cheap'.[164] Economy demanded an agency that would represent the public interest. Despite a range of other duties, a small staff and relative 'ignorance of the requirements of the great spending departments', the Treasury was the appropriate state department to represent the public interest.[165]

However, war, financial reform and the slow grind of investigation, report and implementation hampered its ascent to power. For example,

---

[161] *Ibid.*; TNA, ADM. 114/3, Information of the Attorney General.

[162] Brewer, *The Sinews of Power*, 92, 129, 250.

[163] D. M. Clark, *The Rise of the British Treasury: Colonial Administration of the Eighteenth Century* (Newton Abbot, 1960), 33; I. R. Christie, 'Economical reform and "The influence of the Crown", 1780' in *Myth and Reality in Eighteenth Century British Politics*, ed. Christie (London, 1970), 296–310.

[164] J. Torrance, 'Social class and bureaucratic innovation'.

[165] Roseveare, *The Treasury 1660–1870*, 37–8, 56, 61–6.

so far as the naval departments were concerned, until the end of the Napoleonic War, once the Commons had voted financial supplies, the Treasury did little more to control naval expenditure than to ensure issues from the Exchequer remained within the sums voted by Parliament.[166] Nevertheless, three factors gradually increased the role of the Treasury after 1780.

The first was concern in the House of Commons about the growth in each successive war of the government's debt arising from loans and the issue of bills. Estimates of the National Debt,[167] and statements of the naval debt,[168] were submitted to Parliament each year, their format remaining the same for the whole of the eighteenth century. During the second half of the century, opinion focussed on the growth, decline or 'progress' of the debt.[169] Soon after the beginning of the American War of Independence, a precept of the House of Commons on 21 April 1777 called upon the Admiralty to explain the reasons why the debt of the navy had increased and the Admiralty was obliged to provide similar explanations in subsequent years,[170] even after the war. In a similar vein, in May 1785 the Commons demanded a return of 'monies which have been applied to naval services in the year 1784 over and above the grants of Parliament for the service of that year', and in 1786 it made a point of comparing over-spending in the period of peace twenty years before.[171]

Backed by such concern, the Treasury could make conditions upon which it would sanction expenditure. Hence in 1784 the allowance of an additional salary to the Navy Board commissioners for managing the transport service was annotated: 'P.S. This allowance from the Treasury is only a temporary thing'.[172] To monitor income and expenditure, the Treasury Board received periodic accounts of the state of balances held by naval accountants, and of the Navy, Victualling and Transport bills in course for payment.[173] By the end of the century the Treasury also received annual 'comparisons between the debt of the navy shewing the expense of each year with [the] increase or decrease thereof by the respective naval departments'.[174]

---

[166] NMM, ADM. A/2953, 4 June 1802; Binney, *British Public Finance*, 140.
[167] TNA, ADM. 49/38, estimates of the navy debt 1710–1801.
[168] TNA, ADM. 49/173, statements of debt 1686–1715, 1809.
[169] TNA, ADM. 7/567, the progress of the navy debt 1762–1805.
[170] TNA, ADM. 49/38, estimates of the navy debt 1710–1801.
[171] NMM, ADM. BP/6A, 10 May 1785; ADM. BP/6B, 6 Mar. 1786.
[172] NMM, ADM. BP/5, 24 Mar. 1784.
[173] The loose papers of the Secretary to the Admiralty contain many such statements; one example is in NMM, ADM. BP/5, 18 June 1784.
[174] TNA, ADM. 49/39, comparisons between the debt of the navy, 1801–12.

That these comparisons went to the First Lord of the Treasury is indicated by the annotation of a much-amended account for 1799: 'the debt as sent Mr Pitt was wrong, these corrections not being made till some days after'. Explanatory of changes in the amount of debt, Pitt also received 'particulars of the [annual] expense of all the naval departments . . . as far as can be ascertained from payments and bills issued'. The 'particulars' were made up in the Navy Office from accounts collected from each of the navy departments – the Navy, Victualling, Transport, Sick and Hurt – and from the office of the Paymaster of the Forces.[175]

The second factor making for Treasury control was the growth in the scale and complexity of war finance. Actual costs grew steadily. By the fourth year of each European war in the second half of the eighteenth century, the costs of the army, navy and ordnance had risen to between two-thirds and three-quarters of the amount of government expenditure. During the Napoleonic War, the relative costs of the armed forces declined but only because the cost of servicing debt had risen from between a fifth and a quarter of expenditure to about one-third of expenses. The growth of the debt made more complex the business of servicing it, comprising as it did successive loans to government.[176]

The Younger Pitt presided over the Treasury at perhaps the most critical period in the growth of the funded debt. Loans were raised by the government through the issue of annuities, consols, securities, bonds – all generally known as government stock. By 1786 there were eleven loan stocks in existence, supplied with money from 103 accounts, 77 of them conduits from revenues as they were paid into the Exchequer and then paid out under other heads; payments from the other 26 by-passed the Exchequer and went directly into the loan funds, some receiving supplement from other accounts. The rigid subdivision of revenue accounts meant that each type of taxable material had to have separately valued customs or excise duties attached to it. Management of these finances was a bureaucratic headache.

One of Pitt's major achievements in 1787 was to consolidate the revenues, for example from customs duties. It was a task which demanded an intellectual grasp that extended in his Consolidation bill to 2,537 separate resolutions covering existing duties. These were channelled into 8 new accounts, themselves funnelled into one consolidated fund that would serve all public expenditure. The effect on the bureaucracy of reform

---

[175] The accounts dated from 31 December 1798, and the aggregate version was annotated 'copy to Mr Pitt' 20 Feb. 1799: TNA, ADM. 49/40.

[176] Sinclair, *History of the Public Revenue*, III, 147, lists annual total loans 1793–1802, which together amounted to £232 millions.

was dramatic: 60 or 70 account books were reduced to about a dozen. Pitt claimed that the public accounts were removed from 'obscurity and intricacy' and rendered 'so clear and intelligible' that there was 'no man who may not, with a small degree of attention, become . . . master of the subject'.[177]

Even when the funded debt was under control, during wartime the unfunded debt required regular attention. The navy debt, comprising bills in course issued by all the naval departments, required regular transfers of funds to prevent the course getting out of hand and to maintain the credit of the state. In 1796 the introduction of ninety-day bills cut the duration of the course to three months. In 1797 the increase in rate per man in the Sea Service budget from £4 to £7 enlarged the supply of money available to meet the course. Nevertheless, funds still had to be juggled and generated. The record of the debt of the navy in 1799 includes reference to £110,000 'given by the Treasury to the Navy out of £500,000 voted for the service of Portugal in 1798'; and to £200,000 'given by the Treasury out of the vote of credit of £300,000 voted for the service of 1799'. That for 1800 mentions that 'in the months of October and November 1800 £500,000 was raised by the issuing of navy ninety day bills payable in 1801'; the sum raised was distributed to pay the bills of the Navy, Victualling and Transport departments.[178]

By the early nineteenth century the unfunded debt was much criticised. Sir John Sinclair observed in 1804 that, among other things, it placed 'a load upon the public with hardly any control'. He regretted that the contractor must not only have a profit upon his goods or services but demand a price in proportion to the risk of losing from the delay before he was paid according to the course.[179] Such criticisms permitted the Treasury to attempt the clearance of the debt in bills on a scheduled basis. Hence by 1810 the Admiralty was able to supply monthly estimates of bills that would need to be cleared throughout each year.[180]

The third factor strengthening the role of the Treasury was the series of parliamentary commissions and select committees into financial matters. Most notable among the former were the Commission appointed to inquire into fees, gratuities, perquisites and emoluments, 1785–8, and, specifically on naval administration, the Commission to inquire into irregularities, frauds and abuses practiced in the naval departments, 1802–5. Acting for the Treasury were the parliamentary Select

---

[177] Ehrman, *The Younger Pitt: The Years of Acclaim*, 269–73.
[178] TNA, ADM. 49/39, comparisons between the debt of the navy, 1801–12.
[179] Sinclair, *History of the Public Revenue*, III, 43–8.
[180] NMM, ADM. BP/30A, 15 Jan. 1810, printed in *British Naval Documents*, 476.

Committees on finance, 1797–1803, and the Select Committee on public expenditure appointed in 1810. During the early 1790s, the Admiralty was able to ignore many of the recommendations of the Commission on fees. Naturally enough, therefore, the hints urging implementation from the finance committee of 1797 became more firm, and the recommendations were eventually adopted by 1801.[181]

In 1798 the Select Committee on finance merely observed, of naval estimates amounting to more than £13 million, that 'a charge like this, from its magnitude, most evidently requires the strictest control'.[182] By 1810 the Select Committee on public expenditure was ready to press for the public audit of state expenditure across all government departments. It recommended the strengthening of the small Audit Office which existed within the Treasury.[183]

Hitherto, the Treasury had been little evident in influencing the conduct of financial affairs within the naval departments.[184] However, in response to expectations raised by the Select Committee on public expenditure, in 1811 the Treasury began to take an active interest in the navy's financial procedures. In March it required from the Admiralty reports from the subordinate boards of the manner in which departmental expenditure was authorised, and accounts of this expenditure were examined and passed. Suggestions for improvements in procedure were requested and the Victualling Board proposed an annual return to the Treasury of outstanding imprests and accounts, with an explanation of each.[185]

By 1812 the Treasury may be detected in accounts produced by the Navy Office of the annual average expense of fitting ships of each class, for home and foreign stations. The averaging of expenses from the cumulated costs of individual ships, by class, was an achievement. More important for the Treasury, however, was the knowledge it acquired of the costs of running the navy, which it could employ to check estimates and expenditure.[186] Peace in 1815 provided the Treasury with the opportunity to assert the power this information provided. Its influence was apparent in the restructuring and reduction of the naval departments.[187] The assistant secretary to the Treasury Board and its auditor, George

[181] See also J. Hoppit, 'Checking the Leviathan, 1688–1832' in *The Political Economy of British Historical Experience 1688–1914*, ed. Winch and O'Brien, 267–94.

[182] 31st Report of the Select Committee on Finance, 1798 in *Reports of Committees of the House of Commons*, 1797–1803, 1st series, XIII, 5.

[183] 5th Report from the Committee on the Public Expenditure of the United Kingdom, printed in Roseveare, *The Treasury 1660–1870*, 154–6.

[184] Crimmin, 'Admiralty relations with the Treasury, 1783–1806'.

[185] NMM, ADM. DP/31A, 21 Mar. 1811.

[186] NMM, ADM. BP/32A, 9 Jan. 1812.

[187] NMM, ADM. DP/36B, 28 Oct. 1816.

Harrison, was the principal adviser to the naval departments. His influence was informal, through conversation with the Comptroller of the Navy and the chairman of the Victualling Board.[188] But the resulting rationalisation demonstrated the new power of the Treasury. The Select Committee on finance in 1817 was told that by then 'no measure that involves an expenditure of public money' could be taken without previous communication with the First Lord of the Treasury and the Chancellor of the Exchequer.[189]

The growth in Treasury power over expenditure reduced the degree of independence in managing their finances hitherto possessed by individual departments in the state's bureaucracy. The growth in Treasury control derived from a demand for accountability, revealed in the impeachment of Lord Melville, as well as from the technicality of state finances. It diminished the flexibility the state's bureaucracy hitherto possessed in estimating, accounting and auditing its expenditure. With this relative independence – and spending equivalent to 43 per cent of the navy's budget (about 13 per cent of the state's expenditure in 1797) on stores and provisions – naval bureaucracy helped to shape the state's economy. During wartime, between 1755 and 1815, about three-fifths of state expenditure was on the armed forces. A proportion of this expenditure was met from loans and repaid as long-term debt. Over the whole period, about 15 per cent of national income was consumed by the military forces. In spending the state's money, the naval boards were careful to cultivate contractors who were trustworthy and to close loopholes against those who took advantage of imperfections in the system of contracting. As in the greater world of commerce, this system relied upon credit and gained strength from the maritime economy. This underlying economy also contributed to the capabilities of the state to develop its navy, improve its ordnance, secure provisions and obtain shipping for transport, and it is to these aspects of supply that the following chapters now turn.

---

[188] NMM, ADM. BP/36B, 18 Nov. 1816.
[189] Roseveare, *The Treasury 1660–1870*, 159.

4   Naval growth and infrastructure

Britain's maritime economy, united to the logistical capabilities of the state bureaucracy, facilitated the maintenance of a navy vital to the defence of the nation, its trade and colonies, and to mounting operations overseas. For the maintenance and expansion of the British navy derived from efforts in both the state dockyards and the merchant ship yards and from the purchase and hire of vessels from the private sector. Although better-known for the management of affairs at sea, the Admiralty developed an important role in planning for the supply of ships and resources. Meanwhile the Navy Board, which managed the dockyards, benefited from improvements in its methods of controlling the yards. There the number of docks was fundamental to the size of the naval force that could be maintained at sea, for they determined the number of ships that could be fitted, refitted and repaired for service. The size of the workforce and the efficiency of work incentive schemes were also important, as was the supply of materials in which shortages occasionally threatened crises. Yet foresight, purchasing initiatives and a responsive commercial sector managed to provide all the materials needed by the navy, even when it reached its greatest size during the Napoleonic War, and despite about half coming from sources outside Britain.

The expansion of the British fleet

No two sources agree about the size of the British navy between 1755 and 1815. Suffice to say, the British fleet increased from about 267 ships and vessels to 841 in that period (see table 4.1). At the beginning of 1810, the navy amounted to 976 vessels, over three and a half times its size in 1755. This fleet comprised 200 ships of the line and 776 smaller vessels. Not all were in commission for sea or harbour service: only 143 of the ships of the line (35 for harbour service) and 621 frigates, sloops and

Table 4.1 *Numbers of ships in the British navy, 1755–1815*

	Admiralty / Navy Office figures[a]	C. Derrick's Memoir figures[b]	W. James's Naval History[c]
1755	267	320	
1760	379	412	
1765	357		
1770	337		
1775	308	340	
1780	481	480	
1785	438		
1790	422	478	
1795	524	599	483
1800	768		729
1805	800	949	726
1810			976
1815			841

[a] TNA, ADM.7/567, corrected by the Commissioners for revising and digesting the civil affairs of the navy to allow for double entries.
[b] C. Derrick (Navy Office chief clerk), *Memoirs of the Rise and Progress of the Royal Navy* (London, 1806).
[c] Figures for 1 January of each year: W. James, *Naval History of Great Britain*, tables for selected years. Includes ships being built.

other vessels (37 for harbour service).[1] Even so, the physical demands of this number of ships were immense and the main burden of maintaining them fell on the Royal dockyards. It is a cliché, often repeated, that the state's dockyards were the largest industrial organisation in Britain. But all the materials they used were provided by merchant contractors and an increasing number of naval vessels were built and repaired in merchant shipyards. The private sector thus underpinned the state's resources.

Ships built in the Royal dockyards

Most of the largest warships, the first and second rates, and many of the smaller ones were constructed in the Royal dockyards at Deptford and Woolwich on the River Thames, at Chatham and Sheerness dockyards on the River Medway in Kent, and at Portsmouth and Plymouth on the south coast of England. The number of ships built in the royal yards was governed in part by the number of building slips available. Between 1700

[1] TNA, ADM. 7/567; Morriss, *Royal Dockyards*, 16–17, derived from James, *Naval History of Great Britain*, tables 1–25.

and 1830 the number in the six main yards increased at least fivefold. In 1700 Chatham and Portsmouth each had only one slip, and Plymouth had none.[2] But by 1742 there were twenty-one,[3] and by 1796 there were twenty-six: six at Chatham, five at Deptford and five at Portsmouth, four at Woolwich, four at Plymouth and two at Sheerness. Excluding Sheerness, each of the other five yards had enough large slips to build three first rates; Chatham had four slips of that size. Together the six yards had the capacity to construct simultaneously eleven first rates, five third rates, eight frigates and two sloops.[4]

But the number of ships built also depended on the number of ship-wrights employed. It was calculated that to have completed on each slip a ship of the largest size and completed each vessel in the shortest possible time would have taken the full-time labour of 40 shipwrights for a first rate, 30 for a third rate, 16 to 20 for a frigate depending on its size, and 12 for a sloop. To have fully manned each building slip in each yard in 1805 would have absorbed the labour of 150 shipwrights at Deptford, 140 at Woolwich, 192 at Chatham, 34 at Sheerness, 122 at Portsmouth and 120 at Plymouth – 758 men in total.[5] Yet in wartime, when refitting and repairs took priority, there were never enough shipwrights to per-mit this number to be employed on new construction. It was partly to develop a supply of labour not governed by demand from existing royal and private yards that, during the Napoleonic War, new building began at a new yard at Pater near Pembroke in south-west Wales.[6] Even with this facility, however, only forty-one ships of the line and seventy-eight other vessels were launched from the Royal dockyards in the twenty-two years between 1793 and 1815.[7]

Ships built in the merchant yards

In these circumstances, throughout the eighteenth century the Admi-ralty relied on the merchant yards for new building, especially for the many necessary small warships.[8] The merchant shipyards could build these vessels more cheaply and faster than the royal yards. They were permitted to build ships up to the size of small ships of the line: that

[2] J. G. Coad, *The Royal Dockyards, 1690–1850: Architecture and Engineering Works of the Sailing Navy* (Aldershot, 1989), 109.
[3] Baugh, *British Naval Administration in the Age of Walpole*, 261.
[4] TNA, ADM. 106/2237, 17 Aug. 1805.
[5] NMM, ADM. A/2990, 24 July 1805; TNA, ADM. 106/2237, 17 Aug. 1805.
[6] NMM, ADM. BP/36B, 12 Sept. 1816.
[7] Morriss, *Royal Dockyards*, 26–9.
[8] *Naval Administration, 1715–1750*, ed. Baugh, 192.

is, 60-gun ships before 1748 and 74-gun ships after 1763. For this, they were generally allowed three years. Smaller vessels were built in less time, but all were paid for by a series of imprests or advances from the time of contract signature to launching. Failures to keep to schedule, in particular to complete within contract deadlines, were punished by financial penalties.[9]

Before 1756 contract ship-building yards had mainly been confined to the yards along the Rivers Thames and Medway where there were about fifty merchant docks and launching ways. These yards were most easily kept under observation by overseers appointed from the royal yards, from where some materials were shipped, and to which hulls were floated for fitting and rigging from the naval storehouses.[10] Some building was also put out to yards at Ipswich and Harwich,[11] and in the Solent area, both regions easily accessible from the main dockyards. However, with the growth of the navy after 1756, the Admiralty extended its contracting north and west. This was not undertaken lightly, for there were risks. As Lord Sandwich put it in 1781:

Ships built at a great distance from the dockyards occasion vast delays and expense in getting their stores to them, which must be sent from one of the established yards . . . Engaging persons to build, who are not equal to the undertaking, gains no ground and is liable to every kind of abuse. When a warship is contracted for, a considerable imprest is made to the builder, to the amount of the value of the ninth of the ship; if he is not a man of credit and integrity he will delay your business, and employ your money to other purposes, being secure of your work, which you cannot take out of his hands on account of the imprest advanced.[12]

Nevertheless, during the second half of the eighteenth century, under a system of overseers appointed from the royal yards, contract ship building spread along the south coast, into the West Country, up the east coast to King's Lynn, Hull and Newcastle, and up the west coast from Bristol to Milford and Liverpool.[13] Before 1776 merchant building also took

---

[9] See B. Pool, *Navy Board Contracts 1660–1832: Contract Administration under the Navy Board* (London, 1966), 94–7, 130–5; also B. Pool, 'Some notes on warship building by contract in the eighteenth century', *MM* 49(1963), 105–19. For abatements and premiums for launch after and before contract date, 1801–17, see TNA, ADM. 49/102.

[10] P. Banbury, *Shipbuilders of the Thames and Medway* (Newton Abbot, 1971), 43; Baugh, *British Naval Administration in the Age of Walpole*, 255–7.

[11] A. G. E. Jones, 'Shipbuilding in Ipswich, 1700–1750', *MM* 43(1957), 294–305; and 'Shipbuilding in Ipswich, 1750–1800', *MM* 58(1972), 183–93.

[12] *The Private Papers of John, Fourth Earl of Sandwich, 1771–1782*, ed. G. R. Barnes and J. H. Owen (4 vols., NRS, 1932–8), IV, 293.

[13] See R. J. B. Knight, 'The building and maintenance of the British fleet during the Anglo-French wars, 1688–1815' in *Les marines de guerre européennes XVIIe–XVIIIe siècles*,

place in the colonies, especially in North America. In 1746–9, against Navy Board opinion, two small frigates were built in New England.[14] However, the subordinate board was usually able to build at places in England where there were significant numbers of artificers accustomed to working on large ships and where materials were available. It consistently employed the larger, long-established builders like Barnard, Dudman, Randall, Brent and Wells along the River Thames, Brindley on the Medway, and Hill at Bristol, whose expertise and business methods were trusted.

The merchant yards were an invaluable resource in time of danger. In 1778–81 Lord Sandwich made great use of them to regain parity with the combined forces of France, Spain and Holland, placing orders for 34 new third rates.[15] Between 1782 and 1790 15 new ships were launched in the dockyards, 30 from the merchant yards.[16] This proportion increased the percentage of merchant-built ships in the royal navy. Between 1688 and 1755 only 29 per cent of the Royal navy's line of battle ships came from merchant yards. However, between 1756 and 1815 the latter built 52 per cent.[17] The proportion of ships under 64 guns built by private yards also increased; by June 1805, of 683 frigates and smaller vessels in the navy, 63 per cent came from merchant yards and only 13 per cent from the royal yards.[18]

During the eighteenth century suspicions of profiteering and slovenly work gave rise to prejudice against merchant construction. Indeed, despite the great and growing importance of the private sector, some sea officers persisted in condemning merchant-building for being rushed, using green or otherwise defective timber, and producing weaker ships than those built in the royal yards.[19] Their prejudice had a slight foundation. An examination of all the 74s in the navy between 1801 and 1805

35–50, which lists locations with naval contracts from 1660 to 1832. For an objection to building at Milford Haven in Wales in 1790, see NMM, ADM. BP/10, 27 Aug. 1790.

[14] Pool, *Navy Board Contracts*, 84–6; J. A. Goldenburg, 'An analysis of shipbuilding sites in *Lloyd's Register* of 1776', *MM* 59(1973), 419–35.

[15] Baugh, 'Why did Britain lose command of the sea during the war for America?' in *The British Navy and the Use of Naval Power in the Eighteenth century*, ed. Black and Woodfine, 49–69; R. J. B. Knight, 'The Royal Navy's recovery after the early phase of the American Revolutionary War' in *The Aftermath of Defeat: Societies, Armed Forces and the Challenge of Recovery*, ed. G. J. Andreopoulos and H. E. Selesky (New Haven, CT, 1994), 10–25.

[16] Derrick, *Memoirs of the Rise and Progress of the Royal Navy*, 249; NMM, Middleton Papers, MID.1/141/1.

[17] This percentage excludes those built in India and Canada: Knight, 'The building and maintenance of the British Fleet during the Anglo-French wars, 1688–1815'.

[18] *PP* 1805(192), VIII, 277.

[19] Baugh, *British Naval Administration in the Age of Walpole*, 258–9.

shows that those built in merchant yards lasted on average forty years, those from the dockyards only four years longer.[20]

While he was First Lord of the Admiralty, Lord St Vincent sued one private shipbuilder for alleged defects in the *Ajax*, launched in 1798. In 1802–3 he also reduced the number of contracts for the construction of large ships with the aim of building all the navy's ships of the line in the Royal dockyards. However, even after the war, he refrained from agreeing any new building contracts for ships of the line, a policy which threatened to leave Britain short as decayed vessels were scrapped. It was a policy which Pitt attacked in May 1804 and brought about the resignation of the Addington government.[21]

When he succeeded St Vincent as First Lord, Melville promptly reversed St Vincent's policy and placed orders for 39 large ships with merchant builders, who were offered premiums on their prices for the early completion of their contracts.[22] Under the threat of invasion, enhanced prices per ton were also offered for the early completion of gun and smaller vessels.[23] Orders with merchant builders were maintained by Melville's successors so that between 1804 and 1812 the merchant yards launched 44 ships of the line, compared to 17 launchings in the royal yards.[24] The contrast between merchant and royal yards was even more striking for launchings of frigates and other vessels in the same period; 357 came from the merchant yards against 38 from the royal yards.[25]

The vital contribution made by merchant yards to British naval power is here evident. Between 1801 and 1815 the Navy Board made contracts with 113 yards for at least 520 vessels of all sizes.[26] These yards were also building for private ship owners and repairing a great number of their vessels. Most were merchant vessels engaged in private trade but many were hired as transports. One private yard on the River Thames, Dudman's yard, between 1803 and 1814 built 25 naval vessels and 13 merchant ships while repairing 4 royal ships, 123 transports and

---

[20] Morriss, *Royal Dockyards*, 27–9.   [21] Morriss, 'St Vincent and reform, 1801–04'.

[22] WLC, Melville Papers, 14 June 1804.

[23] DRO, 152M/c1804/ON, supplement to paper 18; TNA, ADM. 49/102, 'An account of the ships of war built in merchant yards since the 1st January 1801 for which bills have passed this office, pointing out those on which a premium has been allowed for being launched before the expiration of the time limited by contract . . .'.

[24] The prejudice against merchant builders did not re-emerge, probably countered by a requirement for the dockyards as well as the merchant builders after 1807 to provide realistic estimates of expense for new construction which revealed much comparability of cost: TNA, ADM. 95/8, estimates 1807–16.

[25] James, *Naval History of Great Britain*, tables 12–20.   [26] TNA, ADM. 49/102.

155 merchant ships.[27] As far as the merchant yards were concerned, naval ship building was only a facet of their business, albeit a lucrative one as long as they observed the terms of their contracts.

Ships purchased and hired

A third source of warships for the navy was by purchase. For example, 5 East India Company ships were purchased in 1797–8 to serve as 64-gun ships; another 7 company vessels were acquired in 1804.[28] However, many purchased ships were prizes from the agents of their captors.[29] During the French Revolutionary War the marked growth in the size of the Royal navy was principally due to the purchase of prize ships. Between 1793 and 1801, 43 ships of the line, 109 frigates and 65 other smaller vessels were added to the strength of the navy from this source. They originated from a succession of captures during the French Revolutionary War – 145 French and 53 Dutch warships, 18 Spanish vessels and 1 Danish.[30]

During the Napoleonic War this source of the navy's growth continued. Figures for 1789–1806 show 58 line battleships taken into service, 3 50-gun ships, 121 frigates and 190 sloops and other small vessels: 372 in total.[31] These acquisitions made a large proportion of the navy foreign-built: 35 per cent by 1801; not less than 25 per cent between 1797 and 1812. Once entered on the list of the navy as effective ships, the Admiralty used these vessels no differently from those built in Britain. Of the 27 British ships that fought at Trafalgar, 3 were captured French vessels.[32]

A fourth source of vessels for the navy was the merchant fleet. Before 1794 ships were conventionally hired by the Treasury, Navy and Victualling Boards for transports, a task taken over in 1794 by the Transport Board. However, gunned and manned brigs and cutters were also hired to supplement the warships available to the Admiralty. They were

---

[27] *PP* 1813–14 (88), VIII, 443, Minutes of evidence on the petitions relating to East India built shipping, in H. Doe, 'Enterprising women: maritime business women, 1780–1880', unpub. University of Exeter Ph.D. thesis, 2007, 216.

[28] Morriss, *Royal Dockyards*, 27; DRO, Addington Papers, 152M/c1804/ON, supplement to account 18.

[29] For the process in the early eighteenth century, see J. S. Bromley, 'Prize Office and Prize Agency at Portsmouth 1689–1748' in *Hampshire Studies*, ed. J. Webb, N. Yates and S. Peacock (Portsmouth, 1981), 169–99.

[30] Clowes, *The Royal Navy*, IV, 552–61.

[31] NMM, CAD. A/10, appendices to 15th report of the Commissioners of Naval Revision, apppendix 18.

[32] Morriss, *Royal Dockyards*, 27.

important during the American War of Independence,[33] and during the French Revolutionary and Napoleonic Wars. Between December 1793 and December 1801 the number grew from 54 to 120.[34] By March 1804 there were 68 gun vessels and cutters under hire, with another 19 hired East India Company vessels stationed around the British coasts. The Admiralty accordingly took account of the growing merchant fleet: in 1804 the Prime Minister, Henry Addington, had with his papers an account indicating that, between 1791 and 1802, an average of 45 vessels, each of about 225 tons, were built along the Thames and registered in the port of London each year.[35]

The role of the Admiralty

The Admiralty was responsible to the King in Council for the conduct of affairs at sea. Within parameters approved by the Treasury, it sanctioned expenses, authorised the commissioning of ships, controlled officer appointments, issued regulations and instructions, maintained discipline through courts martial, and managed the marines. The growth in scale of the British navy, the expansion of its global responsibilities, and the frequency of hostilities steadily increased Admiralty power. This included its influence over policy regarding the acquisition of warships, by new construction in the royal or merchant yards, by purchase and by hire. But the first concern of the Admiralty was always the conduct of affairs at sea.

Communication and coordination

Orientated to the conduct of affairs at sea, the Admiralty was heavily dependent for its information on its communications with sea officers. These took the form of Admiralty orders and directions, with regular, often daily, letters and reports from captains and admirals. Moreover, the official correspondence was often supplemented by personal correspondence. At times Admiralty commissioners, like Sir Hugh Palliser in 1778 and Lord Hugh Seymour in 1798, sailed with the Channel fleet and added their private communications to the correspondence, both public and personal, of the fleet commanders.

[33] For contracts of hire for armed vessels in the American War of Independence, see WLC Shelburne Papers, vol. 151, 24 Oct. 1782; for a brig's contract, see NMM, ADM. BP/3, 10 Oct. 1782.
[34] NMM, ADM. BP/24A, 28 Mar. 1804.
[35] DRO, Addington Papers, 152M/c1804/ON, accounts 5, 8, 19.

The official communications provided the Admiralty with its principal means of control at sea. Their value was enhanced after 1800 when clerks began recording every item and subject of correspondence in huge digests and indices.[36] In turn Admiralty proceedings were more formally recorded. By 1810 Board Room journals recorded the correspondence, with the orders and instructions issued by the Admiralty each day to ships in ports around the British Isles and as far south as Lisbon.[37] By 1812 these were accompanied by daily returns to the First Lord of arrivals, sailings, information from and proceedings of all ships in seas around the British Isles and as far south as Cadiz and Madeira.[38]

While the Admiralty Board conducted the navy at sea, it was the Navy Board which supplied the ships for those operations. It managed the dockyards, arranged with merchant builders for ships to be built, and accounted for naval expenditure. In many respects, the Admiralty was dependent on the information, advice and judgements of the subordinate board with regard to construction, purchase and hire. Trust and communication between the Admiralty and the Navy Board was thus essential, a matter stressed by Sir Charles Middleton and Sir Andrew Hamond, both chairmen of the Navy Board, when their recommendations were not heeded.[39]

The Admiralty was further removed from the practical business of dockyard operations. On the other hand, in some respects the Admiralty had wider horizons than the Navy Board. As a political office, it was closer to public opinion, in Parliament as well as the Cabinet, and Admiralty initiatives tended to reflect contemporary concerns. The Admiralty too was responsible for the performance of the navy at sea; it thus had to ensure the logistical support of those operations was what was required at sea. Inevitably it listened to sea officers, and had to ensure the Navy Board took cognisance of their views. At critical times, moreover, the Admiralty took initiatives in instituting new arrangements affecting the construction and maintenance of ships which it judged in the public and naval interest: for example, the introduction of piecework and new dock construction programmes.

---

[36] TNA, ADM. 12/1–4 index Admirals' dispatches 1711–93; ADM. 12/5–174, index and digest in-letters, 1793–1815.

[37] TNA, ADM. 7/257, journal for Jan.–June 1810.

[38] TNA, ADM. 7/502, 20 Mar.–29 July 1812.

[39] R. J. B. Knight, 'Sandwich, Middleton and dockyard appointments', *MM* 57(1971), 175–92; for Middleton's strictures on trust to Howe, see his 'Observations on the Estimates given into parliament by the Navy Board, 21 March 1786', NMM, MID.14/13; Morriss, 'St. Vincent and reform, 1801–04'.

The Admiralty was not helped by the political nature of the office of First Lord, which ensured the holder of the office usually changed with every alteration of the King's ministers. The other board commissioners, though civilians as well as sea officers, also sometimes changed with membership of the government. During the eighteenth century, despite being members of the House of Commons, the two Admiralty Secretaries tended to survive changes of ministry. But even this altered after the political resignations of 1804 and 1806, when an understanding was reached that the first Secretary was political and liable to change, while the second remained permanent.[40]

With a political board, much Admiralty business fell to the Secretaries. At the height of the Napoleonic War in 1809, on becoming first Secretary, J.W. Croker reported his workload as 'quite terrific'. Later, the Duke of Wellington 'said that he looked on the Admiralty [Secretary's office] to be, after Ireland, the most important place out of the Cabinet'.[41] Experience in the Secretary's office was treasured. Sir Philip Stephens, clerk in the Admiralty 1751–9, then successively second and first Secretary 1759–95, was kept on as a commissioner at the board until 1806. Sir Evan Nepean, Secretary 1795–1804, briefly Chief Secretary in Ireland, returned to the Admiralty as a commissioner in late 1804–6, but subsequently became Governor of Bombay in 1812–19.[42]

The Secretaries were not helped by the small scale of Admiralty bureaucracy. Excluding board commissioners and Admiralty Court officials, the Admiralty office amounted only to about forty staff in 1760, including nineteen clerks, rising to about fifty-nine in 1810, including thirty-two clerks.[43] While the bureaucracy of the Admiralty increased by only 50 per cent, the size of the navy in commission tripled in size. In counterbalance to this disproportionate growth in the commissioned fleet, the Admiralty became more economical in its record keeping, and relied for some information upon records kept in the Navy Office. Yet there were disparities between, for example, the former's List Books of ships in commission and the latter's Progress Books of ships out of commission at the dockyards.[44] In 1760 the disparity amounted to 55 ships; by 1780 it was 62 ships; and by 1800 it was 144 ships. It seriously distorted estimates of strength at sea. By 1805 sixteen 74s were listed at

[40] Sainty, *Admiralty Officials*, 24–7, 34–7, 101–3.
[41] B. Pool, ed., *The Croker Papers* (London, 1967), 14.
[42] Sainty, *Admiralty Officials*, 141, 152.   [43] NMM, ADM. BP/28A, 25 Mar. 1828.
[44] The Admiralty's List Books recorded the details of every ship on each station, while the Navy Office Progress Books recorded work performed on each ship when they visited the dockyards: TNA, ADM.8/35 and 99, List Books for 1760 and 1810; TNA, ADM. 180/6, 9–11, Progress Books of which copies are at NMM.

sea at the Admiralty but recorded as being at a dockyard in the Navy Office.[45]

The discrepancy between the records kept at the Admiralty Office and those in the Navy Office reflected the physical distance of the two departments from one another, and the problem of communication that entailed. While the Admiralty remained in Whitehall, the Navy Office was 3 miles away on Tower Hill until 1785 when it moved to Somerset House in the Strand. Nevertheless it remained over half a mile distant from the Admiralty. Always, dockyard matters affecting ships and decisions regarding finance had to be referred to the Admiralty by letter, a practice that extended the time taken for yard communication by two days, a period of no small importance when sea and yard officers hung on decisions affecting a ship's docking on a particular high tide.

Knowledge and planning

Nevertheless, the Admiralty did not neglect its civil departments. In 1749 the Admiralty began a series of dockyard visitations which continued intermittently into the post-war nineteenth century.[46] These could generate conflict as well as understanding, but in the long term they served to promote common knowledge between the boards and an appreciation at the Admiralty of the problems of managing a large-scale industrial organisation. The 1792 visitation demanded from the yard commissioners, among other things, accounts of the contents of every storehouse, the state and condition of every ship being built and in Ordinary, the numbers of all yard employees and reports on all 'experiments trying'.[47]

Being dependent on the dockyards for the building, fitting, refitting and repair of ships, the Admiralty was keen to increase their productivity. Through fifteen years of service from 1744, both as a civil commissioner and as First Lord, Lord Sandwich acquired knowledge of the dockyards that prompted him in 1775 to attempt an increase in the pace of ship building by the introduction of piecework for shipwrights.[48] He also attempted to ensure ships were built of seasoned timber via the requirement for dockyards to keep a three-year stock of timber. The initiative gave rise to the erection of timber seasoning sheds, some of which still

[45] TNA, ADM. 7/567, 'Abstract of the Monthly List Book and the Navy Progress', made by Charles Derrick, 1806.

[46] J. M. Haas, 'The Royal Dockyards: the earliest visitations and reforms, 1749–1778', *HJ* 13(1970), 191–215.

[47] 'Accounts called for by the Admiralty at their visitation', NMM, GRE/8.

[48] J. M. Haas, 'The introduction of task work into the Royal dockyards, 1775', *JBS* 8(1969), 44–68.

survive at Chatham. Sandwich was castigated in 1779 by Sir Charles Middleton, the Comptroller of the Navy Board, for the 'sad management that prevails' at the Admiralty.[49] But he supported Middleton both in the introduction of copper sheathing which increased the period ships could go without refitting and having their bottoms cleaned, and in the introduction of carronades which enhanced the fighting performance of ships.[50]

St Vincent too aimed to make the dockyards more productive. He was encouraged to eschew merchant building for dockyard building in 1802–4 because he was told that, if classed by ability, forty-six dockyard shipwrights could build a 74-gun ship in one year, and that in this way ten new ships of the line could be launched from the Royal dockyards every year, even during wartime when the whole fleet could still be kept in repair.[51] Two fir frigates were built in this way as an experiment, but no ships of the line, as both the shipwrights and Navy Board opposed the scheme. St Vincent's successors, Melville and Barham, reversed his policies and returned to the employment of contractors. After a decade of war, however, many pre-war ships were worn out. At the same time, there was a shortage of timber, partly the product of St Vincent's distrust of contractors, but equally on account of a growing scarcity of suitable domestic wood.[52] In the circumstances, the Admiralty undertook the vital role of forward planning.

Melville revealed a full appreciation of the problem of maintaining the British navy. In September 1804 the enemy had only seventy ships of the line compared to Britain's eighty-one in commission. However, investigation of the condition of those eighty-one revealed differences in durability. Thirty-seven were estimated fit for service for five years; twenty-six were estimated fit for service for three years; and eighteen were considered fit only for home service. Melville concluded that 'provision must be made not only for the current repairs of all ships in service but for a fresh supply of serviceable ships of the line to fill up the place of those become unfit for service at the end of their specified periods'.

Provision had to be made for new building as well as repair. Ultimately, Melville concluded, 'there can be but one limitation – viz. the powers of the country and the means of carrying into execution the building of large ships, either in the King's or merchant yards'. Steps to increase Britain's

---

[49] Middleton's censure may be found in *Letters of Lord Barham*, II, 2–6.
[50] R. J. B. Knight, 'The introduction of copper sheathing into the Royal Navy 1779–1786', *MM* 59(1973), 299–309.
[51] For full explanation of St Vincent's building policy, see DRO, Sidmouth Papers, 152 M/c.1804/ON, paper 21.
[52] Morriss, 'St Vincent and reform, 1801–04'.

naval strength had to 'be systematically begun and adhered to without intermission for a long track of years'.[53] Melville's recommendation was observed by his successors who maintained a succession of orders that gave rise to forty launchings of new ships of the line from merchant yards between 1807 and 1812, with another thirteen from the royal yards.

Melville was advised by Barham who knew the vital importance of new construction. Having been Comptroller of the Navy during and after the American War of Independence, he appreciated the necessity for employing the merchant yards as well as the Royal dockyards to repair and rebuild the fleet.[54] During the Napoleonic War he was determined to increase the number of ships at sea even if this demanded that some rickety vessels had their hull timbers doubled and braced.[55] By this means he created spare capacity in the fleet to permit the replacement and redeployment of ships as wear, tear and the movements of the enemy demanded.[56] It was this spare capacity which permitted the redeployment of ships preceding the battle of Trafalgar.[57]

Not only did Barham think about the logistics of operations at sea, he ensured the dockyards were 'well supplied with every kind of naval stores'. The advantages, he argued, were manifold: 'it not only enables us to refit the fleet with greater celerity, and to keep it in a constant state of activity, but in the political point of view' reduced Britain's dependence 'on the course which affairs may at any time take on the continent, from whence our principle supplies are drawn'.[58] After the American War, Barham had stressed the necessity for foresight and preparation in the supply of the fleet.[59] To Pitt on 22 May 1805, three weeks after he became First Lord of the Admiralty, Barham revealed that he was still following the same policy and was intent on ensuring that all the dock and victualling yards were complete with stores and provisions, and had left 'nothing to seek' for when sudden demands were made upon them.[60]

During the Napoleonic War, the Admiralty used the resources of the empire to enlarge the power of the British navy. In 1812, faced with another American war on top of the European conflict, Robert Dundas, second Viscount Melville, placed orders for building new ships at Bombay as well as at the new Pembroke yard.[61] By these means, he was able to reverse the balance in naval ship building. While only two ships of the line were launched from merchant yards in 1813–14, eleven were launched

[53] WLC Melville Papers, 3 July 1804.
[54] Webb, 'The rebuilding and repair of the fleet 1783–1793'.
[55] *Letters of Lord Barham*, III, 111–12.   [56] *Ibid.*, III, 110–11.
[57] See pp. 50–3.   [58] *Letters of Lord Barham*, III, 113.
[59] *Ibid.*, II, 77.   [60] *Ibid.*, III, 81.
[61] NMM, ADM. BP/32C, 26 Nov. 1812; ADM. BP/33C, 24 July 1813.

from royal yards in 1813–15.[62] By 1815 the second Viscount Melville was accustomed to taking decisions personally regarding new construction.[63] It was a hands-on method of management that fully involved the Admiralty in planning for the navy and reflected the over-arching problem that faced the British state: how to maintain a navy of unprecedented size at a time when resources were becoming scarce.[64]

Navy Board responsibilities

If the Admiralty took the long view, the Navy Board took the short in resource management. It managed the navy's finances, the dockyards and the recruitment of seamen. When war came, it was the Navy Board that ensured ships were fitted and manned; it then informed the Ordnance Board when they might receive their guns, and the Victualling Board when ships could receive water and victuals. Its commanding position in the finance, equipment and commissioning of ships gave the Navy Board the senior position among the boards subordinate to the Admiralty. Its performance in the mobilisation of the navy was critical to power at sea at the opening of hostilities.[65]

The dockyards, overseas yards and stores depots

Central to the Navy Board's responsibilities was the management of the Royal dockyards. There were also small supply depots around the British coast – for example, at Deal, Leith and Kinsale – and larger refitting and replenishing yards abroad. The latter were essential to the maintenance of squadrons on important stations, and were first established in the Mediterranean and the Caribbean.[66] After the capture of Minorca in 1708, a small yard was established in the bay at Port Mahon. War from 1739 saw both the Navy Board and the Victualling Board improve the staffing, storehouses and refitting facilities there. The base was lost in 1756, regained in 1782, but finally given up in 1802.[67] After its

---

[62] Launch figures from Morriss, *Royal Dockyards*, 28.

[63] See his minute about building fir frigates, NMM, ADM. BP/36B, 12 Aug. 1816.

[64] A. Lambert, *The Last Sailing Battlefleet: Maintaining Naval Mastery 1815–1850* (London, 1991), viii–ix, 13–14, 16–22, 113–14, 164–6, 182–3.

[65] So much so that Charles Derrick, chief clerk in the Navy Office department for stores, produced as his memorial 'a collection of orders, letters and minutes' to serve as precedents for future occasions: TNA, ADM. 106/3063–4.

[66] B. Lavery, 'The British navy and its bases, 1793–1815' in *Français et anglais en Mediterranée de la Revolution française à l'indépendance de la Grèce (1789–1830)* (Vincennes, 1992), 159–67.

[67] *Naval Administration, 1715–1750*, ed. Baugh, 326.

capture in 1704, Gibraltar remained undeveloped as a naval base for three decades. But in 1736 a victualling storehouse was established there, and in 1740 a hospital, even though British ships were still careened at Lisbon in preference to Gibraltar. Its value as a refitting base thus remained limited until the second half of the century, its main value being strategic.

Port Royal was used as a naval base from the time of the capture of Jamaica in 1655. By 1713 there was a naval storekeeper who had a storehouse in Kingston, but in 1726–7 Hosier's squadron found the facilities at Port Royal inadequate in terms of both storage and wharves. The port was vulnerable to storms. Having suffered an earthquake in 1692 and a fire in 1703, Port Royal was hit by hurricanes in 1712 and 1722. Central to the Caribbean, Jamaica nevertheless remained a strategic base, supplemented from 1730 by a small yard at Antigua in the Leeward Islands. Food was delivered in both places by contract. Farther north, a yard was established at Halifax on the coast of Nova Scotia during the Seven Years' War,[68] and was supplemented during the American War of Independence by the availability of facilities and supplies at Rhode Island, Charlestown and New York. A yard developed at Bermuda from 1795 and was especially useful during the War of 1812 with the United States.[69]

Elsewhere during the Napoleonic War, yards developed at Malta[70] and the Cape of Good Hope. Farther east, the East India Company's facilities at Bombay were available to the navy from the 1790s. Stores were also available at Madras from 1796, at Penang from 1798, at Mauritius and Trincomalee during the second half of the Napoleonic War, and at Sydney, New South Wales, from 1812. Overseas depots for stores tended to develop as expedience dictated.[71] But there were long-term schemes to improve the major home yards and some overseas yards. A major scheme was implemented by fits and starts at Portsmouth from the 1760s. This included new storehouses and docks. Between 1796 and 1803 the basin was enlarged and the number of docks increased. At Chatham between 1783 and 1804 two vast storehouses were built and the ropeyard reconstructed. A new eastern yard to replace Deptford, Woolwich and Chatham, all subject to the silting of their river approaches,

---

[68] Gwyn, *Ashore and Afloat*, 3–10.
[69] M. Lester, 'Vice-Admiral George Murray and the origins of the Bermuda naval base', *MM* 94(2008), 285–97.
[70] P. MacDougall, 'The formative years: Malta dockyard 1800–1815', *MM* 76(1990), 205–13.
[71] Morriss, *Royal Dockyards*, 4–5.

was under consideration from about 1800 but, eventually, after 1812, Sheerness yard was enlarged.[72]

The methods of control

The main difficulty for the Navy Board was that the Navy Office until 1785 was on Tower Hill in London, then in Somerset House in the Strand in London. That was 40 miles from Chatham, 70 miles from Portsmouth and 220 miles from Plymouth. The Navy Board had its own couriers who could reach Deptford in half an hour, Woolwich in less than an hour, Chatham in 4 hours, Sheerness in 6, Portsmouth in 8 and Plymouth in 24. Yet distance from the site of dockyard operations rendered misunderstanding and misdirection a constant possibility. The distance of the board from Plymouth was especially challenging, for the development of the western squadron of the Channel fleet demanded that that yard grow.[73] However, the Navy Board employed a range of methods to ensure it knew what work was being performed at each yard.

Fundamental was the necessity to be reminded of the facilities available at each yard. About 1795 Nicholas Pocock, the marine artist, was commissioned to produce a series of bird's-eye views that clearly showed each dock and slip.[74] The ships undergoing work were reported weekly in comprehensive 'progresses'. These briefly outlined when the work might be completed, the nature of the repair, time when taken in hand, the work so far performed, the number and type of artificers employed.[75] At the Navy Office, some of this information was abstracted into the Progress Books which recorded where ships were built or acquired and all subsequent work on them. The books recorded the time when ships arrived, docked, undocked and sailed with the nature of the work performed and the costs incurred.[76] The amount of information given for each visit varied considerably. Nevertheless the Surveyors of the Navy could gather from these books the maintenance record of every ship. Files for individual

---

[72] Coad, *The Royal Dockyards, 1690–1850*, 17, 25–6.

[73] Baugh, *British Naval Administration in the Age of Walpole*, 338–9; R. J. B. Knight, 'The performance of the Royal dockyards in England during the American War of Independence' in *The American Revolution and the Sea* (Basildon, 1974), 139–44; NMM, ADM. BP/3, 21 Nov. 1782.

[74] Four views survive in NMM, for Deptford, Woolwich, Chatham and Plymouth yards. See D. Cordingly, *Nicholas Pocock 1740–1821* (London, 1986), 66–9.

[75] Weekly progress for 3 May 1779, transcribed from NMM, POR/J/2, printed in *Portsmouth Dockyard Papers 1774–1783: The American War*, ed. R. J. B. Knight (Portsmouth, 1987), 23–5.

[76] TNA, ADM. 180/6, 10–11, Navy Office Progress Books. Copies at NMM.

ships were not kept in the eighteenth century but 'ships' covers' exist from 1807 documenting design, construction, alterations and refits.[77]

Stationed at each of the Medway and south coast yards was a yard commissioner. From 1806 Deptford and Woolwich together also received a commissioner. In 1794 there was only one yard commissioner at an overseas yard but by 1814 there were ten.[78] Theoretically each yard commissioner was a member of the Navy Board and acted as the eyes, ears and voice of the board. Occasionally other board commissioners were deputed to visit the yards, which invariably resulted in new regulations.[79] The board issued standing orders for the long term and instructions for work in hand, eliciting from the Admiralty priorities with regard to the ships. Superior direction tended to diminish the apparent authority of the yard commissioner but his role was vital in labour relations and in the regulation of life and work within the perimeters of each yard.[80]

This was all the more important because of the dubious value of the Navy Board standing orders. Issued since the seventeenth century to suit the particular circumstances of each yard, they had accumulated in large unclassified files that defied the efforts of officers to discover the board's precise requirements with regard to any practice or operation. In 1767 only at Portsmouth were the orders read every three months, and only at Plymouth were new or important orders fixed up on gates and in other parts of the yard. Twenty years later, Charles Middleton observed that 'such officers as were inclined to act properly had no fixed instructions to direct their conduct; those who had other views had so many holes to creep out at as put it out of powers of office to detect'.[81]

The trouble was compounded by the failure of the Navy Board to revise copies of orders in the Navy Office which contained contradictory or obsolete instructions going back to 1658. In 1764, wanting itself 'to judge whether it may be expedient to make any alteration therein', the Admiralty Board ordered a revision of the orders. Two years later the Navy Board sent copies of those still in force. But the Admiralty noted that procedure was 'dispersed without order or method in a variety of warrants issued at different times through a long course of years, and they are not well digested so as to afford in a proper and clear manner

[77] Originally TNA, ADM. 138, these files were transferred to the NMM between 1961 and 1977. They contain papers on design, calculations, contracts and correspondence.
[78] Thanks are due to Dr John Day for this information.
[79] For the 1785, 1817 and 1818 board visitations, see NMM, ADM. BP/6A, 15 June 1785; ADM. BP/37B, 22 Nov. 1817; ADM. BP/38B, 7 Nov. 1818.
[80] For the assertion of government authority in Sheerness yard, see R. Morriss, 'Labour relations in the royal dockyards, 1801–1805', *MM* 62(1976), 337–46; Morriss, *Royal Dockyards*, 175–7.
[81] *Letters of Lord Barham*, II, 188, 226.

the information necessary to the conduct of the officers and workmen'. A digest was thus ordered, under proper heads, to compose 'one uniform body and system'.[82]

Nothing was achieved before the American War of Independence. Afterwards the Comptroller of the Navy Board, Charles Middleton, undertook a digest and sent it to the Admiralty in 1786. There it was ignored. Middleton urged on the Commissioners on fees the necessity for new standing orders, arguing to Pitt that

almost every efficient appointment in the yards being in the admiralty, and frequently made an object of interest, officers feel less anxious for their conduct, and particularly when they know that a proper vote would cover a corrupt practice. In particular very improper connections are formed between contractors and the offices; and the public suffers in thousands for a trifling gratuity received by a yard officer.[83]

Yet the recommendations of the Commissioners on fees were not referred to the Admiralty until 1792 and that for the dockyards was not implemented until 1801.

By then the Navy Board had achieved a little. In 1798 Navy Office copies of orders were 'reviewed and arranged'. By 1805 it had a selection and index to those that had been in force since 1792, a necessity as orders continued to grow: over 400 were issued in 1801 alone, almost 850 in 1804.[84] That year the Commission for revising and digesting the civil affairs of the navy was appointed 'to revise the instructions and standing orders for the government of the departments of the navy'. Chaired by Middleton, the Commission completed what he began in the 1780s. By December 1806 the yard officers were all in possession of comprehensive printed instructions which remained the basis of yard operations for half a century.[85] Preceded by the prohibition on the receipt of fees and gratuities, the new instructions commenced a new era in the standards of moral probity and personal performance expected of employees in the naval departments.[86]

Dockyard operations and fleet performance

These efforts to improve administrative control accompanied growth in the scale of dockyard operations. The number of docks and artificers and the amount of materials consumed all grew. At the same time efforts were made to improve the speed, productivity and economy of yard

[82] TNA, ADM. 106/2507, 28 June, 7 Sept. 1764, 25 Apr. 1766, 23 June 1767.
[83] *Letters of Lord Barham*, II, 226.   [84] TNA, ADM. 106/2513, 2514.
[85] TNA, ADM. 106/2534.   [86] Morriss, *Naval Power and British Culture*, 147–74.

Table 4.2 *Docks at the Royal dockyards, 1753–1815*[a]

Yard			S – single	D – double		
	1753–6		1793–6		c.1801–15	
Deptford	1S	1D	1S	1D	1S	1D
Woolwich	1S	1D	1S	1D	1S	1D
Chatham	4S		4S		4S	
Sheerness	2S		2S		2S	
Portsmouth	4S		4S	1D	8S	
Plymouth	1S	1D	3S	1D	3S	1D
TOTALS	13S	3D	15S	4D	19S	3D

[a] Morriss, *Royal Dockyards*, 44.

operations. Their performance was critical to command of the sea, which depended on denial of its use to the enemy. Speed of mobilisation and the maintenance of a force at sea capable of defying and defeating the enemy were critical. In the wars against France, Portsmouth and Plymouth became the two front-line yards. Nevertheless, they could not perform all the work demanded by the growth of the fleet and they were supported by the eastern yards, each of which had specialised roles suited to their locations.

The expansion of dock numbers

At the time of the Seven Year' War, Plymouth yard had one single and one double dock. By 1793 two more single docks had been added. The docks at Portsmouth were similarly increased. Prior to the Seven Years' War, it possessed four single docks. By 1793 another double dock had been added; by 1801 the double dock had been reduced to a single and another three singles added.[87] These two front-line dockyards thus doubled their docking capacity from six to twelve docks. On the other hand the four eastern yards in 1815 retained the capacity they had sixty years before.[88] (See table 4.2.) Work on docks was difficult during wartime, and there other priorities in peace when the desire for economy dominated public priorities. The enlargement of the dock capacity of just the two western yards was thus an important achievement.

[87] For the plans and their implementation, see Coad, *The Royal Dockyards, 1690–1850*, 101–4.
[88] Baugh, *British Naval Administration in the Age of Walpole*, 263–75; BL, Add. MSS 27,884; TNA, ADM. 140/555, parts 14, 18.

With more docks, after 1793 more ships of the line were docked at Portsmouth and Plymouth than at any of the other four yards. In 1794 when ships were still fitting for sea, Portsmouth docked fourteen ships of the line, Plymouth eleven, Chatham nine and Woolwich one.[89] Refitting between May and December 1805 followed a similar pattern with Portsmouth equipping nine ships of the line, Plymouth eight, Chatham four and Sheerness one. In this period, Portsmouth refitted even greater numbers of frigates – twenty-two compared to eight at Plymouth and sixteen at Sheerness. The silting of the rivers Thames and Medway hindered ships of the line and frigates reaching the inland yards but their docks were used for some large repairs. These inland yards also performed more new construction, while Deptford acted as a depot for stores.[90]

Fitting and mobilisation speed

Ships were fitted most rapidly during mobilisation at the beginning of wars. The number of docks available, the size of the workforce and the condition of ships determined the speed at which vessels could be equipped for sea. During the Seven Years' War, after nearly two years of mobilisation, by June 1757, the British navy had ninety-seven ships of the line in commission. After two years mobilisation in the French Revolutionary War, by January 1795, the navy had ninety-two ships of the line at sea.[91] In the Napoleonic War, by January 1805, there were eighty-three ships of this size at sea. In this last war, the initial pace of mobilisation was remarkable: during May 1803 alone, twenty-eight ships of the line were equipped, but then the pace of equipment slowed to about two ships of the line a month, probably because the vessels coming forward were in poor condition.

In the Napoleonic War, the entry of Spain late in 1804 precipitated a second mobilisation in January 1805. In March the Navy Board supplied the senior board with a 'list of the line of battle ships in dock and in the ordinary at the several places which have been inspected and may be brought forward for service at sea'. There were '7 at Chatham, 11 at Portsmouth, 11 at Plymouth'. The Navy Board anticipated a mobilisation schedule that would get twenty-six of these twenty-nine ships to sea by September 1805 (see table 4.3).

This postulated an average rate of mobilisation of four ships of the line a month. In the event, the actual rate of commissioning was slower, but

[89] NMM, ADM. BP/15B, 18 Aug. 1795.   [90] NAS, Melville MSS, GD51/2/847.
[91] These figures exclude vessels commissioned for harbour service.

Table 4.3 *Anticipated rate of mobilisation for ships of the line, March–August 1805[a]*

		To be ready for service...
March	74 guns	2 in number
April	80	1
	74	5
May	74	6
June	74	2
July	98	2
	80	1
	74	3
Aug	100	1
	80	1
	74	2
	Total	26

[a] NMM, ADM. Y/2, 11 Mar. 1805.

not by much. First, second and third rates 'in sea pay' increased from 86 on 1 April 1805 to 103 on 1 October, indicating a rate of equipment of seventeen ships of the line in six months, almost three a month.[92]

Very many more frigates, sloops and other smaller vessels were commissioned. By the end of the eighteenth century, the dockyards achieved as equipment rate of 13 or 14 frigates or smaller vessels a month in the first year of each war. The pace then slowed to between 5 and 8 a month. Ships continued to be commissioned until the requisite number was at sea. In the French Revolutionary War, the progressive commissioning of smaller vessels reached a peak of 372 at sea in January 1801; during the Napoleonic War, the peak was reached with 596 frigates and other vessels at sea in 1809.[93]

The condition of the ships affected how many could be sent to sea. After the ships in the best condition were fitted, those with defects were brought forward and the pace of mobilisation slowed; their bottoms were cleaned, their caulking checked, their copper repaired or renewed, and timbers replaced or reinforced. However, the length of time these processes took was sufficiently predictable for the Navy Board to be able to schedule their completion. Even taking into account ships decommissioned, the Navy Board could predict the naval force that would be available to the Admiralty over two years in advance.[94]

[92] TNA, ADM. 8/89.   [93] *Ibid.*
[94] NMM, ADM. BP/109, 'An abstract of the number of ships (and guns) of the Royal Navy that were in commission the 30th Nov. 1758, compared with the number that

This was affected by the availability of docks, artificers and new parts like iron knees but also by the number of ships returning from sea with defects and the time devoted to those already under repair. The general state of repair of the whole navy thus affected the size of the force at sea. Before 1793 few more than 100 ships of the line could be kept in commission. In 1797 there were 108 at sea; by 1808, 113.[95]

After the American War of Independence, Sir Charles Middleton, the Comptroller of the Navy Board, had the ships laid up in Ordinary arranged in divisions, each with a superintending master. He ordered the collection of these ships' stores and equipment into individual berths in each storehouse. The equipment of each was then applicable as needed to ships of the same class. He also ordered the magazines and store-rooms of ships in good condition to be fitted in readiness for sea. The minor mobilisations of 1787, 1790 and 1791 gave practice in equipping ships for sea, many of which had been built or thoroughly repaired since the American War. Middleton later advised the first Viscount Melville that in 1790 there were 'upwards of 90 sail of the line in good condition and every article of their stores provided and so arranged that when the fleet was armed soon after it was done with such rapidity as was never known before'.[96]

From the many ships in Ordinary, the Admiralty set priorities. The dockyard officers supplied the Navy Board with lists of ships in the best condition, specifying how long each would last without repair. From 12 April 1792, the dockyards were required to provide weekly accounts 'of the state and condition of His Majesty's ships at' the ports. These recorded the condition of ships, such as when their bottoms were last cleaned, but provided most detail about the number of men, officers, stores and provisions on board.[97] Sometimes the junior Surveyor or a committee of the Navy Board visited the yards to decide which ships should be fitted first. The Admiralty then sanctioned the lists of ships proposed by the Navy Board or made its own choice, ordering at the same time the quantities of stores to be put on board – enough for three, four or six months, or 'as much . . . as she can stow'.

After their hulls and sheathing were deemed secure, ships were rigged and stored. But all this work was affected by a shortage of labour. In 1803 the Portsmouth Master Shipwright's department was reinforced with three gangs of shipwrights sent from the Thames yards, while the

---

may be in condition for service at the end of the year 1761'; TNA, ADM. 7/6, weekly accounts of ships fitting, 29 Jan.–26 Mar. 1807.
[95] Figures from James, *Naval History of Great Britain*, tables 1–25.
[96] NMM, MID/2/13/6, draft memorandum 22 June 1804.   [97] TNA, ADM. 49/100.

Master Attendant's departments at all four coastal yards were reinforced with 200 contract riggers. The latter were paid by the ship and allowed eleven to fifteen days for a first rate, eight to ten for a third rate, and six to eight for a frigate. In the view of the Society for the Improvement of Naval Architecture, they could perform their work 'in a slovenly and most indecent manner'. They were not always helped by inclement weather. The scale of operations also sometimes demanded the use of re-laid cordage[98] of which sea officers complained.[99] This work gave them incentive to join their ships as soon as possible after they were commissioned, when the gunner, carpenter and boatswain all had to draw their stores. As guns and provisions were taken on board and their draughts increased, ships often had to move into deeper water. Then there was frequently a shortage of yard lighters and unfavourable winds. Even when these obstacles were overcome, there was always the shortage of seamen. In 1803, thirty-two ships of the line were short of 8,000 men. In May 1805, there were still ships 'laying ready to receive men and none to put aboard'.[100]

Refitting and turn-around speed

Thorough refitting should not be confused with the brief visits of ships to ports and depots to replenish victuals and water. This was common around the coasts of Britain where the small yards at Harwich, Leith, Deal and Kinsale (later Haulbowline Island) supplied stores and victuals. Abroad, the overseas yards provided more facilities and smaller vessels could be careened. Here shipwrights, caulkers or carpenters came on board to assist the ship's artificers. At Malta in 1812 there was an abortive attempt to build a dock.[101] This was because ships of the line and even smaller ships with hull defects generally had to return to a home dockyard.[102] Even the newer yards at Bermuda, Trincomalee and the Cape of Good Hope lacked docks, limiting the work that could be performed there. However, from 1782, ships in the Indian Ocean that

---

[98] Society for the Improvement of Naval Architecture, *A View of the Naval Force of Great Britain* (1791), 17; TNA, ADM. 106/2227, 8 May 1801; ADM. 106/2229, 26 July 1802.

[99] For Nelson's complaints about his ships, see *The Dispatches and Letters of Vice Admiral Lord Viscount Nelson*, ed. Sir N. H. Nicholas (7 vols., 1844–6), V, 299, 306, 307, 319, 334, 354.

[100] DRO, Sidmouth MSS, c.1803/ON3; *Letters of Lord Barham*, III, 83.

[101] MacDougall, 'Malta dockyard 1800–1815'.

[102] R. Morriss, 'Problems affecting the maintenance of the British fleet in the Mediterranean' in *Français et anglais en Méditerranée*, 171–80.

required docking were accommodated at the East India Company's yard at Bombay.[103]

Distant from England, ships in need of refitting often had to be self-sufficient. St Vincent encouraged Nelson in August 1803 by observing that 'caulking and every other refitment which in England requires dock-yard inspection your Lordship knows is much better performed by the artificers of the squadron'.[104] He thought ships replenishing at the Chan-nel yards should not exceed six days unless a mast was to be shifted. Many ships took ten to twelve days. But full refitting took much longer.

In the 1740s, during the American War of Independence and the Napoleonic War, full refitting took three to four months for a ship of the line and about two months for a frigate.[105] Prior to the American War, ships had been refitted every three years, work known as 'triennial trimmings'. After copper sheathing was introduced during the American War, ships were kept at sea until their condition began to impede their performance. Some 74s were then refitted after only two and a half years, yet others were kept at sea more than four years. Ships requiring refitting were reported to the Admiralty by commanders-in-chief on each station and they returned as replacements arrived. At the ports, from 1792 the port admirals reported arrival, the number of men and stores wanting, their condition and when their bottoms were last cleaned.[106] They were then placed in a queue for refitting or repair.

When their time came for refitting, they were surveyed by the dock-yard officers, stripped of topmasts and rigging, which could be done by their crews in a few hours. They heaved out the guns and unloaded gun-ners', boatswains' and carpenters' stores, provisions, cables and ballast. Everything had to be listed and much of the rigging tallied to facilitate replacement. Many items were returned into store but some went to a hulk for reuse. As this work neared completion, crews were drafted in batches to stationary receiving ships or turned over to other vessels ready to sail. Charge of the vessel was then transferred to the dockyard Master Attendant who managed each docking as vacancies, tides and priorities dictated.

---

[103] Thanks are due to Dr John Day for this information. See also Lambert, *The Last Sailing Battlefleet*, 179–80.

[104] *The Letters of Lord St. Vincent 1801–4*, ed. D. Bonner Smith (2 vols., NRS, 1921, 1926), II, 320.

[105] Baugh, *British Naval Administration in the Age of Walpole*, 334–5; Knight, 'The perfor-mance of the Royal dockyards in England during the American War of Independence'; Morriss, *Royal Dockyards*, 288.

[106] TNA, ADM. 49/100, State and condition of ships at each port, return for 28 Aug.–3 Sept. 1796.

By this time, ships of the line had usually been at a dockyard about a month. The docking of each ship was usually closely coordinated with the undocking of another, both operations taking place at the height of a spring or neap tide. Owing to the shallowness of most docks, ships of the line had to be hauled in, with ropes, blocks and tackles to multiply the strength of men, horses and capstans. After inspection of the hull, some ships were hauled out of dock again on the same day or that following, to be put aside for repair. However, most stayed and became the focus of intense activity.

Some sheathing was stripped to expose the caulking, which often had to be renewed. Copper sheathing, extended to the whole fleet in 1779–82, became less coated with weed and *crustacea* than its wooden predecessor.[107] But worn sheets had to be replaced and sometimes ships were completely recoppered; 74s were usually resheathed after four or five years at sea. Often other defects were discovered which, if small, were immediately taken in hand: 74s that simply had some worn sheathing replaced were usually undocked after two weeks, others after a month, while most of those completely recoppered were out in six weeks.

Work on hull timbers and sheathing close to the keels was hastened after 1800 by the introduction of a proposal from Robert Seppings, later Surveyor of the navy, for settling ships on to a row of iron wedges when they came into dock. Hitherto ships had been settled directly on dock bottoms, reducing access to keels. Seppings's invention permitted wedges to be removed one by one, then replaced, with a specifically designed battering ram. The process aided access and reduced strain to hull timber.[108]

Once out of dock and in harbour, further work continued on upper works and fittings. Seamen from the receiving ships were available to help load ballast, and replace guns, stores and equipment. Water and provisions were stowed. There were never enough yard craft or 'extra men' to ferry equipment to ships and most 74s took six or seven weeks to prepare for sea.

The principal reason ships of the line took so long to refit was because they had to be unloaded and loaded before docking. This was partly because dry docks were relatively shallow and ships had to be hauled in and out of dock, often with some damage to their keels or sheathing.

---

[107] Knight, 'The introduction of copper sheathing into the Royal Navy, 1779–1786'.

[108] R. Morriss, 'The administration of the Royal dockyards in England during the Revolutionary and Napoleonic Wars, with special reference to the period 1801–1805', unpub. University of London Ph.D. thesis, 1978, 297.

However, between 1796 and 1801 at Portsmouth, the basin was deepened and enlarged. That, with two deep docks leading from it, could be drained and filled with a steam engine. Thus, once in the basin, ships could be docked and their hulls inspected at any state of the tide. The Inspector General of Naval Works maintained it was unnecessary to unload ships which could sail again within days of being docked. However, this applied to a few frigates rather than ships of the line, which dockyard officers still unloaded, partly to avoid strain on frame timbers when supported by shores and sitting on their keels, and partly to gain access to their interior timbers to check for decay.[109]

The larger number of docks at Portsmouth, some of which were deep, accounted for a greater number of ships of the line, frigates and smaller vessels refitted at that yard than at any other by the time of the Napoleonic War. Between May and December 1805 Portsmouth refitted fifty-one vessels, while Plymouth refitted only twenty-five. Sheerness refitted more vessels than Portsmouth but undertook sloops and gun vessels.[110] Chatham refitted only two third rates and two frigates. Apart from the occasional yacht or storeship, the other inland yards did not refit ships.

The distance of Deptford, Woolwich and Chatham from the sea was a growing problem during the second half of the eighteenth century. These yards were 50, 42 and 20 miles, respectively, from the Nore anchorage in the Thames estuary. Up meandering rivers, ships were dependent on tides and winds and could be delayed for weeks by unfavourable combinations. The difficulty was eased in the Medway in 1800 when transporting buoys and moorings were laid down between Gillingham and the upper end of Long Reach, permitting ships to haul up to the dockyard 'in the course of a few' tides.[111] But, against prevailing westerly winds, the problem remained in the Thames.

The silting of the rivers exacerbated this navigational problem. Most ships had to reduce their draught just to get up the rivers Thames and Medway. By the mid eighteenth century even the smallest ships of the line – the 60-gun ships – had draughts of 19 feet 5 inches,[112] and ships of the line were built progressively bigger. Yet by the end of the century spring tides at Deptford and Woolwich provided only 19 feet of water, and neap tides 2 feet less. Thus, only frigates were sent to Deptford, and

[109] NMM, ADM. Q/3320, 4 Aug. 1795.   [110] NAS, Melville MSS, GD51/2/847.
[111] TNA, ADM. 106/2228, 30 Oct. 1801; *PP* 1805 (193), VIII, 487.
[112] B. Lavery, *The Ship of the Line: The Development of the Battlefleet 1650–1850* (London, 1983), 202.

they had their guns removed at Northfleet or Gravesend. Larger ships went up to Woolwich but also left their guns at Gravesend. The River Medway below Chatham was deeper than the Thames below Deptford but silting still reduced the depth at moorings by $2\frac{1}{2}$ feet between 1724 and 1803. Ships of the line going up to Chatham thus had to remove their guns in Pinup Reach, and smaller vessels no farther up than Gillingham Reach. Likewise, large vessels coming down took on board guns, stores and provisions at the mouth of the Medway.[113]

In the eighteenth century shoals in main channels and near to the dockyards were kept in check by manual shovelling from Trinity House dredgers and yard barges manned by scavelmen. They were only capable of operating at relatively shallow depths and of clearing small quantities of silt quite slowly. However, from 1802 in Portsmouth Harbour and from 1807 in the Thames and Medway, bucket-ladder steam dredgers came into use, developed by the Inspector General of Naval Works. The second of these machines was capable of raising 90 tons of mud every hour from depths up to 21 feet, which permitted it to remain active through the greater part of each tide. These maintained accessibility, at least for the smaller vessels going up to Woolwich and Chatham.[114] A new eastern yard was repeatedly mooted between 1800 and 1808, and at the latter date land was actually purchased for a yard at Northfleet. However, instead, from 1813, a 56-acre extension to Sheerness yard was developed into a new model eastern yard by 1823, when the old yard was also taken in hand.[115]

This new eastern yard was too late to affect the speed of ships being refitted and receiving repairs in the east before 1815. Despite the use of mooring buoys and chains and the bucket-ladder steam dredger, large ships continued to have difficulties reaching the inland yards. This meant there were always queues of large ships waiting to be docked at the south coast yards. For those ships waiting at Plymouth, the construction of the breakwater from 1812 made the Sound a more secure harbour. Over 1,000 yards of breakwater was in place by 1815. It may have increased refitting speed by providing greater security for heavily loaded yard craft

[113] NAS, GD51/2/377; also the report of John Rennie to the Commissioners for revising and digesting the civil affairs of the navy, 14 May 1805, BL Add. MSS 27,884.
[114] A. W. Skempton, *A History of the Steam Dredger, 1797–1830* (London, 1975), 2–7; TNA, ADM.106/2232, 9 July 1803; ADM. 106/1883, 11 May 1805.
[115] 8th Report of the Select Committee on Finance (1818), *CR* 1818(97), III, 153–4; *PP* 1826 (164), XX, 505–11.

and lighters.[116] But it did not diminish the main problem of ships waiting to be docked, which in turn reduced the time ships spent at sea.

Repairs and ship preservation

Many ships coming in to refit were often found in serious need of repair. The Navy Board kept a register of 'ships of the line in commission with their condition and the period of service for which they are fit',[117] but predictions of the time ships could run without repair were not always accurate, for dockyard surveys prior to commissioning did not always penetrate the least accessible timbers where dry rot flourished.

The newest ships were expected to run up to eight years and most ships in commission from three to five years.[118] In 1812 Richard Pering, the Clerk of the Cheque at Plymouth dockyard, claimed that first rates could run no longer than five or six years without repair and that the life of the whole navy without repair was no more than eight years.[119] Potentially, this dramatically reduced the use that could be made of ships. For, once slated for repair, the wait for docking and the repair itself removed ships from service for considerable periods. Between 1801 and 1805, 74-gun ships awaiting repair were out of commission, on average, more than fifteen months: that was over a quarter of their useful life. Of course, repairs varied in size: there were small, middling and large, which on 74s took about ten weeks, ten months and sixteen months, respectively.[120]

Nevertheless, repeated repairs did preserve ships. Repair work therefore tended to alternate with refitting. Twelve ships in commission in 1812 had been built before or during the American War of Independence.[121] The alternation of repairs and refitting was exemplified in that on the *Defence* (74 guns)[122] which was launched in 1763 and attended dockyards eleven times between 1780 and 1805, which was about every two and a quarter years, with significant work about

---

[116] L. H. Merrett, 'A most important undertaking: the building of the Plymouth breakwater', *Transport History* 5(1977), 153–4.

[117] NMM, ADM. Y/1, contents, 18 Jan. 1805.

[118] NMM, ADM. Y/4, 30 Apr. 1806; ADM. BP/28, 10 Feb. 1808; ADM. BP/32B, 6 June 1812.

[119] R. Pering, *A brief inquiry into the causes of the premature decay in our wooden bulwarks* (1812).

[120] Morriss, 'The administration of the Royal dockyards', 299.

[121] NMM, ADM. BP/32B, 6 June 1812.

[122] Making good defects and very small repairs cost between £5,319 and £18,105; small to medium repairs cost £20,000–30,000; middling repairs in the region of £30,000–40,000; large repairs £41,250 to £64,006: Ships' Progress Books, TNA, ADM. 180/6, 10, 11, of which there are copies at NMM.

every ten years.[123] The variety, detail and quality of work performed was remarkable.[124] Yet it took time and absorbed most of the labour that could be spared from refitting.

To get more done, the navy had to employ the resources of merchant yards and this was done in times of crisis. It began for ships of the line during the American War of Independence.[125] The practice was repeated for ten frigates in 1790[126] and nine more in 1793–5. Eight 64s, twelve frigates and thirteen sloops were so repaired in 1805,[127] and four 64s and about twenty frigates in 1806–8, contributing to the great growth of ships in commission by 1809. The use of merchant yards for repair then lapsed until 1815.[128]

The use of the private yards became necessary when ships built under pressure during the Seven Years' War were beginning to decay. After 1778 there was 'little opportunity' to give them 'a permanent repair'.[129] The experience, however, contributed to the determination after the war to repair thoroughly all the ships required by the navy to match the forces of France and Spain. In consequence, between 1783 and 1793 eighty-five ships of the line were repaired. The repair of each ship averaged over two years, and in the six years 1784–9 the Navy Board's Extra estimates provided for on-going repair work on over twenty ships each year.[130]

Preparing the 1791 Extra estimate, Charles Derrick, clerk to the Surveyors of the Navy since 1773 and chief clerk 1784–96, studied this rate of repair. In the seven years from 1783 to 1789, sixty-eight ships had been completed, at a rate of nearly ten a year. On the basis that each ship would require a major repair every ten years, he observed that major repairs or new construction would have to be completed on 10 ships of the line every year to maintain a fleet of 100 ships of the line in active use.[131] However, during the Napoleonic War, as Richard Pering observed

---

[123] TNA, ADM. 180/6, 10, 11.
[124] For the most detailed account of work performed at any dockyard, see that of the Master Shipwright, Joseph Tucker, later a Surveyor of the Navy, 'Account of the works performed by the shipwrights on the hulls of HM Ships at the port of Plymouth 11 March 1803 – 30 September 1803': RNM, Admiralty Library MS, 1/3.
[125] *Letters of Lord Barham*, III, 68.   [126] NMM, ADM. BP/60, 16 July 1790.
[127] TNA, ADM. 7/567, 'summary view of the state of ships and vessels building, repairing or in Ordinary on 30th December 1805'; NMM, ADM. BP/34B, 1 June 1813.
[128] TNA, ADM. 49/102, 'An account of ships of war repaired in merchant yards under contract since 1 January 1801 for which bills have passed this office', listing bills passed 9 Dec. 1807 – 15 July 1815. This account shows that two frigates were repaired in merchant yards in 1815.
[129] Wilkinson, *British Navy and the State in the Eighteenth Century*, 206–10; Derrick, *Memoirs of the Rise and Progress of the Royal Navy*, 180–1.
[130] Webb, 'The rebuilding and repair of the fleet, 1783–93'.
[131] *Ibid.*, based on TNA, ADM. 95/5, 16–53, 'Thoughts on Extra Expenses'.

in 1812, the life without repair even of new ships was no more than eight years; indeed the Navy Board often slated ships for repair after no more than five years. But by then, to maintain the necessary 100 ships of the line in a condition for sea service, a minimum of 20 ships of the line had to be repaired each year.

The workforce and productivity

The labour performed in the Royal dockyards was the human basis upon which the British navy was maintained and grew. However, the workforce did not expand at the rate of the navy in the late eighteenth century. Moreover, recruitment was handicapped by the devaluation of incentives, while the system of piecework, introduced to enhance work output, was riddled with weaknesses. Even so, lessons were learned and applied from 1803, which increased the workforce and the efficiency of its employment through the Napoleonic War.

The deficiency of the workforce

Between 1755 and 1814 the British navy more than tripled in size, growing 315 per cent. However, the manpower recruited to the Royal dockyards only doubled, growing 214 per cent from 7,271 artificers and labourers to 15,598 (see tables 4.4 and 4.5) The shipwright element of that workforce enlarged less than 60 per cent, from 3,126 to 4,936 shipwrights, including 897 apprentices.[132] This disparity between the growth in the size of the navy and that of the workforce, especially the shipwright part, explains why there was never enough labour in wartime to build ships as well as refit and repair vessels. Yet this disparity also indicates the achievements of the period, for not only did enough workers have to be recruited, their efforts had to be employed with the utmost efficiency in order to keep a growing number of ships at sea.

For each of the yards there was a peace establishment and a considerably larger one for war. These establishments were guidelines and, especially in wartime, actual numbers bore little relation to that designated. From 1750 the Navy Board informed the Admiralty of the size of the workforce by sending a quarterly account of the numbers in each trade in every yard.[133] These figures indicate that the recruitment of labour was a slow process.

[132] NMM, ADM. BP/23B, account for Sept. 1755 filed at 2 July 1803; ADM. BP/34A, 2 Apr. 1814.
[133] The quarterly accounts may be found in NMM, ADM. BP/.

Table 4.4 *The workforce in the Royal dockyards on 14 September 1755*[a]

Artificers	Deptford	Woolwich	Chatham	Sheerness	Portsmouth	Plymouth	Total
Shipwrights	536	423	690	154	785	538	3126
Quarter boys	8	7	13	8	17	8	61
Caulkers	37	30	92	22	86	75	342
Oakum boys	8	8	27	8	33	28	112
Joiners	29	30	58	13	51	37	218
House Carpenters	33	40	79	34	63	51	300
Wheelwrights	2	2	1	–	2	–	7
Plumbers	1	1	2	–	1	–	5
Pitch Heaters	1	1	2	1	2	1	8
Bricklayers	10	8	20	11	16	19	84
Do Labourers	7	6	14	5	11	–	43
Sailmakers	32	21	37	15	32	30	167
Scavelmen[b]	20	22	62	28	61	30	223
Riggers	37	23	48	24	60	40	232
Do Labourers	19	12	32	15	58	39	175
Labourers	170	130	151	13	207	135	806
Blockmakers	4	1	2	1	2	2	12
Braziers	1	–	1	1	1	1	5
Locksmiths	2	3	2	–	2	1	10
Teams[c]	5	4	9	3	11	6	38
Sawyers	51	36	80	18	68	48	301
Treenail mooters	1	–	2	–	–	1	4
Oarmakers	1	–	–	–	–	1	2
Cooper	–	–	1	–	–	–	1

(cont.)

Table 4.4 (*cont.*)

Artificers	Deptford	Woolwich	Chatham	Sheerness	Portsmouth	Plymouth	Total
Limeburner	–	–	–	–	–	–	–
Masons	–	–	–	–	6	7	13
Pavior	1	–	–	–	–	–	1
Watermen	1	1	–	–	–	–	2
Armourers	1	–	–	1	–	–	2
Smiths	46	66	75	14	74	68	343
Compass makers	2	–	–	–	–	–	2
Rope yards							
Foremen	–	2	2	–	2	2	8
Spinners	–	88	125	–	123	103	439
Hatchellors	–	10	18	–	24	16	68
Winders up	–	6	–	–	12	8	26
Labourers	–	7	17	–	16	11	51
Boys	–	7	9	–	10	8	34
	1066	995	1671	389	1836	1314	7271

[a] 'An account of the number of artificers borne in His Majesty's several yards on the 14th Sept. 1755', NMM, ADM. BP/23B, at 2 July 1803.
[b] Shovel men.
[c] For horses.

Table 4.5 *The workforce in the Royal dockyards on 26 March 1814[a]*

Account includes inferior officers, cabin keepers and apprentices.

	Deptford	Woolwich	Chatham	Sheerness	Portsmouth	Plymouth	Total
Block makers	4	3	4	4	7	6	28
Braziers	1	–	2	–	2	–	5
Bricklayers	13	23	38	13	38	57	182
" labourers	10	20	27	9	42	22	130
Carvers	2	–	1	–	–	–	3
Caulkers	29	43	67	49	129	116	433
Coopers	–	1	1	–	1	2	5
Engine repairers	–	2	–	–	–	–	2
Founders	–	–	–	–	2	–	2
Glaziers	1	1	1	–	–	–	3
Hair bed makers	18	10	–	–	–	–	28
House carpenters	89	121	110	78	245	245	888
Joiners	47	38	76	39	158	107	465
Locksmiths	1	2	2	1	2	4	12
Masons	2	3	5	2	21	29	62
Messengers	9	10	10	10	16	11	66
Oakum boys	13	13	21	17	44	45	153
Oar makers	1	1	1	1	–	1	5
Painters	13	15	15	10	47	29	129
" labourers	–	12	–	5	14	–	31
Paviers	2	1	–	–	–	–	3
Pitch heaters	1	1	1	1	2	2	8
Plumbers	2	2	4	3	7	5	23
Riggers	75	58	108	33	181	141	596
Sail makers	47	29	52	26	77	68	299
Sawyers	140	135	167	44	240	208	934
Scavelmen	40	40	90	40	120	170	500
Shipwrights	553	584	783	267	1433	1316	4936
Smiths	99	86	120	49	182	234	770

(cont.)

Table 4.5 (*cont.*)

	Deptford	Woolwich	Chatham	Sheerness	Portsmouth	Plymouth	Total
Teams	19	21	22	9	40	31	142
Tin men	–	1	–	–	1	1	3
Trenail mooters	–	2	–	–	–	–	2
Warders	12	13	20	23	36	26	130
Waterman	1	–	–	–	–	–	1
Wheelwrights	2	2	2	2	3	3	14
Yard labourers	640	486	520	153	556	606	2961
Persons employed at the Wood mills	–	–	–	–	94	–	94
Metal mills	–	–	–	–	66	–	66
Millwrights shop	–	–	–	–	72	–	72
Rope yards							
Boys	–	24	48	–	42	42	156
Cordage remanufacturers	–	19	–	–	–	–	19
Foremen	–	3	4	–	3	5	15
Hemp dressers	–	6	13	–	16	14	49
Labourers	–	52	84	–	85	82	303
Layers	–	4	4	–	5	4	17
Line and twine spinners	–	3	6	–	5	16	30
Messengers	–	1	1	–	–	1	3
Overseers	–	1	–	–	–	–	1
Porters	–	2	–	–	–	–	2
Spinners	–	124	210	–	183	189	706
Wheel boys	–	7	13	–	10	18	48
Yarn knotters	–	1	19	–	30	13	63
Total	1886	2026	2672	888	4257	3869	15,598

a NMM, ADM. BP/34A, 2 Apr. 1814.

One reason was that, between 1755 and 1815, Britain's maritime economy was expanding, and with it the ship-building and repair work of the country. In consequence, the size of the national pool of labour was really too small for all the shipwright work, both private and public. The earliest census of the national workforce was made by the Admiralty in April 1804, which revealed that merchant yards in Great Britain contained only 5,100 shipwrights with 3,828 apprentices, and 387 caulkers with 153 apprentices.[134] At this time, the royal yards contained 3,280 shipwrights and apprentices, which was 27 per cent of the total shipwright workforce in Britain.[135]

Recruitment to the royal yards was slow because, for the most part, Britain's ship-building labour was dispersed in small groups around the coastal ports, where it served the local economy. The greatest concentration of merchant labour was along the River Thames where, in 1804, fifteen private yards contained over 2,550 shipwrights and other workers in the ship-building industry. By 1813 that workforce had increased, but only marginally, to 2,607 in eighteen yards.[136] Meanwhile, the royal yards had enlarged its shipwright workforce to 4,376, including 771 apprentices, a growth of 33 per cent over its 1804 size.[137]

Thus, even if the workforce of the Royal dockyards did not keep pace with the growth in the navy, and amounted to only a quarter of the national workforce, the figures for the Napoleonic War indicate that it had some success in recruiting manpower. This was important for the management of the workforce because, when numbers were deficient, the skilled men made use of their scarcity value to press for improvements in their terms of employment.[138] The years at the beginning of wars were thus always difficult. For example in 1739 and 1756, the shipwrights resisted regulations against their perquisite of chips, and in 1775 they resisted the imposition of piecework.[139] The first decade of the nineteenth century was especially difficult, owing to inflation, harsh cuts in the workforce at the peace of Amiens, and the reform of payment

---

[134] *PP* 1805(193), VIII, 485.

[135] Figures 1 Mar. 1804 in DRO, 152M/c1804/ON, account 23; and for mid March in NMM, ADM. B/214, 28 Mar. 1804.

[136] *PP*, 1813–14, VIII, 414, Minutes of the evidence on petitions relating to East India built shipping, cited in Doe, 'Enterprising women', 215.

[137] NMM, ADM. BP/33A, 4 Jan. 1813.

[138] R. J. B. Knight, 'From impressment to task work: strikes and disruption in the Royal dockyards, 1688–1788' in *History of Work and Labour Relations in the Royal Dockyards*, ed. K. Lunn and A. Day (London, 1999), 1–20.

[139] B. M. Ranft, 'Labour relations in the Royal dockyards in 1739', *MM* 47(1961), 281–91; *British Naval Documents*, 528–9; Haas, 'The introduction of taskwork'.

for piecework – factors which threatened to undermine the incentives to recruitment traditionally offered by the Royal dockyards.

Incentives undermined

During the whole of the eighteenth century, the dockyards relied on three forms of incentive to attract and satisfy workers: the money they could earn, the perquisites that added materially to the earned wage, and the less quantifiable advantages of work in a large state establishment.

The basic component of the money they could earn was the wage rate for each trade. For most trades these rates had been established in the 1690s. However, by the second half of the eighteenth century, special efforts in the common working hours of the yards were rewarded by multiples of the basic wage, 'two for one' or 'three for one'. These indeed became ceilings to earnings by the piece in the last quarter of the century which permitted some earnings to keep pace with inflation. In addition, overtime was rewarded at the rate of a single day's pay for a 'night' of five hours, and between a quarter and a third of that rate for a 'tide' of one and a half hours. Men working at Plymouth obtained an extra $2\frac{1}{2}$ pence 'lodging money' dating back to when the yard was established; and all men training apprentices were entitled to their earnings. The scavelmen, who cleared docks and channels, received £2 a year towards their boots, while the smiths each received daily a dozen pints of table beer and three of strong beer on account of the heat of their work.[140]

Yet dockyard wages were always paid late. In 1749 the Deptford yard officers advised the Navy Board that 'the people employed in his Majesty's yards by the day at low wages are seldom paid in less than twelve but often not 'til eighteen months after their work has been performed'. By contrast, it was 'the constant practice of the merchant builders to pay the men they employ on task work every fortnight or three weeks at farthest'.[141] During the second half of the century, improvements in accounting and the size of the clerical force reduced the delay in the royal yards but it remained at about six months. In consequence artificers had to live on credit, resorting to money lenders who demanded 10 to 15 per cent interest. Indeed, they even had to pay a fee to the clerk to the Clerk of the Cheque for a note confirming they would be receiving wages.[142]

---

[140] The best single source for earnings dating back to the mid eighteenth century is the 6th Report of the Commissioners appointed to enquire into Irregularities, Frauds and Abuses practiced in the Naval Departments, *PP* 1803–4 (83), III, 16–21.

[141] *Naval Administration, 1715–1750*, ed. Baugh, 322.

[142] *Naval Chronicle* 14(1805), 284, 341.

Such delays justified perquisites. The timber-working trades took off-cuts; the smiths took old moulds. These chips could be sold for firewood or other purposes and earned as much as 8 to 12 pence a bundle. This was disapproved of by the Navy Board because men were suspected of cutting up good wood when waste was not available, because time at the end of each day was used for getting chips together, and because they provided a cover for embezzlement. To prevent the carriage of large pieces of timber, chips were meant to be carried under one arm, but at times workers defied the regulation and carried them on their shoulders.[143] Attempts were made to regulate and even ban these perquisites but, of necessity, they survived in the dockyards until 1801.

Despite the need to take perquisites, those who managed recruitment to the royal yards always pointed, with some justice, to the advantages of treatment by a surgeon, of sick pay, superannuation and security.[144] There was a surgery and a surgeon with an assistant at each yard. In 1800–1, 4 per cent of the workforce of Chatham dockyard was discharged with injuries – 1 per cent with hernias. On account of such injuries, sick pay was a boon: six weeks of the basic daily rate then a small weekly allowance until an artificer either returned to work or received his discharge. For those injured at work or who were old and infirm, a superannuation allowance existed for shipwrights and caulkers from 1764, for other trades apart from rope makers and sawyers from 1771, and for the two remaining trades from 1802. Among the elderly, it was limited to artificers who had thirty years' government service, and only one pension was available to every forty artificers on the books of each yard.[145] However, community values ensured the young 'carried' their elderly friends and relatives in the expectation they too would be so treated in old age. Meanwhile, those with long service often become part of a yard establishment, preserved from discharge at the end of each war to repair and maintain the navy during peace.

Yet the young, vigorous and healthy were not much concerned with the incentives. In 1792 the Surveyor to the East India Company, Gabriel Snodgrass, warned the Admiralty that it valued shipwrights too little, that government employed too few, and that it was 'probable the navy and of course the nation will receive a severe check'.[146] The navy and the nation survived the 1790s when the workforce was gradually enlarged, but as

---

[143] For example in 1756, for which see *British Naval Documents*, 528–9.
[144] *PP* 1806 (2), XI, 665.
[145] 6th Report of the Commissioners of naval enquiry, appendix 4, 196.
[146] 11th Report of the Commissioners appointed to inquire into the state and condition of the Woods, Forests and Land Revenues of the Crown, 1792, *Commons Journals*, XLVII, 364.

food prices reached their height in 1800–1, a 'combination' occurred and there were riots at two dockyards in March and April 1801, when over 300 men prominent in organising petitions for higher rates of pay were discharged.[147] The riots occurred just as chips were finally abolished and replaced by daily payments ranging from 3 to 6 pence according to trade.[148] Then came the peace of Amiens when St Vincent reduced the workforce by 28 per cent[149] – including many elderly men, to the fury of their families.

The necessity to mobilise the navy in 1803, as well as to repair ships which had been in service for the long Revolutionary War, almost precipitated the check Gabriel Snodgrass had anticipated. Dockyard recruitment had been restricted to men under the age of twenty-eight. At the same time, merchant yard earnings were more than double those in the royal yards.[150] There was, moreover, an understanding between the workers in the different sectors. In 1802 many royal yard workers supported a long strike of merchant yard caulkers against the peacetime reduction of their wage rates.[151] To attract men from the private sector, state yard incentives had to be greatly improved.

Innovations and improvements

The Navy Board achieved improvements in four aspects of employment. The age limit for the entry of artificers was eased upwards – to thirty-five in March 1803, forty-five in June 1804 – and lifted completely in August 1805. It was returned to thirty-five in 1806, where it remained until 1813 when it was again raised to forty.

In view of the abolition of perquisites in 1801, to prevent artificers from going into debt between payments, in 1805 weekly subsistence money was introduced – the payment of three-quarters of usual earnings weekly. In 1813, when wage accountancy was streamlined, all earnings were paid fully each week.

Earnings from piecework were enhanced. Piecework had been introduced for some trades in 1758, for others in 1772. But the system had been adopted piecemeal, it was maladministered and open to fraud. The shipwrights had objected to it and, after a strike, only those in the smaller

---

[147] Morriss, 'Labour relations in the Royal Dockyards, 1801–1805'.
[148] 6th Report of the Commission of naval enquiry, 16–121.
[149] From 11,483 in December 1800 to 8,762 in November 1802: NMM, ADM. BP/21A, 14 Mar. 1801; ADM. BP/22B, 5 Nov. 1802.
[150] NMM, ADM. BP/32A, 9 Mar. 1812.
[151] I. J. Prothero, *Artisans and Politics in Early Nineteenth Century London: John Gast and his Times* (London, 1979), 46–8.

yards along the Thames – Deptford, Woolwich and Sheerness – accepted it in 1775. To avoid their post-war pay dropping, the Chatham and Portsmouth men acceded in 1783, and those in Plymouth in 1788.[152] Yet this piecemeal adoption resulted in differences in its management at every yard. There were two forms of piecework: task work for new work with fixed quantities and values, and job work for repairs where no items were identical. Not surprisingly, the latter caused a range of problems. Propositions for jobs varied from yard to yard; unmeasured work in overtime was mixed with measured work in the day time; and reports of work performed were sometimes passed without check. At Plymouth, creative accounting permitted artificers to earn the highest rates of pay allowed, so that the wages bill rose progressively every year between 1793 and 1800. To limit earnings, ceilings – two or three times the day rates – had been adopted. But, as foremen and leading men of gangs had a vested interest in optimising earnings (their apprentices were paid for work with the men), frauds were alleged. This was one reason St Vincent persuaded the Cabinet to sanction the Commission to inquire into irregularities, frauds and abuses in 1802. He and the Commission took a negative attitude to job work and in 1804 could propose nothing more than a return to payment by the day for repairs.[153]

However, in 1803, the Navy Board removed all ceilings from earnings by the piece, even repair work. Its policy was supported by the new First Lords. Melville obtained the Commission for revising and digesting the civil affairs of the navy, which completely overhauled the system. Barham, who as Comptroller had witnessed the introduction of piecework, took the view that it was 'of such importance towards hurrying up the fleet as cannot be given up without ruin attending it'. He thought 'by working task and job the public receive at least one half more of the labour of the artificers'. Thus in 1805 the value of work performed by task was enhanced by 20–25 per cent and new prices set for both task and job work.[154]

Meanwhile, the system of apprenticeship was reformed. The initial changes deterred the entry of apprentices. With the aim of introducing some theoretical training in schools, in 1802 apprentices were removed from their former masters and rebound to the head of each department. Their former masters were renamed 'instructors' and limited to receiving two-thirds of their earnings, which were limited to a single day's pay. From £70 a year, the value of an apprentice dropped to about £16.

[152] Haas, 'The introduction of taskwork'.
[153] 6th Report of the Commissioners of naval enquiry, 14–48, 498–9.
[154] Letters of Lord Barham, III, 72; Morriss, Royal Dockyards, 110–20.

Parents officially received one-third of this amount and now found this amount too little to keep boys in the royal yards. Many thus left, and few joined. However, in 1804, the situation was reversed: boys who had served half their apprenticeship were allowed to keep all their earnings from piecework, enlarging the shares of instructors and parents.[155]

These measures enhanced recruitment. From 1806 until 1809 shipwright numbers rose; they fell off in 1809–11, but grew again in 1812–14. By March 1814 the yards held 15,598 men, of whom 4,936 were shipwrights.[156] During the Napoleonic War alone, the total workforce grew by 80 per cent, and the shipwright force by 69 per cent. Moreover, on account of the work of the Commission of naval revision, the management of the workforce became more systematic, certainly more economic, and arguably more efficient in enhancing work output.

The supply of materials

In 1788 Charles Middleton observed that no aspect of managing the navy was of more importance than the supply of stores. The scale of the supply necessary is difficult to convey, but some figures for 1801, when the navy amounted to about 750,000 tons,[157] may indicate the quantities of principal materials required to maintain the existing force in operation: nearly 35,000 loads of oak timber, knees and thick stuff; 4,700 loads of 3–4½-inch plank;[158] over 970 masts greater than 21 inches in diameter; 14,935 tons of hemp; 1,431 tons of iron; 95,585 bolts of canvas; 949 tons of copper sheets; 18,826 barrels of tar; and 5,539 barrels of pitch.[159]

For Britain between 1755 and 1815 the greatest challenge was the maintenance of supplies sufficient to meet naval needs at a time when British merchant shipping, as well as the Royal navy, was growing in size, and there were demands from European rivals as well. Under these pressures, prices tended to rise and, for some commodities, there were crises in supply. However, British supplies never gave out, partly on account of confidence in British finance, partly on account of the lengths

---

[155] 6th Report of the Commissioners of naval enquiry, appendix 120, 408–11; 3rd Report of the Commissioners of naval revision, 424.

[156] NMM, ADM. BP/34A, 2 Apr. 1814.

[157] J. Fincham, *A History of Naval Architecture* (London, 1851, repr. 1979), 214. Tentatively, 750,000 tons of naval shipping may be compared to 1,797,000 tons of merchant shipping registered in the UK in 1801. That relationship would suggest a need for naval stores in the order of 5:12.

[158] These were timber wagon loads. One load of rough timber was calculated to produce about half a ton of shipping.

[159] NAS, Melville MSS, GD51/2/793/1; NMM, CAD. A/10, appendices to the unprinted 15th Report of the Commissioners of naval revision, appendix 152.

to which the government went to maintain existing sources and develop new ones, and partly on account of innovations that were made to off-set shortages. For scarcity proved the mother of invention, and developments in the technology of ship building were one response to an environment of growing scarcity.

Methods of supply

Britain's naval stores may be divided into those of foreign origin and those of domestic origin. From within the British Isles came oak, elm, beech and ash timber, copper, pitch, lead and manufactured items like blocks, canvas, cordage, pumps and ironmongery. Consumption naturally expanded in wartime. The amount of canvas purchased for the navy during the American War of Independence rose from 5,274 bolts in 1775 to 84,564 bolts in 1780.[160] There were then sixty-nine suppliers spread throughout England and Scotland, with the greatest numbers in Warrington, Bridport and London. By 1813 the number had declined to thirty, with the greatest number in Dundee.[161]

Stores from abroad included Prussian and Polish oak and fir; Norwegian, Russian, Ukrainian and American masts; Russian hemp; Swedish and Spanish iron; Swedish, Russian and American tar and pitch. During the last half of the eighteenth century, sources of materials like timber became less accessible. The decline in accessibility drove up prices at source, diminishing the proportionate cost of freight and making longer voyages to obtain the material more viable. Thus, although Norway's coastal fir timber was cut away, it was supplemented by increased numbers of masts from Russia through Riga, fir baulks through Memel, and pine boards or deals from Finland and Russia through St Petersburg and Riga.[162]

Because many of Britain's naval stores came from the Baltic, foreign policy was always concerned with the region. To reduce dependence on the Baltic, in the late seventeenth and early eighteenth centuries the board of Trade and Plantations attempted to encourage a supply of naval stores from North America, offering bounties on imports.[163] But this source

---

[160] NMM, ADM. BP/14, 26 Mar. 1794 encl., 'Sail cloth purchased for the navy, 1774–1793'.
[161] RNM, Admiralty Library MS. 1B. 12, 243.
[162] S.-E. Astrom, 'North European timber exports to Great Britain, 1760–1810' in *Shipping, Trade and Commerce: Essays in Memory of Ralph Davies*, ed. P. H. Cottrell and D. H. Aldcroft (Leicester, 1981), 81–97.
[163] Binney, *British Public Finance*, 136; *Naval Administration, 1715–1750*, ed. Baugh, 237–41.

never replaced the Baltic, even during wars in which British trade with that region was threatened.[164]

Naval stores and equipment were purchased for the navy by two methods – by contract and by agents acting on commission. The contracts were made either centrally by the Navy Board or locally by the dockyard commissioners. The Navy Board made contracts annually for the main materials consumed in the yards, especially those imported from abroad like timber, masts, pitch, tar and hemp. These were generally delivered to each yard, or landed at Deptford or Woolwich and distributed to the other yards by storeship.[165] A register of storeship movements for 1816–18 shows that, by then, they carried materials as far as Antigua, St Helena and the Brazil station.[166] In 1813 stores received at Woolwich totalled 12,134 tons.[167] Otherwise, yard commissioners made contracts locally, many of them standing from year to year; in 1782 Chatham had 50, Portsmouth 103, and Plymouth 94. They supplied locally available materials and equipment like sand, ironwork, paving stones, tar brushes, lanterns and ballast baskets. Standing contracts avoided the necessity of re-making numerous contracts but there was a tendency for them to become family concerns or to contain unfavourable terms that went unchecked for years.[168]

Agents were employed on commission to purchase stores when market conditions prevented contractors from obtaining sufficient supplies. The method had been used since at least the mid seventeenth century. Generally it was not favoured by the Navy Board as, once a sample was accepted, the whole purchase had to be received and this invariably included some sub-standard material. Other contractors also resented the special funding of an individual who could mop up a scarce commodity and farther drive up its price.[169] Nevertheless, it was a necessary expedient. The Navy Board usually employed a prominent merchant with whom it was familiar and who acted under Admiralty instructions. This was commonly the case with Russian hemp, the quality of which placed

---

[164] Duffy, 'The foundations of British naval power', 60–1.

[165] TNA, ADM. 49/32, Navy Board abstracts of contracts, 1762–7, records supplies of masts, timber, canvas, train oil, tallow, rosin, bar iron, hemp, pitch, tar and hoops. Names of contractors and places for delivery are recorded, and the former are indexed.

[166] RNM, Admiralty Library MS. 305. The movements of each storeship were registered separately.

[167] NMM, ADM. BP/41A, 22 Jan. 1821.

[168] For contracts made with the yard commissioners about 1780, see WLC, Shelburne Papers, vol. 151, fos. 51–64; similar lists for 1702 may be found in NMM Sergison Papers, SER/126; and for after 1815 in NMM, ADM. BP/39B, 1 May 1819.

[169] Pool, *Navy Board Contracts*, 104–7, 124–5. Pool was Director of Naval Contracts, 1948–60, and surveys eighteenth-century operations with much insight.

it in high demand among warring West European powers, and it was purchased on commission during both the American War of Independence and French Revolutionary War.[170]

On financial grounds, the hemp purchases of the latter war were particularly criticised in 1803–5 by the Commission of naval enquiry. Supplies of hemp from Russia had been poor in quality in 1793. To supply the British navy better and 'by the superiority of our credit [to] forestall the supply of the enemy', in 1795 the merchant Andrew Lindgren was secretly commissioned to buy all the hemp he could obtain in Russia.[171] His commission was extended until 1799. However, the costs were high. In 1795 the commissions were doubled by the employment of merchants in London and Riga to handle Lindgren's shipments. His accounts show the multiplicity of other charges – for freight, import duty, insurance and demurrage – to which imports were also liable.[172]

Purchases of stores were made on the basis of demands for stores received at the Navy Office in the office for examining storekeepers' accounts. A separate Contracts Office was established in 1803. There were two types of demand: occasional demands, when unexpected shortages occurred, which might trigger a shipment of material from one yard to another; and periodical demands – daily, weekly, monthly, quarterly, half-yearly and annual – of which the quarterly and annual demands were the most important, providing information for contract deliveries. To aid central control, in 1784 the Navy Board required each dockyard to fix a formal establishment for each type of store, material or piece of equipment.[173] Calculations of remains could vary according to differences in book-keeping methods employed at each yard, and in 1786 the Commissioners on fees recommended the standardisation of methods according to the system employed at Portsmouth. This was universally adopted by 1798 and annual surveys of remains were ordered in 1801. Meanwhile attempts were made to simplify accounting practices. But only in 1815, when peace resumed, did the Navy Board feel it was 'likely at length to arrive at a correct mode of accounting for the receipt and expenditure of stores'.[174]

---

[170] Beveridge, *Prices and Wages in England*, I, 620; NMM, ADM. BP/15A, 28 Apr. 1795.
[171] NMM, ADM. BP/13, 14 Nov. 1793. Spencer to Hamond, 6 Jan. 1795. Correspondence between Lord Spencer, Sir Andrew Hamond and Lindegren, 6 Jan. 1795–27 Feb. 1799 is in the Hamond papers, WRP.
[172] Lindgren's accounts are in the NMM, MS87/079. The 12th Report of the Commission of naval enquiry is summarised by Pool, *Navy Board Contracts*, 124–5.
[173] NMM, MID.1/43/1.
[174] TNA, ADM. 106/2515, 13 Mar. 1804; NMM, ADM. BP/32B, 27 June 1812; ADM. BP/35B, 19 July 1815.

When the Navy Board wanted to make contracts for the supply of the navy, advertisements inviting tenders were placed in the *London Gazette* and on notices posted in the Stock Exchange. Merchants who submitted tenders were called one by one before the Navy Board which proposed terms below those of the lowest bidder. The merchants were induced to reduce their offers, while the board was induced to modify its terms, until agreements were reached for parts or all of what was needed. Quantities and time limits were settled during negotiations. The system aimed to promote competition and economy. Between 1730 and 1770 there were generally three to eight bidders for a single contract, and ten to twelve bidders for quantities divided between several contracts. Before 1770 all offers were recorded in the minutes of the Navy Board; after 1770 only the lowest, or that agreed, so that the number of bidders is unknown. After 1794, offers to supply stores from abroad were frequently made to the board by merchant houses that already held contracts, and their offers were generally accepted. These offers were probably made by invitation and agreed to maintain working arrangements.[175]

Contracts for foreign stores were made early in the spring, between February and April. The contractors for Baltic stores were usually merchant houses dealing with that region and were given till the end of the year to supply their commodities. This gave them the period during which the Baltic was ice-free to fulfil the terms of their contracts. Prices differed according to the yards at which deliveries were made, and delivery rates could double during war when insurance rates were higher and delays greater.[176]

Prices indicate the ready availability of most materials throughout the period 1755 to 1815. Indeed from the end of the Seven Years' War to the end of the American War of Independence, 1764–84, prices rose barely at all; some even declined.[177] Even through two decades of almost unceasing war between 1793 and 1815, some contract prices were little affected. Copper sheets, for example, dropped in price. Those of train oil, pitch and tallow increased by only 25 per cent; that of tar by 50 per cent. However, the price of rough English oak doubled in price; so too did that of canvas and hemp, while the cost of large Riga masts (over 23 inches in diameter) tripled in price, and that of timber through Riga and Danzig, including fir, increased sevenfold.[178]

---

[175] Beveridge, *Prices and Wages in England*, I, 617–18.   [176] *Ibid.*, I, 618–19.
[177] NMM, ADM. BP/7, 21 Mar. 1787.
[178] 6th Report of the Select Committee on Finance, *PP* 1817(410), IV, 270, appendix 22. For the manuscript account sent to the Admiralty Secretary, see NMM, ADM. BP/37A, 24 May 1817.

Crises and commerce

The rise in prices denotes the increasing depletion of customary sources of timber and masts, the competition for those from the Baltic, and the difficulty of extracting naval stores from the region. The continuing availability of all stores at their source has to be stressed. However, these sources became increasingly remote and were accessed across political boundaries which increased the costs of extraction. These difficulties were evident in the few supposed crises in supply between 1755 and 1815. There were three – in the supply of masts, timber and hemp – but each demonstrates that difficulties were overcome by commercial means.

Britain benefited by trade treaties of 1734 with the Empress Anne of Russia, and of 1766 with Catherine II.[179] By 1778, when 607 ships arrived at Kronstadt, 252 were English, 147 Dutch and 47 Swedish; only 12 were Russian and 1 was French.[180] The British were favoured by being permitted to maintain their own factory at St Petersburg and to avoid Russian legal and financial restrictions. They were thus able to concentrate the export of Russian naval stores in their own hands – in effect, those of a few merchant houses with headquarters in London.[181] The volume of this trade gave them economies of scale and their financial resources facilitated the long-term credit transactions necessary for the order and delivery of timber, masts and hemp from the Russian hinterland. Their dominance in the Russian market prompted foreign buyers – French, Dutch and Spanish – to commission British merchants to buy for them.[182] Hence, when crises arose, the British were the best equipped of all the West European powers in the eighteenth century to surmount them.

The alleged crisis in mast supply during the American War of Independence was a case in point. R. G. Albion in 1926 suggested the Royal navy had become critically dependent on the very large sticks imported from New England and that, when the supply was cut, delays in replacing masts affected the strength of the navy at sea and the course of the war.[183] R. J. B. Knight has shown that, by the time of the Seven Years' War, the British navy was using 'no more than a dozen' large sticks

[179] D. K. Reading, *The Anglo-Russian Commercial Treaty of 1734* (New Haven, CT, 1938), 295.
[180] 'Rise and Progress of the Trade to Russia', printed in *Naval Chronicle* 2(1799), 394–400.
[181] Morison, Spenser, Thornton, Collins, Wale-Pierson, Took, Gom.
[182] P. W. Bamford, *Forests and French Sea Power 1660–1789* (Toronto, 1956), 141–2.
[183] Albion, *Forests and Seapower*; see also his 'The timber problem of the Royal Navy, 1652–1852', *MM* 38(1952), 4–22.

over 32 inches in diameter from America each year. Hence, when war came, the American supply was not seriously missed. Instead the dockyards improved their construction of 'made masts', made more of them, and imported more mast timber from Norway and Riga in the Baltic. From 3,434 sticks in 1771, the navy's average annual consumption of Baltic masts increased to nearly 4,940 in 1778–82. By the time of the following war, owing to remorseless felling near the Norwegian coast, the supply from Riga was the main source of mast timber and provided 73 per cent of Britain's mast imports by 1795.[184] These were supplemented after 1785 by 'American masts' and after 1806 by masts from Canada.[185]

A crisis in the supply of oak timber in 1803–4 also revealed the ability of Britain to secure its needs by commercial means. In 1771 the dockyards consumed an estimated 22,000 loads of oak timber a year. Thirty years later, consumption was estimated at 36,000 loads – two such wagon loads providing a little over one ton of shipping.[186] These quantities may appear large, but are placed in perspective by the amounts needed in 1783 to build a first rate of 100 guns, a third rate of 74 guns, and a frigate of 36 guns: 5,560 loads, 3,212 loads and 1,237 loads respectively.[187] Periodic and occasional returns from the dockyards kept the Navy Board informed of the quantities in stock, and from 1776 the Admiralty was informed of these amounts annually each spring.[188]

By 1786 the Navy Board was convinced there was a 'scarcity of large timber in this island and particularly so where the inland navigations and good roads have reached'. The Admiralty commissioned a survey which indicated in England 551,900 loads 'fit for the navy' with another 52,410 fit 'in a period of years'. The survey was reassuring for it excluded Gloucestershire and Shropshire, 'the principal counties from which the Navy is likely to be supplied'. But, as the Deptford yard purveyor pointed out, the dockyards experienced a scarcity principally because the

---

[184] R. J. B. Knight, 'New England forests and British seapower: Albion revised', *American Neptune* 46(1986), 221–9; Astrom, 'North European timber exports to Great Britain, 1760–1810', 81–97, appendix 3, 'Masts'.

[185] Albion, *Forests and Sea Power*, appendix: 'Imports into Great Britain, 1799–1815 – great and middling masts', 421; NMM, ADM. B/228, 3 Sept. 1806; ADM. BP/48A, 15 Feb. 1828.

[186] R. J. B. Knight, 'The Royal Dockyards in England at the time of the American War', unpub. University of London Ph.D. thesis, 1972, 214; SRO, Melville MSS. GD51/2/793/1.

[187] NMM, MID.8/5/4, quantities given for each class of ships.

[188] During the eighteenth century, the annual statement of remains was sent to the Admiralty in March; after the turn of the century it was in January or early February. See NMM, ADM. BP/.

merchant yards paid higher prices, set fewer conditions, and paid without fail three months after delivery.[189]

During the French Revolutionary War, the yards could not maintain a stock of three years'[190] supply and timber prices began to escalate. The construction of East India Company ships was blamed for the shortage of large timbers suitable for ships of 64 and 74 guns,[191] and early in 1803 the Company was persuaded to reduce the scantlings of its ships. Even so, in March 1803 the dockyards had only 34,562 loads, less than the 39,700 they had consumed in 1801 alone.[192] In 1802 the Navy Board had made new contracts; indeed between 1801 and 1804 the prices allowed to contractors were increased three times. Even so, some merchants who made engagements in 1802 were still unable to fulfil their contracts in 1804.

R. G. Albion dramatised the crisis: 'That long-dreaded failure, more complete than on any previous occasion, materialised just at the time Napoleon was threatening England with invasion.' It was exacerbated by the First Lord of the Admiralty, St Vincent, who thought the merchants were profiteering, and, in March 1804, he refused to make them any further concessions and ordered the Navy Board's junior Surveyor and the dockyard purveyors to make purchases personally in the country sales. In this way, 3,000 loads were secured. But by May 1804 yard stocks had increased only to 36,570 loads.[193]

That month St Vincent was obliged to resign with the Addington government. His successor, Lord Melville, immediately cancelled the 1802 contracts and again raised the prices paid for timber. Hitherto, foreign oak had been regarded with indifference and only small quantities had been used.[194] However, as Melville now put it, if Britain was to have a navy, it would only be by 'recourse to a considerable extent, either to the timber of other countries or to ships built in other countries'.[195] New contracts, including secret deals for imports from parts of Europe occupied by the French,[196] and for Canada Oak and Live Oak[197] from

---

[189] NMM, MID.1/169, B. Slade to C. Middleton, 24 Mar. 1786; ADM. BP/6, 10 Apr. 1786; ADM. BP/7, 10 Dec. 1787 enclosing the survey dated 14 Dec. 1787.

[190] NMM, ADM. BP/24A, 12 Apr. 1804, account of oak timber and knees in store 1775–1803.

[191] NMM, ADM. BP/14, 15 May 1794.

[192] *PP* 1805(152), VIII, 217.   [193] Morriss, 'St Vincent and reform, 1801–04'.

[194] NMM, ADM. BP/24A, 12 Apr. 1804.   [195] WLC, Melville MSS, 5 July 1804.

[196] TNA, ADM. 106/3574, 4 Sept. 1804. One contractor, Jonathan Larking, imported from Holland and Germany, reassured and warned the Comptroller of the Navy: 'I will forfeit my whole Fortune & existence if I do not get this Timber over if any one can, at least I [will if I am] not interrupted by others being encouraged to try at the same time.'

[197] NMM, ADM. BP/26, 14 Apr. 1806.

North America, enlarged oak stocks. By January 1805 there were 41,926 loads in stock; two years later there was double that, including over 13,550 loads of foreign oak. Stocks continued to grow, and necessarily so as average annual consumption rose to more than 53,000 loads between 1803 and 1812. In 1809 it was 64,475 loads; in 1812 it was 74,346 loads. Despite the scale of this consumption, by the summer of 1816 there were 81,740 loads of English oak timber in store.[198]

After 1804 the timber that was purchased simply came from farther afield, both within Britain and overseas. Between 1803 and 1809 three times as much English oak timber was brought by ship around Land's End to the naval dockyards as was cut for them in the Home Counties.[199] The royal forests and Crown estates had never supplied as much as expected of them. But from 2,350 loads in 1801, their supply increased to an annual average of 5,322 between 1804 and 1807.[200]

In Europe new sources of timber were opened in Croatia and Albania, and hopes were entertained of resources in southern Russia that could be tapped through the Black Sea. However, as P. K. Crimmin has shown, these regions lacked both physical infrastructure – adequate roads, navigable rivers, loading quays and shipping – and financial facilities that would permit the Navy Board to purchase on credit. For, on principle, the Navy Board declined to make payments for stores before they were delivered to a dockyard. Hence the new sources provided stores spasmodically and only in a trickle – even that interrupted by the defeats of the Austrian empire at the hands of the French at Austerlitz and Wagram. Had more cash been available in advance, these new sources of supply might have been more forthcoming.[201]

Conversely, once a commercial and physical infrastructure was in place, even the power of Napoleon had difficulty in disrupting an established trade. This was apparent in the supply of naval stores from the Baltic in 1808–13, during which time there was a potential crisis in the supply of hemp. Following the Treaty of Tilsit between France and Russia in 1807 and the implementation of Napoleon's continental system, Russian and north German ports were closed to British ships. In 1805, 11,000 ships had passed through The Sound; two years later the number had dropped to 6,000. The supply of hemp was particularly affected.

[198] NMM, ADM. BP/34B, 20 May 1814; ADM. BP/36B, 26 Sept. 1816.
[199] NMM, ADM. BP/34B, 27 June 1814.   [200] NMM, ADM. BP/31A, 6 Jan. 1811.
[201] P. K. Crimmin, "'A great object with us to procure this timber . . .": the Royal Navy's search for ship timber in the Eastern Mediterranean and Southern Russia, 1803–1815', *IJMH* 4(1992), 83–115.

Hemp imports to Britain in 1807 were worth £639,507; in 1808 they were one-third of that.

Hemp could be obtained from a variety of sources around the world – seventeen were listed for the period 1797–1806 – but St Petersburg and Riga were considered the best. In 1801, when Britain imported 37,428 tons, of which 14,935 tons were consumed by the navy, 86 per cent came from Russia and 6 per cent from north Germany.[202] Efforts were made to develop other supplies of hemp from Canada and India in 1790–2, and from Ireland, Spain, New South Wales, Italy and the Adriatic in 1800–9.[203] But none of these new sources could provide the quantities of the quality supplied by Russia, where it was cultivated widely by peasants on small-holdings and sold to the agents of merchants based in the Baltic ports.[204]

After 1807, despite Napoleon's Milan and Berlin decrees against trade with Britain, these Baltic merchants remained eager to maintain their sales. In consequence, cargoes were shipped from minor ports under minimal scrutiny in vessels that were, or purported to be, neutral. The masters of these vessels often carried false papers and certificates claiming their vessels to be owned in, and their trade between, places in amity with France. American vessels figured prominently. Masters sailing on account of British contractors were furnished with licences, issued by the British government, conferring immunity from detention by the Royal Navy 'notwithstanding all the documents which accompany the ship and cargo may represent the same to be destined to any other neutral or hostile port'. After the destruction of the Danish fleet in 1807, the British navy in the Baltic protected neutral and allied vessels from Danish and Norwegian privateers and small naval vessels by convoying them through The Sound; after 1809, they were convoyed through the Great Belt, virtually to their destinations in the eastern Baltic.[205]

Submission to convoy minimised insurance cover for the ship owner, though for a Baltic voyage it was still three times that for other voyages. It also deterred deviations of ships to ports where other foreign, perhaps enemy, cargoes might be available. Given the quantities of naval

---

[202] 15th Report of the Commissioners for revising and digesting the civil affairs of the navy, appendix 147 (unprinted): NMM, CAD.A/10, hemp imports 1797–1806; A.W. Crosby, *America, Russia, Hemp and Napoleon: American Trade with Russia and the Baltic, 1783–1812* (Ohio, 1965), 110.

[203] TNA, ADM. 106/3575, 16 Jan., 22, 24 Dec. 1800; NMM ADM. BP/28, 2 Jan., 31 May 1808; ADM. BP/29A, 28 Apr. 1809; ADM. BP/29B, 5, 16 Oct. 1809.

[204] Report of Joseph Banks, 30 July 1802, printed as appendix 140 to 15th Report of the Commissioners of naval revision, NMM, CAD.A/10.

[205] *The Saumarez Papers: Selections from the Baltic Correspondence of Vice-Admiral Sir James Saumarez 1808–1812*, ed. A. N. Ryan (NRS, 1968), xiv–xxv.

stores imported from the Baltic into Britain, the scale of the convoys was enormous. Between mid-June and November 1809, 2,210 merchant-men were escorted through the Great Belt. Inward convoys normally comprised over 100 ships but outward often amounted to 300 or 400.[206] Yet the work for the navy was self-sustaining. In 1812, when 12,250 tons of Riga and St Petersburg hemp were ordered, the Navy Board formed agreements with twelve contractors. Some undertook to supply quantities as small as 300 tons, others to supply 1,500 tons, the largest to supply 2,200 tons. The ships chartered by these contractors carried between 150 and 200 tons; the smallest contractor thus employed two ships, the largest as many as fifteen. Together, the contractors to the navy for this one commodity in 1812 required licences for sixty-one ships.[207]

Globalisation and innovation

The Baltic trade demonstrates the durability of trading relations, once formed. They were vital to the supply of masts, timber and hemp despite shortages that have been termed crises. But, as P. K. Crimmin has shown, the development of new sources of supply of naval stores rested on the prior or complementary evolution of a commercial and phys-ical infrastructure. During the Napoleonic War the sources of timber supply became global partly because European trade, finance and ship-ping enlarged the geographical range of their routine operations. Beyond Europe, the search for timber was extended to the Cape of Good Hope, India and New Zealand. After the war, contracts were made for the deliv-ery of an oak substitute from Brazil and of Canada Red Pine from north of Montreal.[208]

Globalisation was driven by the increasing scarcity and cost of materials at their traditional sources. This increase diminished the relative cost of their freight, and encouraged imports from greater distances. Britain's large and growing merchant shipping fleet facilitated those imports, while the size of Britain's navy helped to protect them. Economics, as well as fear of Napoleon, thus contributed to the search for new sources of supply in the early nineteenth century.

Meanwhile, scarcities encouraged innovation. Made-masts and spars supplemented naturally grown sticks. By 1790 Portsmouth yard alone was making over thirty made-masts and bowsprits a year.[209] In 1806

[206] A. N. Ryan, 'The defence of British trade in the Baltic, 1807–1813', *EHR* 74(1959), 443–66.
[207] NMM, ADM. BP/32B, 12 May 1812.
[208] NMM, ADM. BP/36A, 2 Feb. 1816; ADM. BP/43A, 1 Apr. 1823.
[209] NMM, GRE/8.

timber shortages prompted the Navy Board and dockyard officers to consider altering 'the shifts of some of the timbers' and scarphing 'others so as to render them less difficult to obtain', a challenge to which Robert Seppings at Chatham yard responded.[210] Five years later he introduced a system of repair which minimised the use of large timbers. He used shelf pieces to connect beams to frame timbers instead of scarce knees; he used small timbers to reinforce frame timbers by filling the spaces between them to make a solid mass of timber; and he increased latitudinal strength by using diagonal trusses and riders between frame timbers and diagonal planking for the decks. The trusses and in-filling were invariably sound timbers that had already been used. So strong did the system prove that even larger ships could be built; the first on the Seppings principle being the 120-gun *Howe* launched in 1815.[211]

Innovation occurred too in ropemaking. By 1795 the demand for cordage for the British fleet had drained the storehouses and exceeded all that could be made both in the dockyards and in private rope walks. Five productivity improvements were made in government rope manufacturing between 1798 and 1803.[212] Yet in 1804–6 the quantities of cordage made by contract almost equalled that made in the dockyards: in those years, these two sources produced an annual average of 6,824 tons and 7,115 tons respectively.[213]

This heavy dependence on private rope manufacturers was unusual. For the Admiralty still preferred key equipment to be made in the government yards, an objective that lay behind the development of the block mills at Portsmouth between 1803 and 1805. Three years later, the mills had an annual output of 150,000 blocks.[214] A similar motive prompted the development of a metal mill at Portsmouth between 1803 and 1805. Since 1779, copper sheathing had been obtained from contractors, but the mill permitted the navy to recycle its old copper and achieve partial self-sufficiency.

Nevertheless, the raw materials still had to be obtained from the private sector. Ultimately, it was the ability of the latter to make materials available which permitted the navy to build and maintain the largest sailing fleet of any national power at the beginning of the nineteenth century.

[210] NMM, ADM. BP/26, 16 Apr., 9 May 1806.
[211] D. K. Browne, 'The structural improvements to wooden ships instigated by Robert Seppings', *Naval Architect* 3(1979), 103–4.
[212] Mr Watson Fenwick, Master Ropemaker at Chatham, was responsible: TNA, ADM. 106/2232, 30 Dec. 1803.
[213] NMM, ADM. BP/15B, 4 Sept. 1795; NMM, CAD.A/10, 15th Report of the Commissioners of naval revision, appendices 71–2.
[214] K. R. Gilbert, *The Portsmouth Blockmaking Machinery* (London, 1965). The number of blocks used by a 74-gun ship is 922.

As the search for timber indicates, the British state could take initiatives in finding new sources of raw material but the purchase of that material and its shipment to Britain depended on a commercial infrastructure. Similarly, Britain might be able to fund the extraction of the material but that finance was dependent on credit facilities. In this sense, British naval power only existed within the context of a global mercantile system. Largely ignored in naval history, the international connections of merchants and their logistical capabilities hold the key to the supply of materials for Britain's maritime dominance.

Not only did Britain's maritime economy facilitate the supply of raw materials, about half the value of which came from overseas, it provided the shipwrights to work in the Royal dockyards. They still had to be paid and motivated, but piecework was introduced by degrees during the second half of the eighteenth century and payment was reformed early the following century. Nevertheless, it was the number of docks accessible for ships of the line from the sea that determined the size of the navy that could be maintained in commission, and that number was increased with deep, steam-pumped docks at Portsmouth at the turn of the century, and by the introduction of steam dredging of river approaches. These innovations were managed during hostilities; so too was the revision of standing orders that improved control of the dockyards from London. Although more associated with the management of the Sea Service, the Admiralty played an important role in lobbying for these improvements and planning for the future. As the first Lord Melville realised, however, naval power rested on a complete system, in which the work of the merchant yards complemented that in the royal dockyards, and the supply of stores benefited from naval defence and state finance. Nowhere was the importance of Britain's maritime economy and the organising capability of the state's bureaucracy more evident.

5   Ordnance and technology

Like ships in the navy, the ordnance used by the British armed forces was manufactured principally in Britain: about half the gunpowder, most of the small arms, and all of the cannon in private establishments. These munitions were obtained by the board of Ordnance which took orders from the Navy and War Offices and placed them with contractors. Little appreciated during the eighteenth century, the efficiency of the board's technical staff after 1783 ensured the British armed forces had enough ordnance of the best possible quality. Their achievement was the greater because, with the use of coal and steam power, iron founding in Britain underwent a geographical and technological revolution during the third quarter of the century and was virtually a new industry. Scientific knowledge also changed, giving rise to greater knowledge of the combustible properties of gunpowder. But the Ordnance Office was not alone in transforming the state's munitions.[1] Contractors adapted to the greater stringency of ordnance specifications. Pressure from the Navy Board contributed to the introduction of the carronade, while artillery and sea officers, including the navy's gunners, came to appreciate the greater power of the new guns and powder and the challenges of their employment.

The relocation of gun manufacture

The Wealden industry

Before the time of the Seven Years' War, the Ordnance Board contracted for cannon with private gun founders largely found in Kent and Sussex where ironstone was dug out of the Weald.[2] The area was originally well

---

[1] See G. Cole, 'The Office of Ordnance and the arming of the fleet in the French Revolutionary and Napoleonic Wars', unpub. University of Exeter Ph.D. thesis (2008) 244, 362.

[2] B. Lavery, *The Arming and Fitting of English Ships of War 1600–1815* (London, 1987), 80–3.

Table 5.1 *The tonnage and prices paid by the Ordnance Board for guns, 1700–1770*[a]

Period	Years	Tonnage	Per year	£ value	£ per ton
1700 – Apr. 1702	2	746	373	11,581	15.52
May 1702 – Mar. 1713	11	4,200	382	67,517	16.07
Apr. 1713 – Sept. 1739	26	6,043	232	110,139	18.23
Oct. 1739 – Oct. 1748	10	9,966	996	192,511	19.32
Nov. 1748 – Apr. 1756	7	2,637	377	46,990	17.82
May 1756 – Feb. 1763	7	14,342	2,049	285,343	19.90
Mar. 1763–70	7	3,251	464	49,373	15.19

[a] Tomlinson, 'Wealden gunfounding', appendix, table 1.

wooded, and this supplied charcoal for smelting the iron. It was also relatively convenient for dockyards on the rivers Thames and Medway and for Portsmouth dockyard. In 1574, of fifty-eight furnaces in England and Wales, fifty-one were in the Weald, which produced all the guns demanded by the Crown.[3] Between 1700 and 1770, of twenty-six founders receiving payments from the Ordnance Office, seventeen had their furnaces in Kent and Sussex. During the Seven Years' War, 90 per cent of payments for guns went to founders in the Weald.[4]

Between about 1700 and 1770 the tonnage of guns purchased by the Ordnance Board grew markedly in each succeeding war. The prices it paid reflect this growth (see Table 5.1).

During the Seven Years' War the prices per ton of some individual gun contracts rose as high as £22 and £24. Such prices stimulated founding on a larger scale outside the Weald. Indeed, where raw materials and transport were cheap, outsiders undercut the prices required by Wealden manufacturers by large margins. The Carron Company, based at Falkirk in central Scotland from 1759, was able to supply cannon to the Ordnance Board in 1764 for only £14 a ton.[5]

This new source of heavy ordnance broke the monopoly of the south-east. Some founding remained in the Weald, but henceforth gun founding developed on an unprecedented scale in the Midlands of England, in Wales and in Scotland. From the 1760s there was thus a shift in the

[3] For the manufacturing process, and for founders throughout Western Europe, see A. N. Kennard, *Gunfounding and Gunfounders: A Directory of Cannon Founders from Earliest Times to 1850* (London, 1986). For a more limited selection, see Elvin, *British Gunfounders*.

[4] H. C. Tomlinson, 'Wealden gunfounding: an analysis of its demise in the eighteenth century', *EconHR* 29 (1976), 383–400.

[5] *Ibid.*, appendix table 3.

geographical location and quantitative scale of gun founding. Wealden production dwindled, unable to compete with production elsewhere for a number of reasons.

Firstly, the output of Wealden foundries was relatively small. The contracts of the Browne family in the seventeenth century show that only about 7 tons of ordnance could be produced on any one 'found-day' (which lasted a week and of which there were only about thirty in a year). The annual output of the foundry of John Fuller of Heathfield, Sussex, in the 1750s was only about 200 tons.[6]

Secondly, the output of Wealden guns was not dependable. It varied from week to week, and year to year. Output depended on the length of the annual blast which in turn depended on the duration of the local water supply, for the blast of the furnaces was raised by bellows driven by water wheels. Consequently, when the wheels ceased to turn, the production of cast iron and of guns ceased too. The site and situation of individual foundries was critical, for they determined the catchment area which supplied water through the summer months. Climatic variations and differences in water reserves in the chalk or limestone affected output.[7] John Fuller observed of his foundry in 1754:

A furnace is a fickle mistress and must be humoured and her favours not be depended upon. I have known her produce 12 tons per week, and sometimes but 9, nay sometimes but 8. The excellency of a founder is to humour her dispositions but never to force her inclinations.[8]

Under these circumstances, founders dare not take orders for guns that demanded delivery within short time periods; usually they required periods of not less than one year.[9]

Thirdly, the quality of Wealden guns was uncertain for the technology of their production was limited in its precision. The method of casting guns had not changed since the sixteenth century. Guns were cast within three-piece loam moulds. The Ordnance Board issued drafts and dimensions of the guns it required; the founder cut mould boards to the profile of the guns, which were turned to finalise the final form of the mould. This was assembled, greased with butter, and received molten iron standing vertically within a pit. Even if molten iron was cast in a

---

[6] Tomlinson, *Guns and Government*, 108.

[7] Gun production was seriously affected by a long-lasting drought in the 1740s: letter of C. J. N. Trollope to NMM, 16 Feb. 1986, p. 4, filed in NMM Library.

[8] Sussex Archaeological Society, RF 15/25, letter 24 Oct. 1754, quoted in Tomlinson, 'Wealden gunfounding', 390.

[9] H. Blackman, 'Gunfounding at Heathfield in the eighteenth century', *Sussex Archaeological Collections* 67(1926), 39, quoted in Tomlinson, *Guns and Government*, 109.

steady stream to prevent uneven cooling, the quality of the metal could vary according to whether it came from the beginning or the middle of a blast. At the strategic moment, a nowell bar was inserted within the mould to preserve the hollow which became the barrel of the gun. Yet sometimes the angle of the nowell bar could be off-centre. Anyway, when the moulds were broken up after the metal had set, no two guns were ever identical.

Fourthly, inexact knowledge and technology prevailed over precision in production methods. There was some mechanisation. For example, primitive boring machinery was used to drill out the touch-holes, while cast barrels were reamed out with a boring bar. But the location and angle of the bore of these cannon varied and not just because the technology was crude. An earlier John Fuller explained to the Ordnance Boards that touch-holes were bored too high because

Our people no more understand 4 inch 21/100 nor 3 inch 66/100 than they do algebra, nor if they did is it possible in so violent a motion as the boring a gun to come up it not even to a tenth of an inch. And to hit to a point with a drill in the boring of a touch hole through a thickness of 5 or 6 inches of metal is impossible to be sure of though they often luckily hit it. These things are easily done upon paper, but it is quite a different thing when they come to be put in practice.[10]

Ignorance and imprecision ensured many guns were rejected, which could easily bring about the bankruptcy of a gun foundry.

Fifthly, the cost of raw materials and labour was high. Wealden foundries were not self-sufficient and had to pay for both fuel and iron-stone. The fuel was derived from coppice wood of 'middling growth' that was fifteen to eighteen years of age, and had been cut and corded.[11] The price per cord or bundle varied according to demand. During wartime, when most orders for guns were placed, it could be double its peacetime price. The wood then had to be coaled, that is made into charcoal, and transported to the furnace. The ironstone had to be dug from temporary open-cast mines or pits which were refilled after the stone had been extracted. The ironstone too had to be transported to the foundry. At Heathfield, Sussex, in 1745–6, charcoal and ironstone accounted for 65 per cent of the expenses of running a foundry. A large proportion of this

[10] Sussex Archaeological Society, RF 15/25, 5 Sept. 1732, quoted in Tomlinson, 'Wealden gunfounding', 391.
[11] P. Benaerts, *Les origines de la grande industrie allemande* (Paris, 1933), 454, quotes an estimate that 10,000 tons of iron made with charcoal demanded the felling of 40,000 hectares or 100,000 acres of forest. That is, the manufacture of 1 ton of iron demanded the felling of 10 acres of woodland. Cited in E. A. Wrigley, 'The supply of raw materials in the Industrial Revolution', *EconHR* 2nd ser., 15(1962), repr. in *The Causes of the Industrial Revolution in England*, ed. Hartwell, 97–120.

was for labour in digging, cutting and haulage, but at the foundry too wages were expensive, accounting for about 15 per cent of production costs.

Finally, there was the cost of getting the cast guns to the purchaser. This could far exceed that of getting the raw material to the foundry, for the rainfall necessary to fill the mill ponds turned Wealden tracks into quagmire which resisted the passage of heavy wagons. Many tracks were bedded in clay and became impassable in winter, difficult at best in summer. Deliveries took the quickest route to the rivers Medway and Thames but water transport was available for only the second half of a journey to London, or, for the Ordnance Board, to Woolwich.[12]

Heathfield foundry produced other goods but specialised in founding guns which brought a profit of about 25 per cent; that is, £1,181 17s 6d for an outlay of £4,775 in 1745–6. This was typical of other foundries. As long as the Ordnance Board, which was their main customer, purchased the products, the industry was financially secure. However, being dependent on variable supplies of water and charcoal, and prey to imprecise production, neither the supply nor the purchase of the guns could be guaranteed. Indeed, in 1764, when the Carron Company at Falkirk offered guns at £14 a ton, the Heathfield foundry could not match that price. Although Heathfield survived longer, a number of Wealden furnaces closed in the late 1760s. There was a brief revival in 1773–5 when the government declined to take Carron Company guns owing to their low rate of proof, and the American War of Independence preserved the industry another decade. But by 1787 most of the foundries in the Weald were no longer producing guns.[13]

The coalfield industry

Soon after 1800, Arthur Young attributed the shift in cannon founding from the south-east of England to 'some late discoveries' of the Scots who thereby could 'work the manufacture so much cheaper than could be afforded in Sussex'.[14] The Carron Company certainly took advantage of new technology but the 'discoveries' were not confined to the Scots. Indeed, its common use created what, in comparison with the Wealden industry, can only be called a mass production process in south Wales, the West Midlands and central Scotland. Orders for guns and carronades

---

[12] Tomlinson, *Guns and Government*, 109.
[13] Tomlinson, 'Wealden gunfounding', 387–8, 393–4.
[14] A. Young, *General View of the Agriculture of Sussex . . .* (1808), 431–2, quoted in Tomlinson, 'Wealdon gunfounding', 396.

initially enhanced the importance of central Scotland, but the Black Country of the West Midlands grew in importance in the Napoleonic War.

The geographical shift of the industry between 1764 and 1787 was because raw materials, the largest cost in traditional production, were mined and processed in the new localities in large quantities relatively cheaply.[15]

Instead of charcoal, coke coal was used. At Coalbrookdale, Shropshire, in 1708–9, the first Abraham Darby adapted a charcoal blast furnace to the use of coal and converted the latter to coke by burning it in open heaps in the same way as charcoal was made from wood. This drove off most of the sulphur from the coal. Initially he imported 'charde coale' from Bristol but by 1715 he was using 'clod coal' available locally and suited to his blast furnaces.[16] Elsewhere, founders were not so lucky and sulphur remained a problem, which may explain why his process took a long time to spread beyond Shropshire. Nevertheless, rising demand for iron and increasing charcoal prices in the early 1750s favoured the use of coal.[17] By 1788 there were 77 furnaces in England and Wales, of which 53 used coke coal and 24 charcoal. The last of the charcoal furnaces – in Sussex – was to close in 1813. By 1823 there were 237 blast furnaces spread across Britain – all smelting with coal.[18]

Linked to the coalfields were fields of iron ore. The ore was mined and refined in increasingly large quantities. These permitted economies in gun production, especially where transport by river, canal and primitive railway encouraged the adoption of technology which both freed the production process from water power and, after much trial and error, achieved precision of manufacture.

For, instead of dependence upon a supply of water power to blast their furnaces, the new foundries were able to employ steam engines to supply their blast. At the Carron iron works in 1769–70, John Smeaton acted as a consulting engineer[19] to develop a 'blowing engine' for one furnace. This was driven by a water-wheel. But, soon after, John Wilkinson in

---

[15] Additional factors may have been the threat of competition from Russian and American colonial imports, and the ceiling placed on Swedish iron exports from the late 1740s. See C. Evans, A. Eklund and G. Ryden, 'Baltic iron and the organization of the British iron market in the eighteenth century' in *Britain and the Baltic: Studies in Commercial, Political and Cultural Relations 500–2000*, ed. P. Salmon and T. Barrow (Sunderland, 2003), 131–56.

[16] B. Trinder, *The Industrial Revolution in Shropshire* (Chichester, 1973, repr. 2000), 21–4.

[17] J. R. Harris, *The British Iron Industry 1770–1850* (Basingstoke, 1988), 30–7.

[18] W. K. V. Gale, *Ironworking* (Aylesbury, 1981), 6–9.

[19] John Smeaton (1724–92) worked on the Clyde canal, near the end of which stood the Carron iron works: *New DNB*.

Shropshire erected 'blowing cylinders' powered by a Newcomen steam engine. Because the furnace was freed from water power, the blast could be maintained for longer and a larger quantity of uniform quality iron could be produced. In 1769 the Carron Company was producing 40–50 tons of cast iron a week; by April 1771 it was producing about 90 tons a week.

Not only did steam power permit larger quantities of cast iron to be refined, it permitted more guns to be produced more accurately. For they were cast solid, not hollow with nowell bar, and the barrels were bored out using a cannon lathe.[20] John Smeaton at the Carron iron works developed a boring mill driven by water-wheel.[21] Yet again, John Wilkinson in Shropshire improved upon the idea, patenting a cannon lathe in 1774.[22] The solid casting was stronger and the lathe made for accurate boring; indeed, he also developed a cylinder lathe which bored out cylinders for Watt steam engines. In consequence the Ordnance Board could expect barrels hollowed to a fraction of an inch, making for reduced 'windage' and greater power of shot. Not surprisingly, in 1775 the board determined to accept only ordnance bored 'out of the solid'.[23]

Cannon production was transformed between 1770 and 1775. The new fuel and new technology permitted many iron works with good supplies of iron ore, coal and water transport to enter ordnance production. For example, from 1771 the Walker company of Rotherham and Sheffield secured Ordnance Board contracts which were executed at its Holmes foundry by the River Don. By 1781 it had cast 1,220 cannon, some of the largest calibre.[24]

[20] The barrels of earlier hollow cast cannon had been reamed but the tool had a tendency to follow the eccentricities of the rough cast and possess the same deviation from straightness. The earlier reaming lathes had operated on a horizontal plane; the new cannon lathes worked on a vertical plane using the weight of the cannon suspended over the machine tool. Vertical boring had developed in Holland and Germany by 1774: H. H. Jackson and C. de Beer, *Eighteenth Century Gunfounding* (Newton Abbot, 1973), 71–3.

[21] John Smeaton (1724–92) was a partner in Bersham foundry and in the New Willey iron company, and erected a new furnace at Bradley and was accustomed to using coked coal: *New DNB*.

[22] The cannon lathe was similar to the machinery established by Verbruggen in 1770–3 at Woolwich for the board of Ordnance. However, Wilkinson may have seen the horizontal boring machines in France and Holland. See below.

[23] Tomlinson, 'Wealden gunfounding', 388. The board's determination was probably influenced by the opinion and accuracy of the solid boring of Jan Verbruggen, appointed to manage the Royal Foundry, Woolwich, by the Ordnance Board in 1770: Jackson and de Beer, *Eighteenth Century Gunfounding*, 74.

[24] 100 of 102 cannon on board *Victory* at Trafalgar were Walker products. They can be identified by the 'W.Co' that was cut in the round end of the guns' supporting trunnions;

To both smelting and steam power, the supply of coal and iron was vital. Coal was used in a variety of refining operations and its production rose in Britain from 6 million tons in 1770 to 11 million in 1800. Cast iron production rose even more steeply from 17 million tons in 1740 to 68 million in 1796 and 258 million in 1806. By 1800 ordnance manufacture was consuming about 25 per cent of all the iron produced in Britain, and demand required much be imported from Sweden and America.[25]

Sea, river and canal transport were critical to the development of this new industry. Owing to the great bulk of the raw materials and the weight of the finished products, foundry sites were developed with availability of water transport in mind.[26] To supply its customers the Carron Company had its own small fleet of ships.[27] So too did Crawley's foundry near Newcastle. As early as 1768, Crawley employed three ships to bring 2,100 tons of iron ore a year from the Baltic, with another 500 tons imported in other vessels.[28] Such founders were able to ship their products from Scotland and the rivers of north-east England to the River Thames where cannon for the Ordnance Board were landed at Woolwich for testing, with the minimum of land transport. By the 1790s, while placing small orders with gun-founders in London, the Ordnance Board placed its largest orders with contractors close to the raw materials and distant from London (see table 5.2).

Ordnance Board responsibilities

The Ordnance Board had its origins in late medieval times, and had become an important organ of government under the Tudor monarchs. With its headquarters still in the Tower of London, it also had officers in Palace Yard in Westminster, with a staff which grew from 117 to 227 between 1797 and 1815.[29] The Ordnance department was headed by the Master General but he was usually a statesman or soldier and was little involved in the running of the department and often away from London. The real work was done by the board which was composed of principal officers including the lieutenant general, surveyor general, clerk of the ordnance, the storekeeper and the clerk of deliveries.[30]

---

160 Rotherham products survive and have been located: J. L. Ferns, 'Missing cannons: the Walker Company of Rotherham', *The Local Historian* 17 (1986), 236–41.
25 K. Dawson, *The Industrial Revolution* (London, 1972), 42.
26 Wrigley, 'The supply of raw materials'.   27 Lavery, *Arming and Fitting*, 83.
28 Dawson, *The Industrial Revolution*, 43.
29 J. West, *Gunpowder, Government and War in the Mid-Eighteenth Century* (London, 1991), 13; Cole, 'The Office of Ordnance', 38.
30 West, *Gunpowder, Government and War*, 9–10; Cole, 'The Office of Ordnance', 39.

Table 5.2 *Guns and carronades ordered by the Ordnance Board, June 1796 – November 1798*[a]

	Guns	Carronades
Carron Company, Falkirk	1,200	1,650
Walker & Co., Rotherham	1,250	–
Alexander Raby, London	–	500
Alexander Brodie, Shrops.	784	–
Clyde & Co.	–	400
Dawson & Co., Bradford	1,050	–
Wiggins & Graham, London	–	300
John Sturges & Co., Bradford	–	950
Francis Kinman, London	–	150

[a] Figures in H. A. Baker, *The Crisis in Naval Ordnance* (NMM Monograph 56, Greenwich, 1983), 38–40, derived from Proof Books, in Armouries Library, Tower of London.

Purchase, proofing and distribution

The Ordnance Board did not itself undertake the casting of iron cannon. In 1716 the board had established a foundry to cast brass cannon at Woolwich where some prestigious guns were still cast in the mid eighteenth century. But the Brass Foundry became increasingly a centre for testing and the maintenance of expertise. It complemented the Royal Laboratory for testing gunpowder at Woolwich. For the main supply of both the army and the navy, the Ordnance Board took orders from the War and Navy Offices and placed them with appropriate contractors.[31] It remained accountable for the proofing and issue of guns but had no immediate concern in their manufacture other than to place orders and set specifications. However, this task was central to the quality of the state's ordnance. Indeed, it was critical also to the quality of the small arms, powder and other equipment used by the armed forces.

The board ordered gunners' stores from a host of ironmongers, armourers, wheelwrights, cutlers, braziers and other tradesmen. Small arms had long been made in the east end of London, in the Minories and East Smithfield, conveniently near to the Tower of London where deliveries were tested or proofed.[32] By the mid eighteenth century, small arms were also being produced in the Black Country of the West Midlands. The industry was highly skilled but quite dispersed. Hand guns were

[31] For guns needed for new ships being built, see lists sent by Navy Board to the Ordnance Office, 'in pursuance of letters from the Admiralty Secretary', TNA, ADM. 106/3067.
[32] Tomlinson, *Guns and Government*, 107–17.

made in parts by a variety of craftsmen and assembled at the finishing stage. The board thus dealt with powerful merchant-manufacturers.

However, foreign states like Portugal, as well as the East India Company, also placed large orders for arms. Despite having to pay with bills, delaying cash for six months or more, the Ordnance Board fought this competition and by 1800 it established a near monopoly of the supply in Britain. Nevertheless, national defence and allies on the continent called for unprecedented quantities of small arms and there were fears of shortages. In 1804 the Ordnance Board thus began assembling small arms at the Tower and formed its own factory for parts at Lewisham, near Woolwich. The latter did not start effective production till 1808 and the Ordnance Board was forced to place large orders for arms with continental producers. By the middle of the Napoleonic War, however, the board had developed a surplus of small arms which served to arm Spain, Russia, Prussia and Sweden.[33]

While small arms were tested or proofed at the Tower of London, heavy ordnance was proofed in the Royal Arsenal at Woolwich. Here there were storehouses, proving grounds and facilities for making, testing and storing guns, shot and powder – although the main magazine for powder was nearby at Greenwich. Satisfactory ordnance was then shipped to the depots for issue to the army and navy. Ordnance storage capacity for the navy at Chatham, Portsmouth and Plymouth was substantial, largely the product of expansion in the eighteenth century.[34] In addition there were separate powder magazines: at Upnor Castle opposite Chatham;[35] at Priddy's Hard on the Gosport side of Portsmouth harbour from 1773; and to the north of Plymouth where Morice Yard operated from 1720.[36] Ships proceeding up the River Thames had to deposit their powder at Purfleet or Gravesend.[37] By the end of the French Revolutionary war, these magazines could not all cope with the amount of powder returned from ships and they were extended by the use of powder hulks.[38] Guns

---

[33] Glover, *Peninsular Preparation*, 47–62.

[34] Coad, *The Royal Dockyards 1690–1850*, 245–70.

[35] For the out- and in-letters of the Upnor magazine to and from the board of Ordnance, 1745–1840, 1761–1850, see TNA, ADM. 160/1–40, 56–111.

[36] For these establishments in the late seventeenth to early eighteenth century, see Tomlinson, *Guns and Government*, 118–23; also Tomlinson, 'The Ordnance Office and the Navy', *EHR* 90(1975), 35–6. In 1702–13, supplies of ordnance were also available at Harwich, Hull, Kinsale, Dover, Berwick, Tynemouth and Leith. For maps, pictures and developmental history, see Coad, *The Royal Dockyards 1690–1850*, 177–86, 245–70. The period after 1770 is covered by D. Evans, *Arming the Fleet: The Development of the Royal Ordnance Yards 1770–1945* (Gosport, 2006), 8–49.

[37] TNA, WO. 55/1745, fo. 123, 9 Oct. 1765.

[38] A. Saunders, 'Upnor castle and gunpowder supply to the navy 1801–4', *MM* 91(2005), 160–74.

too were stored on board ships in Ordinary in the keeping of seamen but under the eye of the Ordnance officials.[39]

As the Ordnance Board also served the army and military garrisons, the number of its employees and outstations grew during the late eighteenth century, especially in wartime. By 1812 the three inspectorates of artillery, gunpowder and carriages at the Royal Arsenal at Woolwich reached over 3,500 men. There must have been well over 1,000 more at the powder mills, laboratories and port yards and magazines. By 1809 there were forty-four outstations submitting accounts in the British Isles, with another thirty depots and yards abroad.[40] These figures exclude depots in Ireland, which before 1801 had its own separate Board of Ordnance. The work of the latter was transferred to London in 1801 but its accounts were still kept separately.

Reputation and achievement

During the early eighteenth century ordnance and gunners' stores were regularly the cause of complaints about poor quality and deficient quantities.[41] A defence then had been that the Ordnance Board served the army as well as the navy and that there were differences of opinion. Howard Tomlinson, who examined the management of the Ordnance Office in the period before 1714, admits the board had particular difficulties owing to the time contracts took to fulfil, the challenges of standardisation, and ill equipment with wharves and stores.[42] He argues that, because the structure of government remained the same until the nineteenth century, the conflicts of opinion would persist.[43]

They did.[44] Significantly, the first precedent of a series maintained at the Admiralty until 1810 for the regulation of the supply of 'the guns and small arms' was an 'order in council of 10 July 1679 requiring the Master General of the Ordnance to comply with all directions issuing from the Admiralty'.[45] That this order was still necessary was evident between

---

[39] Lavery, *Arming and Fitting*, 80.

[40] Cole, 'The Office of Ordnance', 139, 272, 296–7, 359. For overseas establishments, Cole cites TNA, WO.46/156, pp. 235–9, which lists Anholt, Annapolis, Antigua, Bahamas, Barbados, Bermuda, New Brunswick, Cape of Good Hope, Ceylon, St Christopher, St Croix, Curaçao, St Thomas's, St Lucia, Demerara, Dominica, Gibraltar, Grenada, Halifax, Jamaica, Malta, Minorca, Newfoundland, Placentia, Quebec, Quebec Field Train, Tobago, Trinidad, St Vincent and Surinam.

[41] Tomlinson, 'The Ordnance Office and the Navy, 1660–1714'.

[42] Tomlinson, *Guns and Government*, 136–9, 144–61.   [43] *Ibid.*, 162–6.

[44] Comment from Dr Gareth Cole, to whom many thanks are due for reading and commenting on this chapter.

[45] TNA, ADM. 7/677.

1779 and 1781 when the Admiralty, driven by the Navy Board, was obliged to enforce compliance of the Ordnance Board in the introduction of the carronade, to which the officers of the Ordnance were initially opposed, doing all they could to obstruct it. Conflicts of this nature reinforced distrust of the Ordnance department, which was unfortunate, since it was partly based on prejudice, evident in the reputations of succeeding master generals.

The Duke of Richmond, Master General between 1782 and 1795, had a remarkable ability to inspire dislike.[46] Yet the Ordnance department under Richmond took many initiatives to improve its service and the quality of its supplies. Ordnance department finances were reformed (the estimates offered to the House of Commons were incomprehensible), it was one of the first government departments to abolish fees and perquisites, and it began the topographical survey of the south coast that became the Ordnance Survey.[47] Richmond strove for efficiency, even at the lowest level of yard operations,[48] and his immediate subordinates worked hard to avoid misunderstandings and mis-management.[49] Indeed, the main achievement of Richmond's administration has gone virtually unrecognised. For, after the relocation of the gun casting industry to new sources of supply of raw materials, and with rising standards of precision in engineering and chemistry, the Ordnance department was responsible for establishing and enforcing higher standards of gun and powder production.

In 1780 a post of Inspector of Artillery was created, and filled by Thomas Blomefield who served the public until his death in 1822. In 1789 another inspectorate was established, that of Gunpowder Manufactories, and filled by General Sir William Congreve who, since 1783, had been Deputy Comptroller of the Royal Laboratory at Woolwich. He would serve the public until 1814 when he was succeeded by his son and namesake.[50] Richmond's support for the work of these two men was vital to the qualitative development of ordnance in Britain.

[46] For his difficult personality, see Cole, 'The Office of Ordnance', 51–5.

[47] Richmond also organised the fore-runner of the Royal Horse Artillery and fostered improvements in small arms.

[48] Evident in his reluctance to oblige Elizabeth Wickham who requested her son be entered as a labourer at Portsmouth. He 'Having been twice discharged from the King's [dock] yard for idleness, I cannot think of taking him (at least for the present) into the Gun Wharf which is a place of some trust, but I have written to Lt Col. Phipps to employ him on the Works, where we shall see how he behaves': HRO, 109M91/CG23, Richmond to Ordnance officers, Portsmouth, 26 Oct. 1786.

[49] NMM, MID.1/110, Lewis to Nepean, 11 Oct. 1787.

[50] Cole, 'The Office of Ordnance', 137, 183.

Lord Cornwallis, who succeeded Richmond in 1795, has been branded 'too much a colonial soldier' but, in a difficult period, was also Lord Lieutenant for Dublin while Master General. Lord Chatham, who followed between 1801 and 1810 has been branded as lazy. Yet his administration was characterised by expansion and reform: innovations included the establishment of laboratories at Portsmouth and Plymouth to restore ageing gunpowder, and the Royal Artillery Academy at Woolwich.[51]

Indeed, when personality and opinion are ignored, the Ordnance department may be regarded as responsive to technological change. Equally important was the adoption of contemporary administrative ideology to achieve efficiency. Individual responsibility characterised the inspectorates of Artillery and Gunpowder Manufactories. And the principle of these posts was extended with Inspectors of Small Arms, of Carriages and of Barracks. Such individual responsibility, also applied at board level, made the Ordnance Department a model for post-war reformers.[52]

Quantity control for guns

The establishments

In theory, provision of the number of guns wanted by the Royal navy was straightforward. Ships were classed according to the number of guns they carried and joined a class or rate in which all the ships, more or less, carried the same number of guns. To facilitate equipment, manning and payment, each of these classes had established numbers of guns.[53] Hence, to calculate the total number of guns required by the navy, those demanded by each class might be calculated by multiplying the number of ships in that class by the number of guns each carried. This was, however, to impose theory on practice. Experience revealed that gun establishments did not determine the guns wanted by the navy and that the demands of the navy were in fact determined by those already available to the Ordnance Board, those condemned as defective and the structure of individual ships.

The establishments had originated after the Dutch Wars when in 1674 the Admiralty attempted to impose uniformity on the fleet and permit the Ordnance Board to keep ships supplied with the appropriate ordnance.

---

[51] Glover, *Peninsular Preparation*, 38–9; Cole, 'The Office of Ordnance', 56–7.

[52] Cole, 'The Office of Ordnance', 22–3, 40, 183, 359; P. Burroughs, 'The Ordnance Department and colonial defence, 1821–1855', *Journal of Imperial and Commonwealth History* 10(1981), 125–49.

[53] Establishments of masts, yards, ropes and guns to 1745, see TNA, ADM. 106/3067.

However, the Ordnance Board was not even involved in forming the first establishment. Moreover, this was for new ships. The great variety of sizes and number of guns in existing ships could only be moderated. Ship construction, such as the number of their ports, still dictated the armament they could carry. Practicality and pragmatism prevailed over theory.[54]

This was acknowledged in 1685 when consultation over a new establishment included the Ordnance Board and took into account the number of guns 'on hand' and noted the sizes of which there was an 'excess' and a 'shortage'. Yet, afterwards, new sizes of ship and revivals of old forms produced new sizes of gun. In consequence there were numerous deviations from the establishment and some confusion in the gunning of ships. Moreover, war between 1689 and 1698 revealed some ships were too heavily armed for their structure. A new establishment of 1703 thus reduced the weight of metal in ships of the line, which were intended to conform to class at least in the calibres of their guns, if not exactly in their number. Indeed, another new establishment in 1716 used the weights rather than their traditional names to describe the guns. Still, however, a common response of the Ordnance Board to requests for guns of a particular size was that 'the guns in store could not comply with the same'.[55]

Peace until 1739 permitted the Ordnance Board to continue to arm ships as best it could. Ships in existence tend to have been armed according to the establishment which prevailed when they were built.[56] However, war brought shocks from the size of the Spanish and French ships, and the Admiralty Board of 1744, which included Anson, enlarged the establishment of small ships of the line, fostering 74- and 64-gun ships. Practical sailing performance required their hulls to be enlarged too, beginning a period in which ships were designed around their guns rather than having their guns imposed on them. Design theory developed, with lines of flotation and centres of gravity naturally affected by the weight of armament. The Seven Years' War enhanced the pursuit of a balance between sailing performance and fighting power. Thereafter ships were built on the lines of previously successful ships.[57]

During the American War of Independence, the introduction of the carronade added a further form of armament which really finished the establishments as realistic reflections of weaponnes. There remained much consistency between the main deck armament of ships of the same

---

[54] Lavery, *Arming and Fitting*, 115–16.   [55] *Ibid.*, 117–19.
[56] Hence in 1743 the Navy Board looked back to 1716, 1733 and 'how gunn'd in 1741'; NMM, ADM. BP/105, 'Establishments of guns for the Royal Navy'.
[57] Lavery, *Arming and Fitting*, 119–20; for contemporary discussion of these matters, see *British Naval Documents*, 486–95.

Table 5.3 *The establishment of iron cannon for a ship of each rate c. 1794[a]*

	Ships of the line (no. of guns)			
	100	98	74	64
32 pounders	30	28	28	–
24	28	–	–	26
18	–	30	28	26
12	42	40	–	–
9	–	–	18	12
6	–	–	–	–
4	–	–	–	–
3	–	–	–	–
½	–	–	12	12

	Frigates and sloops (no. of guns)								
	50	44	38	36	32	28	24	20	18
32 pounders	–	–	–	–	–	–	–	–	–
24	22	–	–	–	–	–	–	–	–
18	–	20	28	26	–	–	–	–	–
12	22	22	–	–	26	–	–	–	–
9	–	–	10	10	–	24	22	20	–
6	6	6	–	–	6	4	2	–	18
4	–	–	–	–	–	–	–	–	–
3	–	–	–	–	–	–	–	–	–
½	12	12	12	12	12	12	12	12	12

[a] TNA, ADM.160/150, 'A proportion of the Ordnance and Stores for a ship of each rate in the Royal Navy'.

rate (see table 5.3). But by 1794 carronades added between six and twelve guns to the main deck establishment.[58] By the time of the Napoleonic Wars, the establishments were recognised as but a guide to ship power rather than a means of logistical calculation.

Returns and contracts

How then did the Ordnance Board provide the quantity of guns and ordnance equipment required by the navy? Pragmatically, as it had always done. It used the guns in store and, should there be insufficient there, it ordered more from contractors. For this purpose, and for the Ordnance estimates, it used the establishments to develop comprehensive lists of the ordnance stores necessary for the equipment of ships of each class (see

[58] NMM, ADM. BP/8, 15 May 1788; MID.9/2/27.

table 5.4), distinguishing ships by rate or class and by whether employed on foreign or Channel service.[59] When ships were decommissioned, most of these were returned into store. The Ordnance Board was able to demand annual, later monthly, returns of stores 'in immediate readiness' for service from its depots at the Tower, Woolwich, Sheerness, Chatham, Portsmouth and Plymouth.[60]

The Ordnance Board thus focussed on immediate needs. It maintained stocks of stores at its magazines and yards, and ensured that those issued to ships were cared for by the issue of warrants to ships' gunners.[61] As such, its task was relatively uncomplicated, especially during peace. However, in wartime the number of ships being built or needing replacement guns increased. Demands for guns and stores therefore escalated and the Ordnance Board could run short.

After 1755 each war saw an increase of about twenty ships of the line, demanding between 1,280 and 2,000 cannon. Growth in the whole navy was far greater. In the Seven Years' War, the increase between 1755 and 1762 was 136 ships. During the French Revolutionary and Napoleonic Wars the growth between 1792 and 1809 was 504 ships. It has been calculated that the total number of cannon needed for the navy between 1793 and 1798 grew from 9,518 to 16,004. During the first year of the Revolutionary War, 2,292 extra cannon were needed; during the second, 2,552; during the third, 948; in the fourth, 386; and in the fifth, 308.[62]

With so many cannon wanting for the navy in the first two years of the French Revolutionary War, the period has been characterised as one of crisis. Yet supply to ships at that time was aided by the series of preceding mobilisations which prompted the Ordnance Board to ensure all ships were complete in their ordnance. Returns in May 1790 specified all ships incomplete, without either guns or carriage or both, and showed that thirty-eight ships (including fourteen 74s and nine frigates) lacked any

---

[59] TNA, WO. 55/1745, fols. 8–13.   [60] *Ibid.*, fols. 120–1, 1 Oct. 1765.
[61] TNA, ADM. 6/3–32, Gunners' warrants, 1695–1815.
[62] Dr. G. Cole, presentation at the Greenwich Maritime Institute, University of Greenwich, 13 Apr. 2007. The number of cannon needed by the navy increased as follows:

1793	9,518
1794	11,810
1795	14,362
1796	15,310
1797	15,696
1798	16,004

Table 5.4 *The ordnance stores and equipment
supplied to each ship in the Royal navy c. 1795*[a]

Iron ordnance with carriages
Axletrees
Trucks, pairs
Beds and coins
Ladles and sponges
Rope sponges
Wadhooks
Heads and rammers
Coins for coining guns
Spikes
Round shot
Grape shot
Boxes for grape shot
Double headed shot
Paper cartridges
Tallow
Marline
Junk
Hand grenadoes
Boxes for above
Melting ladle
Copper powder measures
Guns with seven barrels
Musquets
Musquet rods
Bayonets
Scabbards for bayonets
Slings for musquets
Musquetoons
Pistols
Cartouch boxes
Belts for above
Frogs for bayonets
Boxes for cartridges (musquets & pistols)
Flints for above
Shot for above
Fine paper
Funnel of plate
Sweet oil

[a] TNA, ADM.160/150, 'A Proportion of Ordnance and
Stores for a ship of each rate in the Royal Navy'.

ordnance, while eleven wanted 292 guns and 41 carriages.[63] To make good these deficiencies, the Ordnance Board simply enlarged the scale of its contracts and the number of its contractors. By this time, contracts were large. In 1793 one contract, recommended for the London contractor Alexander Raby, consisted of 500 32-pounders, 150 24-pounders, 250 18-pounders and 100 12-pounders. Up to 1,000 carronades were ordered from individual suppliers. But contracts varied in their terms: in 1796 Alexander Brodie in Shropshire agreed to deliver 30 tons a month, while another, the Walker brothers of Rotherham, agreed to supply 2,000 tons a year.[64]

The scale of these deliveries may be contrasted with the earlier output of foundries in the Weald of southern England which produced about 200 tons a year. The new industry permitted the Ordnance Board to meet the needs of the navy relatively easily. This is not to say that improvisation was not occasionally necessary. Sometimes the Ordnance Board fitted new ships with guns taken from other ships, though cautiously, from fear of mixing old and new guns which were thought to have a different resistance. Later it also employed foreign guns that were 'accurately constructed' and for which there was shot of an appropriate size.[65]

The focus here has been on the ordnance needed by the navy. It is important to recall that the Ordnance department also supplied the army and that its demands also grew. Yet the available evidence indicates efficiency in meeting both the detail and the scale of demands.

All the military equipment needed by the army overseas was shipped either from the River Thames or from the yards at the ports. The employment of Ordnance officers as artillery commissaries overseas ensured familiarity and understanding between those despatching stores and those receiving them in foreign ports.[66] Almost daily communication between offices in England ensured attention to the smallest detail. In the Seven Years' War, the Portsmouth officers had to assemble the stores needed for the army's field trains and ship stores to Germany and America, including clothing for the artillery. Some came from the Tower by ship, smaller quantities by wagon overland. In July 1756 one wagon load included a 'small canvas bundle of cloathing for a Drummer at St John's Newfoundland'.[67]

[63] Baker, *The Crisis in Naval Ordnance*, 4–5.   [64] *Ibid.*, 9, 11.
[65] *Ibid.*, 4–6; TNA, ADM. 1/4015, fo. 508, 31 July 1801.
[66] For the services of Richard Veale and William Bache in Germany and elsewhere, see the dissertation of L. A. Burton in Hampshire Record Office, HRO, 109M91/MIS2, and Veale's papers, 109M91/CG19 & CG82.
[67] HRO, 109M91/CO16, 28 Jan., 30 July 1756.

During subsequent wars the scale of demands grew. In the American War of Independence, the Portsmouth officers were not only shipping stores to America and the East Indies but also supplying the militia stationed in Hampshire.[68] During the French Revolutionary War, they shipped the stores and equipment for the St Domingue and Leeward Islands expeditions, while also despatching arms for the use of soldiers in New South Wales. At this time the demands of the expeditions eclipsed those of the navy: in December 1795, 25,000 stands of arms were issued to the agent for transports for the army under Sir Ralph Abercromby.[69]

Quality control of guns

After guns were delivered by contractors, but before they were accepted by the Ordnance Board and distributed to ships, they were measured against contract specifications and they were proved. This was done at the board's main foundry, store and testing grounds at Woolwich. The estate there, known as the Warren, had been used for testing guns since the sixteenth century, and the construction of a foundry in 1716 established it as the board's centre of technical expertise. In the following half-century there was little further development, but the Seven Years' War indicated the necessity for the board of Ordnance to equip ships with ordnance equal in quality to, if not better than, that of Britain's continental rivals. This demanded the recruitment of experts in production of cannon, the installation of machinery to set the standards required in manufacture, and the imposition of categorical tests on cannon offered to the board by contractors.

The recruitment of expertise

By 1770 the board had recruited Jan Verbruggen from Holland. Having become a master founder in Enkhuizen by the age of thirty-four, in 1755 he had been appointed head founder in the Dutch heavy ordnance foundry at the Hague. There, within three years, Verbruggen had introduced machinery for horizontally boring gun barrels from solid castings. The machinery was similar to that developed by Johann Maritz in 1715 in Switzerland. A colleague at that time was Johan Jacob Siegler who, prior to employment at The Hague, had experience of such machinery from work in France's Douai gun foundry. Siegler seems to have resented

---

[68] HRO, 109M91/CO43, 5 Feb., 9 Apr., 16, 27 Jun., 1 Aug. 1778.
[69] HRO, 109M91/CO67, 1, 5, 6, 20, 24, 29 Oct., 5, 19, 24, 26, 28 Nov., 28 Dec. 1795; 109M91/CG55, 12 Oct. 1796.

Verbruggen's appointment and later denounced the quality of his castings. Though Verbruggen appears to have learned much from his efforts to eradicate flaws, the difficulties of work at The Hague encouraged him to accept an offer from the Ordnance Board to move to Woolwich.

As at The Hague, Verbruggen's first task was to install horizontal boring machinery. Casting and boring began at Woolwich under Verbruggen in 1773. This was used principally to bore brass cannon, all of those needed by the navy being cast at Woolwich. But the precision of the boring prompted the Ordnance Board in 1776 to lay down to all contractors that in future it would purchase only bored cannon.[70] At the same time, between 1775 and 1780 the facilities around the foundry were improved with new workshops for smiths and carpenters, new storehouses and a new laboratory. Alongside this complex were firing ranges for proofing guns delivered to the wharves facing onto the River Thames. Significantly too, artillery cadets at the Royal Military Academy were barracked there for instruction.[71]

Jan Verbruggen died in 1780. By 1782, his son Peter Verbruggen had become master founder at Woolwich.[72] The experience of the Verbruggens, father and son, contributed to the quality of the knowledge with which cannon offered to the board of Ordnance were assessed. The rigour of this assessment was established between 1783 and 1789 under Captain Thomas Blomefield. He had joined the navy in 1755 but three years later entered the Military Academy at Woolwich as a cadet. Though rising as an artillery officer, Blomefield served in a bomb vessel at the end of the Seven Years' War, when he also took part in the capture of Martinique and Havana. During the early and late 1770s he was aide-de-camp to General Conway, Lieutenant General of the Ordnance, and to Lord Townsend, Master General. During the American War he was severely wounded serving in Burgoyne's army before its surrender at Saratoga. Invalided home, in 1780 he was appointed Inspector of Artillery and Superintendant of the Royal Brass Foundry at Woolwich.[73]

The new standards

Under Blomefield the quality control of cannon offered to the armed services by contractors reached its zenith. Having been made responsible

[70] Jackson and de Beer, *Eighteenth Century Gunfounding*, 46–51.
[71] O. F. G. Hogg, *The Royal Arsenal: Its Background, Origin and Subsequent History* (Oxford, 1983), 451.
[72] To Peter Verbruggen are attributed the pictures of the interior of the Woolwich Brass Foundry, printed in Jackson and de Beer, *Eighteenth Century Gunfounding*.
[73] Baker, *The Crisis in Naval Ordnance*, iv.

Table 5.5 *Dimension tolerances permitted in Ordnance Board gun contract, 1786*

	42–18 pounders	12–4 pounders
Bore diameter	1/30 inch	1/40 inch
Position of central axis	1/2 inch	1/3 inch
Deviation from exact cylinder	1/10 inch	1/10 inch
Vent diameter – no tolerance	1/20 inch	1/20 inch
Vent position – forward	3/10 inch	3/10 inch
backwards	1/10 inch	1/10 inch
to side	2/10 inch	2/10 inch

Source: J. G. D. Elvin, *British Gunfounders 1700–1855* (privately printed, 1983, copy in NMM), appendix C.

for the proof and examination of all guns, his first task was to survey and re-proof all reserve guns at the ordnance yards. Undertaken with the help of his assistant, Henry Careless, and a working party, the task took until 1789.[74] Guns were classified according to whether they were serviceable, repairable or condemned. All had to be cleaned of rust and tallow. Those condemned amounted to 1,387 over the three years. At Portsmouth alone, 436 were condemned, and 449 classed as repairable. The most common repair was to bouch worn vents, which involved replacing and re-drilling the vents. With only a small number of smiths available from Woolwich, this work took a long time: only 175 of the 449 guns at Portsmouth were repaired by re-bouching in one year between 1786 and 1787. However, local smiths were trained and the overall effect was to produce a smaller but sound stock of guns in store.[75]

Blomefield's second task was to ensure the guns received from contractors were of the standard wanted by the navy. Their guns were examined, then measured. With solid boring of barrels and developing machine tools, a range of accurate measuring instruments were also developed to reveal any deviations from specifications. The tolerances contained in a contract awarded to Walkers of Rotherham in February 1786 extended to no more than one-thirtieth of an inch for guns firing 18- to 42-pound shot, and no more than one-fortieth of an inch for those firing 4- to 12-pound shot (see table 5.5).

---

[74] Careless completed work on the guns at Portsmouth in April 1789, when he went on to Plymouth. The guns condemned at Portsmouth were placed in store: HRO, 109M91/CO57, 11 Feb., 2, 15 Apr., 8 Dec. 1789.

[75] Baker, *The Crisis in Naval Ordnance*, 1–2.

Carronades were not permitted to deviate more than one-twentieth of an inch in any respect from their specifications.[76] An equal challenge for contractors was to achieve a satisfactory metal quality. Poor-quality materials produced pitting in the metal. Blomefield laid down that, in carronades, pits one-fifth of an inch deep in the barrel or one-tenth of an inch in the chamber were sufficient to condemn them.

The late 1780s saw a marked increase in founders offering guns to the navy.[77] The iron available to these new founders was invariably contaminated with sulphur. Under heat the sulphur liquefied, which meant it remained undetected during casting but could liquefy again when cannon heated on being repeatedly fired. Suspecting the contamination, in 1787 Walkers successfully cast guns with 50 per cent of their metal derived from old Wealden guns. The Wealden metal contained lime and phosphorous – which cancelled one another – but no sulphur. However, Walkers also used iron from other sources and produced brittle iron unsuitable for guns. So too did other gun founders, resulting in a rising failure rate on proof.[78] A recent study indicates that the number failing proof was 12.5 per cent in 1780, 14.5 per cent by 1794, and an alarming 25.6 per cent in 1795.[79]

The gun founders blamed the extreme rigor of proving methods laid down by Blomefield. Water pressure proofing had been introduced in 1780.[80] Guns were then fired. Blomefield retained the size of powder charges that had been used since 1719 but, instead of each gun firing a single shot, introduced a two-round proof. If any gun failed from firing two rounds, another two guns from the same batch were selected to fire thirty rounds double-shotted in simulation of a naval action. The guns were inspected after every round. If any of the guns failed in the course of the thirty shots, another two guns were selected for thirty rounds, and so on until all doubtful cannon had been tried from the batch. Should any gun fail the two-round proof, there was a tendency for the whole batch to be rejected. The exhaustive test caused much distress to the gun founders. In 1789 they managed to have the board of Ordnance reduce the thirty-round test to twenty rounds. Yet by then they also had to ensure their guns complied with a new design.

---

[76] Elvin, *British Gunfounders*, appendix C; Baker, *The Crisis in Naval Ordance*, 13–14.

[77] Cookson in Co. Durham, Crawshay in south Wales, Hird Dawson and Hardy in Yorkshire, Sturgess & Co. at Bradford: A. B. Caruana, *The History of English Sea Ordnance 1523–1875* (2 vols., Rotherfield, Sussex, 1994–7), II, 11, based on Kennard, *Gunfounding and Gunfounders*.

[78] Baker, *The Crisis in Naval Ordnance*, 28–9, 33–4; Caruana, *The History of English Sea Ordnance*, II, 12; Trollope letter, 1.

[79] Greenwich Maritime Institute workshop, 13 Apr. 2007.

[80] Hogg, *The Royal Arsenal*, 464–5.

The new design

Blomefield's final measure was to redesign the British naval cannon. With machine boring, and a decline in the windage or waste of explosive gas around cannon balls, the internal pressures upon barrels increased. The subsequent introduction of cylinder gun powder also added to that pressure. Blomefield's new design was intended to strengthen the breeches of cannon. He began experiments with the assistance of Walkers in 1786. Extra thickness was given to the sides of the cannon, adding to their weight. This was counteracted by thinning the chase of the barrel and, although slightly lighter than French guns, the final design was little different in tonnage from the earlier one. Problems were met in, for example, achieving mould dimensions that permitted the easy flow of molten metal. But by 1790 new pattern designs were being distributed to contractors.

In November 1794 two guns of the old design burst under proof and it prompted the termination of all further casts in the old design. Henceforward, the stronger design ensured fewer cannon failed their test. Failures at proof fell to 16 per cent of guns in 1796, then to 5.6 per cent in 1797. A decade later, in 1809–10, the failure rate remained at 3.5 per cent; in 1810–11 it was 4.2 per cent. Many cannon of the old design remained in ships; some were used at Trafalgar. But by 1810 most were phased out.[81]

The Blomefield cannon was thus the standard replacement of the Revolutionary and Napoleonic Wars. Production was aided by the gradual improvement in the manufacture of cast iron. New methods took time to spread. During the late 1780s Henry Cort took five years to demonstrate, prove and introduce his method of making wrought iron on an industrial scale.[82] The communication of methods of production among iron founders and their trial probably took twice that time. It has been suggested that the quality of iron used by gun founders rose on account of the higher furnace temperatures produced by the stronger blast of air driven by Watt's double-acting steam engine.[83] Certainly one contractor, Clyde and Co., is recorded as introducing Neilson's hot blast process in 1798–9.[84] The contracts on a large scale offered by the Ordnance Board must have given incentive to innovate and modernise production processes.

---

[81] Baker, *The Crisis in Naval Ordnance*, 12; Caruana, *The History of English Sea Ordnance*, 11–12; Lavery, *Arming and Fitting*, 94; Cole, 'The Office of Ordnance', 218, 243.

[82] R. A. Mott, 'Dry and wet puddling', *Transactions of the Newcomen Society* 49(1977–8), 153–8.

[83] Caruana, *The History of English Sea Ordnance*, 12.

[84] Elvin, *British Gunfounders*, 15.

These improvements came together in the 1790s. There can be no doubt that the quality of cannon rose significantly, making for greater reliability. Gareth Cole notes that fewer seamen were liable to be killed from the explosion of their own weapons.[85] The credit for this improvement may be given to the board of Ordnance, but more particularly to its Inspector of Artillery, Thomas Blomefield. He promoted a standard and vision which lifted the whole of British gun manufacture to a new level. Although not considered here, army cannon too must have benefited from the same process of improvement. During the French Revolutionary and Napoleonic Wars, British military ordnance entered a new era.

The amplification of gun power

For the navy, this was true not only for the main armament of ships, but for their supplementary guns. For in November 1794 carronades were established as additional guns in virtually every warship. These small guns were mounted on the upper decks of large ships and, owing to their terrific power at close range, influenced the tactics employed by the British in action. During the French Revolutionary War, the navy successfully returned to mêlée tactics which exploited the qualities of the carronade and gave the Royal navy 'a massive advantage in combat'.[86]

The carronade trials

The development of the carronade was part of the history of the Carron Company established in 1760. The company sold its first long guns to the Ordnance Board in 1765. But these fell into disrepute in 1771 after a high proportion burst on proof. Detailed testing of their quality in 1773, including an assay of the metal, resulted in the rejection of those on order, the removal of Carron guns from ships, and termination of the Company's contract to supply long guns to the navy. That supply was not to be renewed until 1795.

To restore its business, the Carron Company improved the quality of its metal and experimented with lighter guns with similarities to army mortars. It was advised by General Robert Melville, an infantry officer, who wanted a light gun of large calibre. Under the impetus of war with the American colonists, then France, production of the carronade began in 1778. The Company armed its own ships sailing to London with them,

---

[85] Cole, 'The Office of Ordnance', 242.   [86] Ibid., 223.

and in 1779 fitted a Liverpool privateer, the *Spitfire*. Her performance in
action was successfully advertised to the King. More importantly, it was
communicated to Sir Charles Middleton, chairman of the Navy Board,
who came from Leith and was acquainted with Charles Gascoigne, a
partner in the Carron Company and responsible for its gun founding.[87]

Probably developed for land carriage as much as naval use, the
carronade capitalised upon the advantages offered by the new boring
machinery. First, the chamber, where the powder charge was placed,
was of a narrower bore than the calibre of the barrel. Because of this,
the metal around the chamber remained thicker, permitting a reduction
in the outside diameter of the barrel and thus the weight of the weapon.
Weight was further reduced to about one quarter the weight of long
cannon of comparable calibre because the barrels were much shorter: a
carronade of 1780 that fired an 18-pound shot was only 2 feet 4 inches
long while the cannon of similar calibre was 9 feet. Shortness of barrel
made for less accuracy at a distance, but precise boring made for reduced
windage, adequate accuracy and great power at short range.[88]

Sir Charles Middleton seems to have been instrumental in obtaining
trials. The first used weapons of small calibre were mounted on 6 July
1779 at Woolwich in the presence of the Master and Surveyor General
of the Ordnance. The trials compared the performance of an iron 12-
pounder gun, length 8 ft 6 in, weight 32 cwt 0 qtr 18 lb, mounted on a
ship carriage, and a 12-pounder carronade, length 1 ft 10 in, mounted
on a newly constructed sliding carriage, firing at two bulk heads, each
about 2 feet thick, placed one behind another 15 yards apart. After one
of these first trials, Middleton noted perceptively:

At 350 yards distance, the common 12 pounder put a shot through the upper
part of the bulkhead where the plank was four inches thick. It took off a piece
of the timber and after going through the lining of three inches went almost
through the second bulk head. The carronade at about the same distance with
one pound of powder buried itself only seven inches in the wale. But at 200
yards distance with 1½ pound of powder, it put the ball through the first bulkhead where the plank was five inches and the lining four and through a four
inch plank of the second, so that I think the carronade at 200 yards went with
more velocity than the common gun at 350, but in no degree equal at the same
distance.[89]

Two days later a trial compared the effect of

sea service grape shot, each grape containing 9 balls of one pound; case shot each
containing 23 balls of half a pound weight; langrege shot, each case containing

[87] Lavery, *Arming and Fitting*, 104–5; Talbott, *The Pen and Ink Sailor*, 62–3.
[88] Lavery, *Arming and Fitting*, 105.   [89] NMM, MID.9/2/5 & 17.

21 pieces of iron 1 inch square and 3 inches long, medium weight of each piece of iron 12oz; fired from a new 12 pounder iron gun mounted on a ship carriage, length 8 feet 6 inches, weight 32 cwt. 0qtr. 18lb., with 4lb. of powder. The target fired at was of 1/2 inch deal 9 feet high and 30 yards wide.

Part of the trial was to measure the 'spread from the centre' and effect on sails and rigging.

Each grape left a round smooth hole while the langrege cut and tore.[90] Middleton was impressed. He came away, convinced two men could work each carronade instead of the ten needed for long guns; that, being so short, they became less hot and could be fired twice in the time taken for one long gun's shot; and that 'such guns from their lightness may be used to great advantage on the poops of all ships and on the quarter decks of others'. To Lord Sandwich, First Lord at the Admiralty, he concluded that, if used along with howitzers in the tops, they would soon 'give us a great superiority over the enemy in all naval actions and particularly within musquet shot'.[91] Sandwich referred his observations to George III on 8 July 1779.

The persuasion of opponents

Under the impetus of its Comptroller, the Navy Board acted promptly. On 15 July the Admiralty requested the Ordnance Board to acquire carronades and fit them to ships as they came in for refitting. The following day the Navy Board provided an establishment of 12- and 18-pound carronades for different types of ship. The initial response of captains varied, opposition arising principally because carronades got in the way of rigging. By the end of 1779 the Navy Board was ready to relax the establishment, and from 9 March 1780 carronades were fitted on the application of a captain.[92]

By then a growing body of naval opinion favoured more and bigger carronades. One reason was the favourable reports that resulted from Rodney's Moonlight Battle with the Spanish on 16 January 1780. Captain John Elliot of the 74 Edgar reported his carronades must have done more damage than musketry and would have been 'still more serviceable' if they had been of the same calibre as his upper deck guns.[93] On 20 March 1780 the Navy Board therefore requested experiments on the force and range of 24-, 16- and 12-pounder carronades compared to 6-, 4- and 3-pounder long guns, commonly mounted in small vessels. Not until

[90] NMM, MID.9/2/18 & 23.   [91] NMM, MID.9/2/5 & 23.
[92] Lavery, *Arming and Fitting*, 105–6; Talbott, *The Pen and Ink Sailor*, 64–5.
[93] NMM, MID.9/2/3.

21 October 1780 did the Ordnance Board report to the Admiralty and then it reported its view that 'the [long] guns have greatly the preference' and that 'carronades are of little use in the Royal Navy'. The Admiralty did not send this negative report to the Navy Board until 15 December. It gave rise to immediate confutation.

Three days later Middleton retorted the Ordnance Board must have mistaken the intention of the trial: it was limited to seeing if carronades might replace long guns in brigs and sloops. On the broader principle of their general use, he contended:

Unless therefore it can be proved that shot fired from carronades of 12 and 18 pounders are unequal to the cutting of rigging and sails, to the disabling masts and yards, to the killing men and occasioning of splinters at the common distances in which ships generally begin to engage; that common guns are as manageable as carronades when ships fall on board each other in action; that they can be fired as often and to equal advantage in such situations with as few men; that the guns of the common construction can be placed with safety on the extreme parts of a ship's frame and added to the guns already allowed, without being obliged to increase the number of men, we cannot agree with the opinion of the board of Ordnance that carronades are of little use to the Royal Navy.

The Navy Board backed its Comptroller with reports from captains – such as that from Captain MacBride who had fought the *Conte de Artois* in the *Bienfaisant* – and by reference to talk 'that the boatswain of the Flora, assisted only by his boy made a surprising number of discharges with a forecastle 18 pounder carronade when alongside the Nymphe in action'.[94]

The Ordnance Board did not press its opinion. On the contrary, for events moved against it. The enemy learned of the carronade. In July 1781 the *Tartar* privateer of Glasgow, smaller than a naval sloop but armed with sixteen 32-pounder and six 18-pounder carronades, engaged six Dutch East Indiamen, each four times the tonnage of the privateer. Five escaped in the night, but the one principally engaged sank with one survivor. Then in October 1781 an American frigate built at Amsterdam for Congress was reported to be armed with forty-eight 42-pounder carronades.[95]

Meanwhile, the Carron Company began casting larger carronades. In September 1781 it experimented with a 100-pounder weighing 48–50 cwt; the barrel was $9\frac{1}{8}$ inches in diameter, the chamber 8 inches in diameter; and its length, including the chamber, was six calibres, that is,

[94] NMM, ADM. BP/1, 20 Mar., 18 Dec. 1780; MID.9/2/1, 4, 5, 6 & 7.
[95] NMM, MID.9/2/9 and 12, Reports to the Lord Advocate of Scotland, 13 July, 16 Oct. 1781.

4 feet 9½ inches. With 11 pounds of powder, it fired balls of 100 pounds between 2,400 and 2,705 yards and canisters packed with balls between ½ and 4 pounds to a gross weight of 97 or 98 pounds. One canister first grazed the water at 270 yards, then scattered in a column for 1,300 yards; it contained 32 balls of 2 pounds and 27 of 1 pound. With ten men, the 100-pounder carronade could be 'wrought, pointed and fired in three minutes'.[96] An awed observer informed the Lord Advocate of Scotland on 6 September,

It is altogether impossible that any ship can resist the shock given by the 100 pound ball. If one of our line of battle ships or 50 gun ships has only four of these guns on board there is not a ship of the enemy of the first rate . . . which would dare approach her. To come within two or three hundred yards of that distance at which ships engage and to continue there half an hour would be attended with inevitable destruction to the enemy.

The observer continued that, if the French had these guns 'now on board their fleet we are undone for ever'. Indeed, he foresaw the new gun would soon be 'universally received . . . by every maritime state in the world . . . £50,000 will turn the scales of the war at sea . . . no ship is safe within a mile of it'.[97]

At Leith the experiments continued, comparing a 100-pound carronade to a standard long 24-pounder and a 68-pound carronade to a long 12-pounder. The results favoured the former. Calculations even considered the complete substitution of carronades for long guns and proved the dominant power of the former in a 74. They not only delivered more shot, they employed for fewer men and less powder (see table 5.7).

Indeed, since the carronades could be fired at least twice for every discharge of the conventional cannon, if against an enemy 74, they could be reckoned to deliver 66 cwt of shot to the 14 received. That was an extra 52 cwt – more than 2½ tons – of shot at every complete discharge of the carronades.[98]

By December 1781 further trials at sea reinforced favourable opinion. Captain MacBride had had three trials of them in the *Artois*: 'one case shot from the large carronade almost entirely unrigg'd one of the privateers and brought down every sail but his jib . . . No small guns in close action can stand against them nor has any defect or failure now or on former occasions happened to our carronades, and they were fired three

---

[96] NMM, MID.9/2/19 and 28.
[97] NMM, MID.9/2/10, 'Copies of two letters from a gentleman in Edinburgh to Lord Advocate dates 6th and 7th Sept 1781'.
[98] NMM, MID.9/2/16, 20.

Table 5.6 The establishment of carronades for a ship of every class in the Royal navy in 1794[a]

Rate of guns	Quarter deck		Fore castle		Round house		Total
	Number nature	Pounders	Number nature	Pounders	Number nature	Pounders	
1 100	–	–	2	32	6	24	8
2 90	–	–	2	32	6	18	8
3 74/80	–	–	2	32	6	18	8
4 64	–	–	2	24	6	18	8
4 50	4	24	2	24	6	12	12
5 44	6	18	2	18	–	–	8
5 38/36	4	32	2	32	–	–	8
5 32	4	24	2	24	–	–	6
6 28	6	18	2	24	–	–	6
6 24	6	12	2	18	–	–	8
6 20	6	12	2	12	–	–	8
Sloops	6	12	2	12	–	–	8
Brigs	6	12	–	–	–	–	6
Cutters	4	12	–	–	–	–	4
For the French ships							
3 84	6	32	2	24	8	24	16
3 74	4	32	2	24	6	24	12

[a] TNA, ADM. 106/150, 'Establishment of carronades for a ship of every class in the Royal Navy', dated 24 Nov. 1794.

Table 5.7 *The differences in ordnance resources and shot fired between a 74-gun ship armed with long guns and one equipped completely with carronades*[a]

Standard 74 with long guns	Mounted instead with carronades
30 × 32 pounders	30 × 68 pounders 8 inch diameter
30 × 18 pounders	30 × 42 pounders
14 × 9 pounders	14 × 32 pounders
weigh 158 tons	weigh 94 tons
employ 744 men	employ 368 men
consume 5 barrels powder	consume 3 barrels powder
discharge 14 cwt shot	discharge 33 cwt shot
'Difference in favour of the carronades'	
64 tons less weight	
376 fewer men	
2 barrels less powder	
discharge 19 cwt more shot	

[a] NMM, MID.9/2/16.

to two at least oftner than the great guns.' Having discovered how best to restrain and incline the carronades, he also found 'our seamen's prejudices are removed and they prefer much being quartered at a carronade to a great gun'. To remove the prejudice held at the board of Ordnance, and 'to justify our persevering even to an appearance of obstinacy in recommending these guns', the Navy Board requested the Admiralty send a copy of MacBride's report to the Ordnance Board. MacBride had anyway recommended the carronades be lengthened by one to two calibres and the Navy Board supported the proposal.[99]

The carronade in service

Heartened by this support, on 28 December 1781 the Navy Board recommended two 68-pounder carronades be allowed to all classes of ship capable of supporting them, and 42- and 32-pounders to those of smaller rates – on the request of their captains. Moreover, 'as we wish the merits of these guns to be tried on a larger scale, and to increase if possible by their means the number of ships capable of acting in a line of battle', the Navy Board proposed the *Rainbow* of 44 guns be fitted completely with carronades.[100] Both the Admiralty and the Ordnance Board complied. A list of ships supplied with carronades of 22 July 1782 records the *Rainbow* equipped as requested, the only naval warship at that time

[99] NMM, MID.9/2/13; ADM. BP/2, 12 Dec. 1781.
[100] NMM, MID.9/2/21; ADM. BP/2, 28 Dec. 1781.

fitted completely with these guns. Her armament was tested at the Nore on 31 July 1782 with eight naval captains present, before she was sent for trial at sea.[101] There her success was dramatic. On 11 September 1782 the *Rainbow*'s master, Charles Duncan, reported from Plymouth that she had 'taken and brought in here the Hebe French frigate of 40 guns 300 men, without any action more than firing the starboard carronade on the forecastle . . . 26 times'.[102]

By July 1782, 36 ships of the line had been fitted with 166 carronades, 77 frigates with 530 carronades, and 45 smaller vessels with 278 carronades. In all, the Ordnance department had issued at least 978 carronades to 158 vessels. To supply the guns, in April 1782 the Carron Company was casting and finishing between 60 and 80 carronades a week, devoting nearly its entire production to the gun.[103] Carronades continued to be supplied, but hostilities with the European powers arising from the American war were terminated by order on 13 February 1783.[104]

By that time the reputation of the carronade was established, and the Ordnance Board won over. Technical refinements followed. The critical reduction of windage gave rise to the proposal for an experienced gunner on each gun wharf to inspect shot and ensure they were rust free and rubbed with grease.[105] To accustom new-raised men to using the guns, the Admiralty had carronades placed on receiving ships and the gunners of ships in Ordinary attend to exercise them daily. It submitted instructions for the guidance of these gunners and proposed that gun locks, suggested in January 1781 by Captain Sir Charles Douglas,[106] be used for carronades as well as long guns.[107] Other modifications included a nozzle of a few inches to project their blast more clear of a ship's upper works; a mounting ring under the gun to obviate the need for trunnions; a vertical screw thread through the rear button to facilitate elevation and depression; and sights along the barrel. In consequence the carronade became slightly longer and heavier, but more manageable and accurate.[108]

The carronade received greater use in the wars from 1793. A new establishment of carronades for ships, superseding that of 1779, was adopted on 19 November 1794 (see table 5.6). Significantly, Sir Charles Middleton was an Admiralty commissioner from May 1794, becoming

---

[101] NMM, MID.9/2/21 & 22.   [102] NMM, MID.9/2/14.

[103] Talbott, *The Pen and Ink Sailor*, 67.

[104] Syrett, *The Royal Navy in European Waters*, 162.

[105] NMM, ADM. BP/2, 28 Dec. 1781; MID.1/108, W. Langton to Middleton, 31 Mar. 1805.

[106] NMM, ADM. BP/2, 17 Jan. 1781.   [107] NMM, ADM. BP/3, 13 Apr. 1782.

[108] Lavery, *Arming and Fitting*, 106–7.

the senior naval lord in March 1795.[109] From August 1795 every ship bigger than a 16-gun brig was ordered to be supplied with a carronade for her launch. From March 1798 every line of battle ship was fitted to receive carronades on the quarter-deck and forecastle. In 1799 the carronade was made the general quarter-deck and forecastle gun of frigates. From 21 February 1800, 20- and 24-gun frigates were fitted for 32-pounder carronades on their main decks rather than long 9-pounders. Almost inevitably they spread too to French and Spanish vessels, which carried 36-, 32- and 24-pounder carronades.[110]

Some prejudice among sea officers remained. Lord St Vincent observed to Captain George Murray in April 1799: 'the rage for carronades in ships of the line passes my understanding; they are a great strain upon the decks and sides, extremely dangerous in action, and create so much noise and confusion, where silence and attendance on the braces should preside, that I never suffer one to be mounted in a ship I serve on board of'.[111] Yet others were devotees. From 1793 Nelson kept two 68-pounder carronades with him, transferring them as he moved from ship to ship. Their ability to clear decks of men and damage rigging tailor-made them for Nelson's preferred close-range battle. Hence he asserted 'no captain can do very wrong if he places his ship alongside that of an enemy'.[112]

Chemistry and gunpowder

Improvements in the manufacture of guns after 1785 were linked to an improvement in the quality of gunpowder, of which large quantities were consumed by each weapon (see table 5.8). Manufacturing improvements began in 1750 but came rapidly after 1783, deriving from a range of scientific investigations into the explosive behaviour of powder. Without question, the combined effects of these improvements by 1795 affected the quality of British gunnery for both the army and the navy.[113]

The supply of powder

The Ordnance Board had become responsible for supplying gunpowder to both the army and the navy in 1664.[114] It obtained these supplies by

[109] Middleton served as an Admiralty commissioner 12 May 1794 – 20 Nov. 1795.
[110] Clowes, *The Royal Navy*, IV, 155, 544.
[111] Typescript copies of letters received by Sir George Murray: NMM, MS84/057.
[112] Knight, *The Pursuit of Victory*, 139–40.
[113] Caruana, *The History of English Sea Ordnance*, 25.
[114] Tomlinson, *Guns and Government*, 111–17.

Table 5.8 *Proportions of powder for sea guns, carronades and small arms c. 1800*[a]

	Proof		Service		Saluting		Scaling	
Guns								
	lb	oz	lb	oz	lb	oz	lb	oz
42	25	0	14	0	10	0	3	0
32	21	0	10	11	8	0	2	12
24	18	0	8	0	6	0	2	0
18	15	0	6	0	4	8	1	8
12	12	0	4	0	3	0	1	0
9	9	0	3	0	2	4	0	12
6	6	0	2	0	2	0	0	8
4	4	0	1	5	1	5	0	6
3	3	0	1	0	1	0	0	4
½	0	8	0	3	0	3	0	1
Carronades								
	lb	oz	lb	oz	lb	oz	lb	oz
42	9	0	4	8	4	8	1	8
32	8	0	4	0	4	0	1	4
24	6	0	3	0	3	0	1	0
18	4	0	2	0	2	0	1	0
12	3	0	1	8	1	8	0	12
Small arms								
	oz	drams	oz	drams	oz	drams	oz	drams
Wall–piece	2	8	0	10	–	–	–	–
Musquet	0	12	0	6	–	–	–	–
Pistol	0	6	0	3	–	–	–	–

'*NB.* These proportions are with Powder in good condition – if it is damp, or damaged, a greater quantity will be necessary.'

[a] TNA, ADM.160/150, 'Proportion of Powder for Sea Guns etc'.

contract until the time of the Seven Years' War when nine mills, all in Kent, Surrey, Essex and Middlesex, supplied the board.[115] During this war, difficulties in obtaining a quantity of sufficient quality, compounded by an inability to import from Holland, prompted the board to purchase and operate its own mills. It thus purchased a gunpowder manufactory at Faversham in 1759, following that in 1789 with the purchase of another at Waltham Abbey, and the purchase of a third at Ballincollig, County Cork, in 1805. However, there were still many, indeed an increasing number of, private gunpowder manufactories around the country and

[115] West, *Gunpowder, Government and War in the Mid-eighteenth century*, 197–211.

a large proportion of the gunpowder required by the armed services was still purchased from contractors.[116] In 1809, about two-fifths was obtained in this way.[117]

Gunpowder was, and is, composed of three raw materials: saltpetre which releases oxygen, charcoal which serves as fuel, and sulphur which ignites at relatively low temperatures. In the eighteenth century, sulphur (or brimstone as it was called) was imported from Italy or Sicily. Charcoal was produced locally from coppice wood.[118] Saltpetre was made in northern Europe from organic matter, or imported from regions with a hot dry season where it forms naturally. In Bengal, India, it was reduced to a crystalline state with a single washing and the English East India Company began importing supplies as a profitable ballast cargo from the seventeenth century. Cheaper than northern saltpetre, imports by the Company gradually grew until, between 1793 and 1809, they amounted to 24,752 tons. Legislation required the Company to sell a proportion on the open market and a quantity to the government. In 1791 the latter purchased 500 tons; by 1808 its purchase was 6,000 tons; and in 1810, 12,500 tons.[119]

The Ordnance Board stock of saltpetre was stored at Rotherhithe on the River Thames, and issued to contractors in the quantities required for them to fulfil their agreements with government. By 1810 contractors required 2,300 to 2,500 tons annually. To meet these issues, the board was required to keep a five-year stock of saltpetre and sulphur. Quantities in 1810 were premised on the necessity for the board of Ordnance to provide 50,000 barrels of gunpowder to the British armed forces and 32,000 barrels to the Spanish forces, and to keep a stock equivalent to two years' consumption.[120]

In the mid eighteenth century, for every 100 lb barrel ordered, the board expected a consumption of $80\frac{1}{4}$ lb of double refined saltpetre, 15 lb of charcoal, and $12\frac{3}{4}$ lb of refined sulphur.[121] This total of 108 lb was reduced to 100 lb in the process of manufacture. Generally

---

[116] A. G. Crocker, G. M. Crocker, K. R. Fairclough and M. J. Wilks, *Gunpowder Mills: Documents of the Seventeenth and Eighteenth Centuries* (Guildford, 2000), 1–5.

[117] Glover, *Peninsular Preparation*, 67, citing 16th Report of the Commission of Military Inquiry, 1810, Appendix 9; 36,623 barrels of 90 pounds came from the state's yards, 24,433 from merchants.

[118] By 1789 wood for charcoal was scarce around London. Alder wood was thus purchased in Hampshire and shipped from Portsmouth to Faversham by the officers of the Ordnance: HRO, 109M91/CO57, 27 Mar., 2, 10 Sept. 1789.

[119] C. N. Parkinson, *Trade in the Eastern Seas 1793–1813* (London, 1966), 78, 83–4.

[120] Library of Congress, Melville Papers, Col. Hadden to Lord Mulgrave, 17 May 1810.

[121] Proportions of saltpetre to charcoal and sulphur varied from 5: 4: 3 to 8: 2: 1, and were commonly 75: 15: 15 by the mid eighteenth century.

20 tons of saltpetre were issued to contractors for the manufacture of 485 barrels of gunpowder, and issued on a deposit of £1,200. The contractors weighed and ground the ingredients, dry mixed them, dampened and incorporated or closely blended them, corned or granulated the blend, removed any loose dust, dried the grains and packed the powder for despatch.

The finished gunpowder was shipped to Greenwich for proof by the Ordnance Board at Purfleet. Batches were accepted or rejected according to whether they achieved the board's standards. During the Seven Years' War, between 72 and 80 per cent of all powder submitted for proof was accepted.[122] Yet up to 60 per cent of a contractor's supply could be rejected and for reasons that were not well understood. The process of production had to be followed with minute attention to detail. By the mid eighteenth century, pestle mills had been replaced by wheel edge runner mills for grinding, and the necessity for the ingredients to be thoroughly blended was appreciated. However, there was no scientific basis to production, and little attention was paid to the fineness of the grind or the thoroughness of the dry mix. The fact that charcoal could absorb moisture from the atmosphere up to one-eighth of its own weight was not known until 1797.

Innovations in manufacture

Greater understanding of the fundamental necessities in the manufacture of gunpowder followed the appointment in 1783 of Major William Congreve to the Royal Laboratory, Woolwich. He was made Comptroller and Inspector of Gunpowder Manufactories in 1789. Responsible for proofing powder at Purfleet, he was able to implement lessons from experiments at the Faversham and Waltham Abbey mills. He was also able to employ the findings of Charles Hutton, professor of mathematics at the Royal Military Academy at Woolwich, and of the scientists Ingenhousz and Benjamin Thompson, all of whom were contributing papers to the Royal Society in the late 1770s and 1780s. Until that time, there had been little detailed analysis of the constituents and activity of gunpowder: existing publications dated from the seventeenth century, with the sole addition of work by B. Robins published in 1742.[123]

Robins had thrown doubt on contemporary methods of measuring the strength of gunpowder and prepared the way for experiment and new

---

[122] West, *Gunpowder, Government and War in the Mid-eighteenth Century*, 218; R. D. Crozier, *Guns, Gunpowder and Saltpetre: A Short History* (Faversham, 1998), 58.

[123] B. Robins, *New Principles of Gunnery* (1742).

thinking. Hutton related the velocity of projectiles to the strengths and qualities of powder, the types and weight of shot, and the dimensions of guns. He, for example, demonstrated that momentum increased with the weight of shot, and diminished with the windage of the barrel.[124] Ingen-housz revealed that in explosions 'the quickness of this propogation of fire depends in great measure upon the interval or interstices which remain among the grains of gunpowder', and that the grains were best proportionate to the size of the firearm.[125] Benjamin Thompson noticed that the degree of ramming affected the force of powder, as did the heat of the weapon, and atmospheric conditions – that is, the temperature and humidity. He went on to analyse the factors affecting the pace of powder combustion.[126] His work established the parameters of knowledge for the next century.[127]

Congreve was able to use these ideas to improve the quality of gun-powder. For this, however, the board of Ordnance needed manufactories under its own control and Congreve's first task was to defend reten-tion of the Faversham mills, under threat of sale in 1783. His second was to ensure that the increased output of these mills was cheaper as well as better than that supplied by contractors – serving to check the prices they asked. His achievement regarding Faversham in fact reversed government policy and resulted in the purchase of the Waltham Abbey manufactory.

The enlargement of government facilities permitted Congreve to undertake the 'recovery' of damp and lumpy powder, restoring it more quickly than new could be made. This became the primary purpose of 'laboratories' founded at Portsmouth and Plymouth. He also improved the methods of extracting saltpetre from powder beyond restoration and of refining saltpetre before it was issued to contractors, which, on being refined twice more by them, became more pure and contributed to higher-quality gunpowder. With heightened purity of content, durability increased. Tests in 1809 and 1810 at Marlborough Downs revealed that Faversham-made powder dating from 1785 fired 9-pound balls 4,319 yards and was still more powerful than any recent product of contrac-tors. Indeed, it was exceeded only by 41 and 111 yards by Faversham

---

[124] C. Hutton, 'The force of fired gunpowder and the initial velocities of cannon balls', *Philosophical Transactions of the Royal Society* 68(1778).

[125] J. Ingen-housz, 'An account of a new kind of inflammable air or gas and a new theory on gunpowder', *Philosophical Transactions of the Royal Society* 69(1779).

[126] B. Thompson, 'New experiments on gunpowder' and 'Experiments to determine the force of fired gunpowder', *Philosophical Transactions of the Royal Society* 71(1781) and 87(1797).

[127] For the contributions of Hutton, Ingenhousz and Thompson, see West, *Gunpowder, Government and War in the Mid-eighteenth Century*, 175–84.

and Waltham Abbey powder made in 1809. Different-sized grains of powder were produced for weapons of a different size and its strength was markedly increased.[128]

For naval combat, this last achievement would seem to have been the most important. In 1785 Congreve experimented with the preparation of charcoal, corresponding about alternative methods with the board of Ordnance and with Richard Watson, Bishop of Llandaff and professor of divinity and chemistry at the University of Cambridge.[129] Watson designed cylinders for the preparation of charcoal which, by evenly baking it, permitted the production of fine, evenly ground charcoal. When made into gunpowder, the product was startling. Test firing at Hythe showed that, while standard gunpowder projected a ball 171 feet, cylinder powder sent it 273 feet, an astonishing 58 per cent further. The Ordnance Board responded promptly, setting up cylinders to make charcoal in Sussex and Cumbria.[130] The Board's own tests were less dramatic in result.[131] Nevertheless, cylinder baking came into operation at Waltham Abbey in 1794 and at Faversham in 1798.

The new powders

Hitherto shot had been fired with powder equivalent to half their weight. A trial in 1796 fired shot with one-third their weight of cylinder powder. With 2 degrees elevation, the shot first touched ground at 1,200 yards. Even with one quarter their weight, shot went 1,000 yards. Two shot fired with the same charge – that is, with powder equal to one-eighth the weight of each shot – still went 1,000 yards. The Board of Admiralty was convinced of the power of the new cylinder powder by the end of 1800 and directed the board of Ordnance to accumulate a stock ready to exchange it for the powder on board ships. The exchange began with ships of the line by an order of 29 April 1801. However, such was the 'superior strength' of the new powder, had it been used in the same quantity as standard powder, there was a danger guns would burst. The Ordnance Board thus saw fit to reduce the charges for guns and the proportions issued to ships, requesting captains be cautioned to observe the instructions for

---

[128] W. Congreve, *A statement of facts relative to the savings which have arisen from manufacturing gunpowder at the royal powder mills; and of the improvements which have been made in its strength and durability since the year 1783* (London, 1811).

[129] C. Russell, 'Richard Watson, gaiters and gunpowder' in *The 1702 Chair of Chemistry at Cambridge: Transformation and Change*, ed. M. D. Archer and C. D. Haley (Cambridge, 2005), 57–83.

[130] Private letter from Professor C. A. Russell, 8 Jan. 2003.

[131] Information from Dr G. Cole, who examined TNA, SUPP.5/117–18.

its use, and especially to deter double shotting owing to the overheating of guns fired that way.[132] The increased strength of cylinder powder was consequently used to achieve economy of consumption rather than increased power of shot.

The use of cylinder powder to strengthen old powder prompted some sceptical comment. One anonymous critic claimed in 1801 that all powder was so mixed and that this reduced the power of English shot compared to that of the French. The board of Ordnance denied the allegation, claiming that 'frequent examination' of the powder captured in French ships revealed the latter was less strong and less good in quality, while old power, restored and mixed with cylinder powder, was perfectly fit for service.[133] Examination of test results confirms the relative weakness of French and Dutch powders compared to the British. Moreover, as more cylinder powder was manufactured, from 1805 it became the predominating powder in use.[134]

The restoration and mixing of powders gave rise to various denominations of powder. Private manufacturers letter-coded powders by the mid eighteenth century.[135] The Ordnance Board colour-coded them as well by the end of the century. Blue powder was made with traditional pit charcoal; white was old that had been restored and mixed with cylinder powder; red was cylinder powder in pure form.[136] In case gunners should forget the colour-coding, the board of Ordnance continued to instruct them in their appropriate usage. In October 1804 ships' gunners were informed:

when Red LG powder is fired no more than one third of one shot's weight is to be used in a charge for any gun and only 1/12 of one shot's weight is to be fired out of any carronade. The Red LG gunpowder is for distant shooting, the White LG & Returned powder or White L is for close fight and none but returned or White L is to be fired in salutes and in sealing pieces of ordnance.[137]

The necessity for these distinctions was evident in the dangers that arose from over-charging guns. For the new powder was so powerful that it could endanger gun crews if used indiscriminately. A trial of ships' guns in November 1810 demonstrated the effect of different charges. With a 4-pound charge and double shot in a 24-pounder, 'the effect of the gun's

---

[132] TNA, ADM. 7/677, fo. 16; ADM. 1/4015, fos. 440–4; Cole, 'The Office of Ordnance', 166–7.

[133] TNA, ADM. 1/4015, 2 May 1801.

[134] Cole, 'The Office of Ordnance', 162–5.

[135] Crocker *et al.*, *Gunpowder Mills*, 19.   [136] Information from Dr Gareth Cole.

[137] Saunders, 'Upnor Castle and gunpowder supply' based on Board of Ordnance letters to the officers of Upnor Castle powder magazine, that of 22 Oct. 1804 quoted here. An order to the same effect of 20 Aug. 1801 is given in TNA, ADM.160/150.

recoil on the breeching was in every respect equal and moderate, and without any strain on the tackle'. But with a 6-pound charge, 'the effect of the recoil on the breeching was so violent as to prove its being unsafe to fire with so great a charge between decks'.[138]

The trial in 1810 demonstrated the power of British powder. The smaller charge in the 24-pounder was sufficient to penetrate timber which was comparable in resistance to a ship's hull.

A butt of wood was constructed at 100 yards distance from the guns, 9 feet long, 6 feet high, and 5 feet 2 inches thick; it was made of sleepers of fir, 9 inches square, placed alternately horizontal and perpendicular, except two thicknesses of deal planks of 3 inches thick by 1 foot broad, placed one horizontally and the other perpendicularly; the butt was connected by five iron bars passing from front to rear, and by four bars passing from one end to the other.

After firing:

On examining the butt it appeared much broken and ruined; ten of the balls had gone quite through to the average distance of 50 yards . . . to which distance considerable splinters of wood were driven. The iron bars which connected the butt were twisted and bent as if they had been wire. The average penetration of such balls as did not go through the butt was about four feet, that is, one foot eight inches more than the thickness of a 74 gun ship near her lower-deck ports.[139]

Such being the effect of 24-pounder guns, the power of 32-pounders was equal to any challenge British ships were likely to face, and during the Napoleonic War 42-pounders were phased out.[140] Too little is known of the improvements made by the French, Spanish and Dutch in their powder and ordnance but they were unlikely to have been able to match the achievements of the board of Ordnance, backed by the British scientific community and an expanding iron founding industry using cheap coal as fuel and steam engines to pump air and bore cylinders.

The Ordnance Board was fortunate in appointing two men, Congreve and Blomefield, who were alive to the scientific and technological developments of their day. It was also fortunate in possessing at Woolwich a centre of expertise in gun manufacture and in being able to develop its own powder manufactories which set the standards to which the British munitions industry had to conform. Certainly by 1810 the Ordnance department of the British state was no longer the conservative establishment it appeared in 1780. The bureaucracy managing the supply of guns and powder had been transformed in response not only to the demand

---

[138] TNA, ADM. 7/677, 21 Nov. 1810.
[139] *Ibid.*, 21 Nov. 1810.   [140] Cole, 'The Office of Ordnance', 204, 242.

for expertise, but to the new organisational ideas for bureaucracy and to the geographical reorganisation of the industry. The latter was heavily dependent on water transport, an aspect of the industry which once again reflected the importance of the maritime economy. Britain's ordnance industry could hardly have provided the same service to the state without sea, river and canal transport, while the state would not have responded in the same way to the new industrial environment without individual responsibility.

6   Manpower and motivation

As a maritime nation, the British should not have lacked for manpower to serve its navy. But the latter had to compete for men with the merchant service, with the army and militia, and with civil employments. Moreover, the state's armed forces had to grow rapidly at the beginning of hostilities and demands for manpower grew in every war between 1755 and 1815.[1] Before 1763 only about 5 per cent of the male population was mobilised for war.[2] By 1811, about 6 per cent were in the regular armed forces, the army and the navy, and another 4 per cent in the militia and volunteers.[3] The state took this 10 per cent from agriculture, manufacturing, construction and commerce, giving Britain a higher ratio of men in the military forces than any other European nation.[4] However, recruitment was a challenge. Methods for the army increased in variety but those for the navy did not. The proportion impressed for the navy was persistently high. But so also was the rate of desertion. The navy's death rate declined, not only due to medical and dietary improvements but because rates of discharge increased far more. This made for a more efficient workforce, yet one that demanded management and motivation, for recruits naturally had their own personal interests and prejudices.[5] Methods of management thus mattered, not least because they formed attitudes to the state.[6] Yet after 1793 methods and attitudes were each subject to utilitarian ideas that pervaded contemporary thinking even in the navy's officer corps.

[1] The term 'manpower' does not exclude women, for whose service see S. J. Stark, *Female Tars: Women Aboard Ship in the Age of Sail* (London, 1998).
[2] John, 'War and the English economy'.
[3] Colley, *Britons: Forging the Nation*, 286–7, fn. 8.
[4] G. Hueckel, 'War and the British economy, 1793–1815: a general equilibrium analysis', *Explorations in Economic History* 10(1973), 365–96.
[5] J. E. Cookson, *The British Armed Nation, 1793–1815* (Oxford, 1997), 8–9, 111, 120.
[6] Colley, *Britons: Forging the Nation*, 283–319. Colley has talked of 'war patriotism' and an 'attachment' to the state that had to be learned. See also Colley, 'The reach of the state, the appeal of the nation: mass arming and political culture in the Napoleonic Wars' in *An Imperial State at War*, ed. Stone, 165–84; see also Introduction by Stone, 22–3.

The state's military manpower

The growth of land and sea forces

The British regular army had an average size of 62,000 men in the War of Austrian Succession. It grew to more than 90,000 men during the Seven Years' War, and 100,000 men in the American War of Independence. In 1807 it reached 205,000, and in 1815 it stood at 233,852 men.

At the same time the army developed a non-professional branch upon which it was able to call for domestic duties. An amateur militia, subject to training and martial law following the Militia Act of 1757, reached 28,000 men by the end of the Seven Years' War. It grew to 40,000 men under the threat of invasion in 1778–9, and 198,500 by 1809. From the time of the Seven Years' War, the militia was backed by fencibles – regular soldiers enlisted for home service for the duration of the war. Legislation of 1794 added Volunteer infantry and Yeomanry cavalry, which placed another 189,000 men in uniform by 1809. Some militia were embodied into regiments that reinforced the regular army for service around the country. Militia and volunteers were raised in Ireland as well during the Napoleonic War,[7] serving the needs of defence and permitting the regular army to perform garrison and expeditionary duties overseas.[8]

The army and militia together grew from about 115,000 men in the Seven Years' War to more than 610,000 men by the last half of the Napoleonic War. By that time, when the manpower required by the navy is added, the state employed at least three-quarters of a million men from the domestic population. From a male population of about 6 million, approximately 1 man in every 9 or 10 of military age was serving in the army, navy or regular militia; if volunteers and amateur militia are added, the ratio becomes 1 man in every 6. In 1805 the ratio was even higher, perhaps 1 in 5, which at the time was compared to the proportion of men under arms in France, Russia and Austria (1 in 14) and in Prussia (1 in 10).[9]

However, outside Britain men were also recruited to the military forces of the empire. By 1815 there were over 160,000 East India Company troops recruited in India, 30,000 'foreign corps' and 25,000 militia and fencibles in the colonies. In 1815 they supplemented the forces raised in

---

[7] By 1815 there were 80,000 yeomanry and militia in Ireland.

[8] Gates, 'The transformation of the army, 1783–1815', 132; J. W. Western, *The English Militia in the Eighteenth Century* (London, 1965), ch. 6; Holmes, *Redcoat*, 99–101; R. Glover, *Peninsular Preparation: The Reform of the British Army 1795–1809* (Cambridge, 1970, repr. 1988), 6; Glover, *Britain at Bay*, 39–47.

[9] Glover, *Britain at Bay*, 43–5, 140–5; Emsley, *British Society and the French Wars 1793–1815*, 132–3; Cookson, *The British Armed Nation*, 95.

Britain and took the total number of men under arms within the empire to more than a million.[10]

The navy was a relatively small proportion of this total (see table 6.1). Its size peaked towards the end of each war. It reached a peak of 84,797 men in 1762 during the Seven Years' War and another of 142,098 in 1810 at the height of the Napoleonic War.[11] During the Seven Years' War, the navy employed little more than two-thirds the number of soldiers retained by the army. By the second half of the Napoleonic War, the navy employed only about one quarter the number of men in the state's land forces.

For the mid eighteenth century, P. J. Marshall considers the navy was more successful in meeting its manpower requirements than the army because it enjoyed more competitive advantages of employment and better-developed powers to impress men. He quotes Rodger on the navy in the Seven Years' War suffering a recruitment problem of 'protracted difficulty rather than of failure'.[12] The same may be said of naval recruitment in the two decades before 1815. After some experimentation in 1795–6, schemes for recruitment to the navy did not change despite the steady output of pamphlet suggestions for improvement.[13] By comparison, the proliferation of schemes for raising soldiers was a symptom of desperation.

The means of growth

Growth in the size of Britain's military forces benefited from the increased efficiency in agriculture and social restructuring. Between the sixteenth and early nineteenth centuries net yields – the quantity available for consumption – in English agriculture increased about 135 per cent. This had no parallel in continental agriculture and released people from the land and fed growing towns where industrialisation supported a workforce and urban poor. Towns in England in the eighteenth century grew at a rate of 1.7 per cent per annum, compared to 0.4 per cent per annum in France. Mobility was aided by a system of poor relief, which supported wives and children, and facilitated both internal migration and emigration.[14]

Excluding Ireland and the American colonies, population figures for Britain rose from an estimated 8 million in 1760 to 10.5 million at the first

[10] Colquhoun, *Treatise on the Wealth, Power and Resources of the British Empire*, 47.
[11] From C. Lloyd, *The British Seaman 1220–1860: A Social Survey* (London, 1968), table 3, 261–4.
[12] Marshall, *The Making and Unmaking of Empires*, 61.
[13] *The Manning of the Royal Navy: Selected Public Pamphlets 1693–1873*, ed. J. S. Bromley (NRS, 1974), 95–172.
[14] Wrigley, 'Society and economy in the eighteenth century', 72–95.

Table 6.1 *Seamen in British merchant and Royal naval ships, 1750–1820*

Year	Paid 6d duty in merchant service[a]	Borne for wages in Royal Navy[b]	Total in merchant and Royal Navy ships
1750	33,040	11,691	44,731
1751	34,080	9,972	44,052
1752	32,513	9,971	42,484
1753	34,441	8,346	42,787
1754	34,193	10,149	44,342
1755	38,710	33,612	72,322
1756	31,789	52,809	84,598
1757	34,761	63,259	98,020
1758	29,171	70,518	99,689
1759	36,449	84,464	120,913
1760	36,693	85,658	122,351
1761	38,377	80,675	119,052
1762	37,625	84,797	122,422
1763	43,441	75,988	119,429
1764	42,034	17,424	59,458
1765	38,272	15,863	54,135
1766	44,599	15,863	60,462
1767	44,658	13,513	58,171
1768	39,951	13,424	53,357
1769	43,530	13,738	57,268
1770	49,062	14,744	63,806
1771	49,926	26,416	76,342
1772	49,213	27,165	76,378
1773	49,669	22,018	71,687
1774	–	18,372	–
1775	–	15,230	–
1776	–	23,914	–
1777	–	46,231	–
1778	–	62,719	–
1779	–	80,275	–
1780	–	91,566	–
1781	–	98,269	–
1782	–	93,168	–
	Seamen in English & Welsh ships		
1783	59,004	107,446	166,450
1784	65,880	39,268	105,148
1785	71,372	22,826	94,198
1786	74,835	13,737	88,572
1787	81,745	14,514	96,259
	Seamen in ships of Britain and empire[c]		
1788	107,925	15,964	123,889
1789	108,962	18,397	127,359

Table 6.1 *(cont.)*

Year	Paid 6d duty in merchant service	Borne for wages in Royal Navy	Total in merchant and Royal Navy ships
1790	112,556	20,025	132,581
1791	117,044	38,801	155,845
1792	118,286	16,613	134,899
1793	118,952	69,868	188,820
1794	119,629	87,331	206,960
1795	116,467	96,001	212,468
1796	124,394	114,365	238,759
1797	–	118,788	–
1798	129,546	122,687	252,333
1799	135,237	128,930	264,167
1800	138,721	126,192	264,913
1801	149,766	125,061	304,827
1802	154,530	129,340	283,870
1803	153,828	49,430	203,258
1804	153,774	84,431	238,205
1805	157,712	109,205	266,917
1806	156,031	111,237	267,268
1807	157,875	119,855	277,730
1808	157,105	140,822	297,927
1809	160,598	141,989	302,587
1810	164,195	142,098	306,293
1811	162,547	130,866	293,413
1812	165,030	131,087	296,117
1813	165,537	130,127	295,664
1814	172,786	126,414	299,200
1815	177,309	78,891	256,200
1816	178,820	35,196	214,016
1817	171,013	22,944	193,957
1818	173,609	23,026	196,635
1819	174,318	23,230	197,548
1820	174,514	23,985	198,499

[a] 'An account of the number of men paying the 6d duty to Greenwich Hospital, voted, borne and mustered in the navy', WLC Shelburne Papers, vol. 137, fo. 1; also account of 27 Jan. 1774 from the Receiver's Office for Greenwich Hospital in NMM, Sandwich Papers, F5/2.

[b] Figures for men borne and mustered in the Royal Navy to 1813 principally from Table 3 in Lloyd, *British Seaman*, 262–3; figures 1814–20 from NMM, Milne Papers, MLN.153/3/41.

[c] The official registration of ships began in 1787. These figures for seamen were a product. They come from NMM, Milne Papers, MLN.153/3/41, and their coverage is deduced from table 2 in Lloyd, *British Seaman*, 260.

census in 1801, to 12 million in 1811 and 14 million in 1821. Between 1760 and 1811 Britain's population grew by about 50 per cent. Over the same period seamen mustered in the navy rose by 67 per cent,[15] soldiers in the regular army by 115 per cent, and the army and militia combined by 430 per cent. While expanding less than Britain's land forces, the navy was favoured by the rapid rate of expansion of naval towns and mercantile ports. Between 1801 and 1811 the suburban populations of Chatham, Portsmouth and Plymouth grew by 25–28 per cent, Bristol by 17 per cent, but those of Liverpool by a remarkable 931 per cent.[16]

Naval recruitment on shore especially benefited from the growth of London, which by the beginning of the eighteenth century contained about 10 per cent of England's population. From London, many boys were supplied to the navy by the Marine Society, founded in 1756 by Jonas Hanway. The charity took destitute boys off the streets of London, gave them clothes and some basic education and sent them to sea either in merchant ships or in the navy. During the Seven Years' War the society sent the navy 10,625 boys and men and continued to perform the same work during subsequent wars; 22,973 adult volunteers reached the navy through the Marine Society between 1793 and 1815.[17]

But it was the size of the shipping industry which really determined the number of skilled men available. Registration of British shipping did not start until 1787; before that date, ship numbers have to be calculated from port books. Ralph Davis has provided the most authoritative estimate of the tonnage of English merchant shipping. He calculated there were 473,000 tons involved in foreign trade in 1755, and 752,000 tons in 1786.[18] Registration thereafter combines ships in coastal and foreign trade, for which there were 1,278,000 tons in 1788 and 2,478,000 tons in 1815.[19] Between 1755 and 1786 there may thus have been a 60 per cent increase in ocean-going tonnage; while between 1788 and 1815 there was a 94 per cent increase in ships in foreign and coastal trade.

This increase produced an abundance of skilled seamen who could be recruited into the navy. According to accounts of seamen paying their

[15] Calculations based on figures in Lloyd, *British Seaman*, 261–4.
[16] J. Marshall, *Account of the Population in each of Six Thousand of the Principal Towns and Parishes in England and Wales . . . at each of the three periods 1801, 1811 and 1821* (London, 1831).
[17] Rodger, *Wooden World*, 162; Lloyd, *British Seaman*, 179, citing NMM, MS57/031.
[18] Davis, *Rise of the English Shipping Industry*, 27, 70.
[19] Mitchell and Deane, *Abstract of British Historical Statistics*, 217, 'Shipping registered in the United Kingdom 1788–1938'.

6 pence a month to the pension fund of Greenwich Hospital (see table 6.1), the number of seamen employed in merchant shipping rose from 38,710 in 1755 to 74,835 in 1786. That represented a 92 per cent increase in paying seamen, exceeding the growth in ship tonnage. After 1787 the registration of seamen included those within the British empire. Between 1788 and 1815 the number of seamen registered rose from 107,925 to 177,309, a 64 per cent increase, which was less than the growth in merchant ship tonnage.[20] Nevertheless this pressure on mercantile manpower encouraged the training of new seamen and improvements in rig and steerage that made for reductions in the average number of seamen required to manage individual merchant ships.[21]

The colonies were an important and often overlooked source of maritime labour. The thirteen American colonies contained over $1\frac{1}{2}$ million people by 1755 and their maritime labour force, like that of the mother country, grew under the protection of the Navigation Laws. However, these colonies did not cooperate willingly in the manning of British naval vessels. In 1708 the 'American Act' gave them virtual immunity from impressment. The Admiralty maintained that the Act expired with the peace of 1714 but the colonists claimed the opposite and prosecuted captains impressing men.[22] During the 1740s and 1750s the British navy persisted in impressing seamen on shore in North America, but by the 1760s the strength of American opinion forced the Admiralty to desist. Instead, naval captains in American waters impressed men when necessary from ships arriving from Europe. Local opinion mattered too in the West Indies, where seamen were impressed from 1746 only with the consent of governors.[23]

After their War of Independence, Americans continued to serve in the British navy. In 1811 3,685, about 2.6 per cent of all seamen and marines, were real or 'pretended' Americans. These pretenders needed a certificate of American citizenship to obtain their discharge.[24] Meanwhile, they served among a far greater number of 'foreigners'. Of these, the Irish were probably the largest contingent. In 1797–8 there were as many as 10,000 Irishmen serving in the Royal navy, nearly 7 per cent of the

---

[20] WLC, Shelburne Papers, vol. 137, fo. 1; NMM, Sandwich Papers, F5/2; NMM, Milne Papers, MLN.153/3/41.

[21] Davis, *Rise of the English Shipping Industry*, 71–3.

[22] *Naval Administration, 1715–1750*, ed. Baugh, 92–3.

[23] N. R. Stout, 'Manning the Royal Navy in North America, 1763–1775', *The American Neptune* 23(1963), 174–85; R. Pares, 'The manning of the navy in the West Indies, 1702–63', *TRHS* 4th ser. 20(1937), 54–60.

[24] For a certificate issued to Edward Phillips, see TNA, ADM. 1/3664, 3 Sept. 1811.

whole.[25] Between 1811 and 1813 there were still around 13,420 so-called 'foreigners', almost 10 per cent of the total naval workforce.[26]

Supplemented by colonists and foreigners, the British population provided both skilled and unskilled seamen, for most went to sea young, remaining at sea for fifteen to twenty years. A 1762 list of neutral seamen made prisoner of war indicates that many had been at sea from the age of five or six, and virtually all from the age of ten or twelve. Early census information indicates seamen typically to be in their early twenties with an average length of service of seven years. The navy classified its seamen according to experience and qualification. Analysis of naval muster books between 1764 and 1782 indicates that most ordinary seamen and landsmen were in their late teens, most able seamen in their early twenties, and petty officers in their late twenties. Yet the sea was a demanding employer and most were leaving the naval service by their early thirties.[27]

The navy preferred to enlist volunteers, and N. A. M. Rodger indicates that during the Seven Years' War they came from all over Britain. All counties were represented but with greater numbers from the maritime counties and from the county in which the captain had personal connections. In 1755 and 1756 the Cornishman Vice-Admiral Edward Boscawen had contingents of Cornish men join his ship, one party of 'stout fellows' from Penzance numbering as many as 55 men. Later in 1793 Captain Edward Pellew also raised large numbers of volunteers from Cornwall, the *Nymphe* in 1793 bearing 104, one-third of the men whose place of birth were identified. Rodger makes the point that the Cornish connection had its parallel in every other county of England: 'Those in particular who came from the poorer and remoter parts of the three kingdoms seem to have found there the most recruits.'[28]

These naval examples had their army parallels. Despite the recent 1745 rebellion, Scottish highland regiments were employed by the British government during the Seven Years' War, and on an even greater scale by the end of the century through the management of their clan chiefs. To this end, Lord North promoted Catholic relief, and the reconciliation was completed by the Scot Henry Dundas who, as Secretary for War and

---

[25] Doorne, 'Mutiny and sedition in the Home commands of the Royal Navy'.

[26] In 1811 there were 11,693 foreign seamen and 3,039 foreign marines, a total of 14,732. In 1813 the figures were 11,727 seamen and 1,478 marines, totalling 13,205: NMM, MLN.153/3/41.

[27] R. Omer and G. Panting (eds.), *Working Men Who Got Wet* (Newfoundland, 1980), 3; Rodger, *Wooden World*, 114, 360–1.

[28] N. A. M. Rodger, ' "A little navy of your own making": Admiral Boscawen and the Cornish connection in the Royal Navy' in *Parameters of British Naval Power 1650–1850*, ed. Duffy, 82–92.

Colonies, raised as much as one-fifth of Britain's military manpower in Scotland, despite that country having only one-tenth of the population. Ireland too contributed soldiers to the British army, though Catholic nationalists were not subdued even after the Union of Ireland with Great Britain in 1801.[29]

Poverty rather than patriotism undoubtedly motivated most Scottish and Irish soldiers. But when invasion threatened in 1803–4, it was proximity to the enemy which stimulated volunteers. Volunteer forces recruited best in the rural southern and western counties of England – Kent, Sussex, Hampshire, Wiltshire, Devon, Somerset and Gloucestershire – where 50 per cent of all men aged between seventeen and forty-five volunteered. Rural counties farther north and east supplied an average of 22 per cent, while counties containing industry farther north averaged 35 per cent, the one exception being Yorkshire which supplied only 20 per cent. In 1804 a return to the House of Commons reported that 44 per cent of eligible males in Scotland were willing to serve in volunteer regiments, compared to 28 per cent of Welsh men in counties making returns, and 37 per cent in England.[30]

Recruitment to the naval service

There were few reports of competition between the army and the navy for the available men. This was probably because the navy was principally looking for trained men. Competition was thus confined to what the navy called landsmen and those who could be employed as marines. But rates of pay were poor for both services and few new recruits could have had the option of choice. Rather, the competition was between the service of the state and the private sector, and in this the state struggled and for three reasons. Rates of state payment were always lower than in the private sector; public money was always short in wartime; and the documentation required for the payment of men outside their ships often denied some their earnings.

Disincentives for seamen

Low rates of pay were a long-standing feature of naval employment. Rates of pay for seamen remained unchanged between 1653 and 1797. They

---

[29] Cookson, *The British Armed Nation*, 11–13, 28.
[30] Colley, 'The reach of the state', 165–84.

were embodied in parliamentary legislation which allowed 24 shillings a lunar month for an able seaman, 19 shillings for an ordinary seaman and (when the rate was established) 18 shillings for a landsman.[31] There were supplements from prize money but these made few fortunes, for the proportions in which it was distributed really brought the seamen little.[32] In consequence the seaman relied on his basic rates of pay. It was supplemented by accommodation and victualling. But deductions were made for the payment of chaplains and surgeons, for slops, breakages, tobacco and 'venereals' (that is, cures for venereal disease) and these left the seaman with wages that could not compete with rates earned in merchant shipping.

What made the situation of the seaman in the navy worse was the infrequency of payment. The naval estimates included payment for a specific number of seamen in employment. Yet the navy estimates voted by Parliament buried seamen's wages within the Sea Service vote which included money for victualling, naval stores and ship repairs. The ability of administrators to use these financial supplies flexibly sometimes meant payments for other essentials could take priority over the payment of wages.

In 1728 the House of Commons had passed an Act for encouraging seamen to enter into His Majesty's service by ensuring there were regular and more frequent 'general pays'. By that time, ships' voyages were no longer intermittent according to season and the state of hostilities. There were always some in commission and on long-term voyages overseas. Meanwhile, seamen were still paid in cash every two or three years or paid with tickets if they were turned over from one ship to another, sent to sick quarters or discharged as unfit. To obtain money, seamen thus took advances from, and sold their tickets to, money-lenders, including landladies, tavern keepers and prostitutes, who were permitted to collect the pay of seamen with 'authorities' – letters of attorney – that possessed no official format or formality.

In 1728 the latter were declared invalid unless signed by the seaman's captain, commander, a Clerk of the Cheque or magistrate, while wages were to be paid 'constantly, regularly and punctually'. Men on ships in commission in home waters for six months were to be paid two months' wages; those on ships in commission for eighteen months were to be paid wages for twelve months. However, seamen still went without pay for long

---

[31] Lloyd, *British Seaman*, 226.

[32] Eighths of any prize money were distributed in the following ratio within each ship: 1 to the flag officer if part of a fleet or squadron, 2 to the captain, 1 between all lieutenants, the master and marine captain, 1 to the warrant officers, 1 among the petty officers, and 2 divided between all the rest of the crew, including marines.

periods, for captains preferred to defer pay days to prevent drunkenness and deter desertion, and convenience delayed payments to the end of commissions.[33] The 1728 Act was thus ignored.

During the war of 1739–48 there were only two general pay orders and 'frequent complaints from poor sailors of the delays and difficulties they meet with' in obtaining payment. Between 1755 and 1757 there was again only one general pay day each year. In 1758 there was thus a new Act for 'establishing a regular method for the punctual, frequent and certain payment' of seamen. It required arrears of wages to be paid up to April 1758, then six months' wages after every twelve months served.[34] Yet during the American War, the Navy Board had 'daily ... solicitations from seamen invalided abroad for payment of their wages, who have no documents to produce to enable us to relieve them'.[35] The board gave them every attention and even made a point of removing obstacles when it could. Yet the board's hands were tied when documentation of earnings had not been received or was incomplete, or if the seamen had no proof of identity or certificate of earnings.[36] Afterwards Captain Robert Tomlinson, who observed the workings of the 1758 Act, wrote 'we found seamen <u>full as averse</u> to enter voluntarily into the navy, as if no such Act had passed, or been provided'.[37]

The Admiralty and Navy Boards did what they could. There was an increasing stream of parliamentary legislation to prevent frauds and improve the delivery of wages.[38] In August 1786 a new Inspector's Branch was established at the Navy Pay Office to handle wills, petitions and powers of attorney. Of these, by January 1791 – after fifty-three months – 10,817 had been registered. In March 1791, when mobilisation had increased the navy from 20,000 to 38,000 men, powers of attorney and wills were being received at the rate of 300 a month. Allotments of pay to relatives were also registered in great numbers. In 1813 alone, when there were over 130,000 men borne on ships' books, 6,645

---

[33] *Naval Administration, 1715–1750*, ed. Baugh, 159–65.

[34] Additional funds were voted to permit payment of arrears. Volunteers had to be paid in advance as soon as they boarded a ship, supernumeraries after ten days on muster books. State officials like customs and excise officers and collectors of the land tax were authorised to act as attornies for the purpose of allotting or remitting money to seamen's dependents: Gradish, *The Manning of the British Navy during the Seven Years War*, 87–101.

[35] NMM, ADM. BP/3, 13 Apr. 1782.   [36] NMM, ADM. BP/4, 25 Apr. 1783.

[37] *The Tomlinson Papers: Selected from the correspondence and pamphlets of Captain Robert Tomlinson, R.N. and Vice Admiral Nicholas Tomlinson*, ed. J. G. Bullocke (NRS, 1935), 121.

[38] For a list of statutes relating to naval pay, see G. L. Green, *The Royal Navy and Anglo-Jewry 1740–1820* (Ealing, 1989), appendix 10.

allotments were registered.[39] The finances of the Chatham Chest (established in 1590) and Greenwich Hospital (funded from 1690) which provided pensions and accommodation for elderly, infirm and invalid seamen were also reformed. In 1784 the Chatham Chest was almost bankrupt so the fund was united with Greenwich Hospital, which by 1805 had 3,243 out-pensioners and 2,410 residents.[40]

However, these provisions did not address the fundamental deficiency of wage rates. During the 1790s inflation of prices and the demand for seamen drove up rates in merchant shipping. Until 1786 the North Sea coal trade had paid its seamen 50 shillings a month; between 1793 and 1795 it paid £6, rising to £9, a month; and in 1796 the rate was still £5 pounds a month. This compares to 18 to 24 shillings in the Royal navy. By 1797 the army, militia and naval lieutenants had just received pay rises. Indeed, the pay of the common soldier, established at 8d a day at the time of the Commonwealth, was raised by 50 per cent.[41] As a result, even soldiers received 25 per cent more than able seamen. A dockyard shipwright and caulker could earn in five days what an able seaman earned in a month, a dockyard rigger took six days, and a rigger's labourer eleven days.[42] In 1797 the infrequency of payment also remained a major issue. Seamen on board the *Nassau*, participants in the Nore mutiny, had not been paid for nineteen months.[43]

In their rapid response to the mutiny at Spithead, the Board of Admiralty and the House of Commons acknowledged the problem. In 1797 pay rates for naval seamen were immediately raised: by 5s 6d for an able seaman; by 4s 6d for an ordinary seaman; and by 3s 6d for a landman. Further increases followed in 1806: of, respectively, 4s 0d, 2s 0d and 1s 0d.[44] By then the monthly pay rates of seamen ranged between 33s 6d and 22s 6d. It would be tempting to suggest these increases in pay rates contributed to the numbers of seamen recruited into the navy during the Napoleonic War. But between 1798 and 1813 monthly wages in the North Sea coal trade only fell below £6 in 1801 and 1802 when peace appeared in prospect and the navy was partially demobilised.[45]

---

[39] NMM, ADM. BP/11, 20 May 1791; ADM. DP/201B, 27 Apr. 1821.
[40] NMM, MID.1/154, C. Proby to C. Middleton, 28 Feb. 1784. See also MID.5/1, letters of Chatham Chest officials.
[41] Barnett, *Britain and her Army*, 241.
[42] 6th Report of the Commissioners into irregularities, 16, 111.
[43] C. Lloyd, 'New light on the mutiny at the Nore', *MM* 46(1960), 285.
[44] Lloyd, *British Seaman*, 226–7.
[45] Ville, *English Shipowning during the Industrial Revolution*, 164. Ville provides wage rates for the North Sea coal trade, the Mediterranean, the Baltic, the Canadian timber trade, the West Indies trade, the Honduras mahogany trade, and trade to South America.

The impressment of seamen

Owing to the inability to recruit as many professional seamen as it needed, the British state was forced to make up its deficiency by impressing these men. The practice of impressing seamen was of long standing and justified on grounds of national defence. Naval officers pressed seamen from merchant ships at sea, sometimes putting substitute men on board, and port admirals pressed around the main dockyard towns. But the main supplier of men to the navy from on shore was the Impress Service, which raised volunteers as well as pressed men. During the eighteenth century this service progressively threw its net wider until it eventually covered most parts of the British Isles.

During the 1740s recruitment arrangements on shore were relatively small-scale in organisation and in geographical extent. During the war of Austrian Succession an account of 4 July 1743 suggests that lieutenants were sent to form 'rendezvous' in only twenty locations outside London and the main naval towns. All twenty places were in the south of the country, the most northerly being Peterborough, the farthest west Totnes in Cornwall and Newport in south Wales.[46] The west of England, Wales and the whole of Scotland were virtually ignored. During the Seven Years' War rendezvous were established farther west and north. Liverpool, Lancaster, Greenock, Glasgow, Edinburgh and Aberdeen each received recruiting lieutenants; gangs were posted in Belfast, Dublin and Cork and paid visits to Limerick and Galway.[47]

By the time of the American War of Independence arrangements on shore had become more organised.[48] (See table 6.2.) In 1775 rendezvous were already located in twelve of the principal ports, including Greenock in Scotland and Cork in Ireland. Once war began, rendezvous were extended to most of the middling ports and some substantial inland towns like Norwich and Bridgewater. Until 1778 rendezvous were concentrated predominately in the south of England and confined to four principal cities in Ireland – Cork, Dublin, Belfast and Waterford. However, with the European war, rendezvous were extended to the smaller inland ports and towns, including industrial towns like Stockton and Wigan, as far west as Limerick and Galway in Ireland, and north to Ayr, and the

---

[46] TNA, ADM. 1/3663, 4 July 1743. Lieutenants were located at Faversham, Canterbury, Dover, Deal, Guildford, Maidstone, Winchester, Salisbury, Shoreham, Chichester, Southampton, Lymington, Weymouth, Poole, Exeter, Plymouth, Totnes, Newport, Yarmouth and Petersfield.

[47] Gradish, *The Manning of the British Navy during the Seven Years War*, 62–3.

[48] *The Manning of the Royal Navy*, 124. The 'Impress Service' first received its title in 1780.

Table 6.2 *Men recruited on shore by the Impress Service, 1775–1783*[a]

	1775	1776	1777	1778	1779	1780	1781	1782	1783	Total
London	2,172	6,936	4,645	5,673	6,443	3,311	3,878	5,287	372	38,717
Poole	44	200	236	172	221	167	78	67	24	1,209
Exeter	72	422	265	321	326	398	274	279	36	2,393
Penzance	1	108	178	74	55	143	125	106	4	794
Bristol	102	543	310	544	692	512	481	537	117	3,838
Liverpool	252	547	518	551	696	731	567	578	75	4,515
Whitehaven	66	131	68	39	72	64	51	36	–	527
Shields	43	161	296	372	268	117	163	146	–	1,566
Hull	84	304	250	258	418	418	428	359	16	2,535
Yarmouth	54	93	166	127	137	89	115	91	5	877
Greenock	52	111	134	450	331	227	309	228	35	1,877
Cork	81	597	662	799	637	473	474	521	8	4,252
Gravesend	–	326	408	392	269	180	213	266	20	2,078
Faversham	–	21	254	158	218	138	147	103	4	1,043
Margate	–	11	136	90	47	22	36	51	–	393
Deal	–	261	557	381	489	398	274	97	22	2,478
Dover	–	161	471	366	455	406	275	105	16	2,255
Folkestone	–	35	265	161	222	125	98	28	–	934
Chichester	–	25	132	121	128	66	58	30	–	560
Southampton	–	130	109	97	294	208	181	151	9	1,179
Dartmouth	–	2–4	217	95	113	218	236	226	30	1,159
Barnstaple	–	10	105	62	120	84	27	59	29	496
Falmouth	–	48	305	238	272	470	230	218	13	1,794
Bridgewater	–	150	140	110	101	88	30	77	22	718
Haverfordwest	–	38	210	163	121	115	107	66	3	823
Berwick	–	34	–	59	56	41	46	41	–	277

Newcastle	—	164	728	608	597	306	320	377	36	3,136
Lynn	—	11	122	105	131	175	118	125	5	792
Harwich	—	38	94	78	79	38	36	53	2	418
Salisbury	—	4	77	73	47	51	50	31	—	333
Norwich	—	18	106	78	165	87	84	75	—	613
I. of Wight	—	80	133	184	360	263	265	319	41	1,645
Jersey	—	18	46	19	19	89	30	36	2	259
Edinburgh	—	36	425	309	918	709	763	400	43	3,603
Dublin	—	635	981	1584	1594	1229	652	1979	230	8,824
Waterford	—	59	406	448	527	418	388	476	38	2,760
Belfast	—	122	186	548	441	272	234	617	138	2,558
Weymouth	—	—	10	81	181	93	137	89	7	518
Saltash	—	—	16	14	44	47	62	19	—	202
Chester	—	—	139	22	87	116	83	104	9	560
Gloucester	—	—	71	94	63	68	36	49	3	384
Guernsey	—	—	51	51	17	51	33	49	2	254
Fareham	—	—	—	113	123	49	52	32	—	369
Fowey	—	—	—	7	—	22	72	26	—	127
Stockton	—	—	—	108	143	148	112	73	—	584
Wivenhoe	—	—	—	94	131	47	38	41	2	353
Newry	—	—	—	86	176	161	121	257	—	801
Hastings	—	—	—	—	138	117	228	219	14	716
Oakhampton	—	—	—	—	134	119	132	123	22	530
Swansea	—	—	—	—	113	56	33	19	1	221
Carnarvon	—	—	—	—	18	27	11	18	—	74
Lancaster	—	—	—	—	33	46	21	7	—	107
Maidstone	—	—	—	—	21	36	82	57	—	196

(cont.)

Table 6.2 (cont.)

	1775	1776	1777	1778	1779	1780	1781	1782	1783	Total
Godalming	–	–	–	–	141	41	44	26	–	252
Limerick	–	–	–	–	54	–	–	–	–	54
Gosport	–	–	–	–	–	36	376	697	86	1,195
Bewdley	–	–	–	–	–	136	230	236	16	618
Wigan	–	–	–	–	–	89	135	99	1	323
Ramsgate	–	–	–	–	–	22	78	–	–	161
Stranraer	–	–	–	–	–	72	–	89	–	336
Ayr	–	–	–	–	–	45	135	149	7	158
Orkneys & Shetland	–	–	–	–	–	58	19	64	17	
Oraston	–	–	–	–	–	–	130	143	14	287
Galway	–	–	–	–	–	–	102	22	124	–
Wexford	–	–	–	–	–	–	–	–	–	–
Londonderry	–	–	–	–	–	–	–	–	–	–
Total No. procured	3023	12611	14628	16577	19695	14552	14241	17062	1616	113,925
Proportion raised in London	72%	55%	32%	34%	33%	23%	27%	31%	23%	34%
Bounties paid to volunteers	–	5899	11617	13217	13601	9590	8467	8590	3251	74,241

a NMM, MID.7/3/2, 'An account of the expense of the Impress Service during the last war, distinguishing... the number of men procured annually... prepared pursuant to a precept from the Honorable House of Commons dated the 19th May 1786'.

Shetland and Orkney islands. By 1782 recruits had been raised in at least sixty-four locations around the British Isles.[49]

The Impress Service was a large service and demanded an administrative bureaucracy. Many recruits involved payments for entry, conduct and subsistence. The same arose from the deployment of inland gangs for whom expenses had to be paid.[50] But the service could be highly productive. The London rendezvous provided 34 per cent of all men entering through rendezvous during the American War. In 1775 it produced nearly three-quarters of the men raised by the Impress Service. In 1776 it raised over half, and thereafter it continued to supply between a third and a fifth of the men raised on shore.

In London, the regulating captains, who supervised the lieutenants in charge of rendezvous, had the assistance of local constables and of Watermans' Hall. The latter, which drew upon the watermen who worked the River Thames, raised very few men. The constables were more productive. They were appointed for each parish under the influence of local magistrates and justices of the peace and, though in principal not supposed to sell alcohol as publicans, seem frequently to have done so.[51] This appears to have promoted their familiarity with those who were regarded as dissolute. In 1780 press warrants were issued to 100 constables all within 20 miles of the City of London. Yet the number of pressed men they provided was still relatively small compared to those raised by naval gangs under lieutenants. Of these, according to an account of October 1778, there were thirteen in London, six sent by individual ships with seven attached to no particular ship. The account indicates that, since the last report, while Watermans' Hall had produced only 2 men, the constables had procured 42 men, while the lieutenants had pressed 2,095 men. In addition, outside London, 5,072 volunteer seamen and watermen had been raised, with 5,643 volunteer landsmen.[52]

The ability for rendezvous to raise volunteer landsmen as well as press seamen gave local authorities the opportunity to dispose of its undesirable inhabitants. Legally, the navy was empowered only to press seamen and watermen. Yet this did not prevent local magistrates and constables offering others who were a burden on a parish, a nuisance or threat to the security of local property – in the hope that if they could not

---

[49] NMM, MID.7/3/2, 'An account of the expense of the Impress Service during the last war, distinguishing ... the number of men procured annually ... prepared pursuant to a precept of the Honorable House of Commons dated the 19th May 1786'.

[50] NMM, ADM. BP/10, 1 June 1790; for the costs of the Impress Service during 1775–83, see NMM, MID.7/3/2.

[51] D. George, *London Life in the Eighteenth Century* (London, 1925, repr. 1965), 292.

[52] TNA, ADM. 1/5117/9.

be impressed they could be pressured into volunteering. The Mayor of London in October 1787 ordered the marshals and constables of the city

to make diligent search and apprehend all persons who have no visible means of obtaining a livelihood or cannot give a good account of themselves, and bring them before his Lordship or one of the Aldermen in order to their being examined and if found fit or proper to serve his Majesty either in the capacity of sailors or soldiers that they may be forthwith sent upon that service or otherwise dealt with according to the law.[53]

The Act for manning the navy in 1795 reiterated these directions, authorising justices of the peace to levy 'all able-bodied, idle and disorderly persons who cannot upon examination prove themselves to exercise and industriously follow some lawful trade or employment, or to have some substance sufficient for their support and maintenance'.[54]

Yet it was trained seamen who were most wanted. No stone was left unturned to secure them. Admirals of the maritime counties made use of rewards 'to any person who shall discover any seaman or seamen who may secrete themselves so that they be taken' by the press.[55] But after the first few years of war, supplies of seamen on shore gradually dwindled. In 1779–81, the total number of recruits, numbers of bounties paid, and the supply of men from London fell away, suggesting the productivity of existing methods of recruitment was limited. During the French Revolutionary War, the Impress Service was resurrected but there was certainly awareness that existing methods had to be supplemented.[56]

Sir Charles Middleton became a sea lord at the Admiralty in May 1794, having, as Comptroller of the Navy Board, been responsible for the Impress Service from 1778 until 1790. Answering the Prime Minister's queries in 1786, he reflected on the numbers raised during the first two years of the Seven Years' War and American War:

Here then in two wars are 37,000 men raised in the first two years, which with a peace establishment of 12,000 is sufficient for 50 sail of the line and 35 frigates. But suppose by adopting some system, or intermixing a certain proportion of landsmen to be raised parochially or otherwise, we could command the first year 30,000, and 20,000 the second, we shall then have the first year 50 sail of the line, and 20 more the second with a proportion of frigates.[57]

[53] TNA, ADM. 1/5118/10, 4 Oct. 1787.
[54] *The Law and Working of the Constitution: Documents 1660–1914*, ed. W. C. Costin and J. Steven Watson (2 vols., London, 1952), II, 8–10.
[55] TNA, ADM. 1/5119, 16 Feb. 1793; NMM, ADM. BP/7, 16 Oct. 1787.
[56] For officers in the Impress Service by place of service 1793–1800, see TNA, ADM. 30/34.
[57] NMM, MID.2/40/9, Answer to Questions, Middleton to Pitt, 12 Feb. 1786.

By the first Quota Act of 5 March 1795, each county of England and
Wales was given a specific number of men to raise for the navy. Rutland
was given 23, West Riding 609. The Act provided for the supply of 9,766
men. The second Act of 16 March 1795 set quotas for the sea ports
of England, Wales and Scotland. London was given 5,704, Liverpool
1,711, Bristol 666. A third Act required that the counties and burghs
of Scotland find 1,814 men. Target numbers were modelled on those
each place had provided during the American War, with allowance for
growth, for example, in the port of Liverpool. The burden of executing
these Acts fell upon local authorities. The local justices had to divide the
county quotas between the parishes, where the overseers of the poor and
churchwardens had to find the men and the bounties to pay them. In the
ports, mayors, magistrates and customs officers became commissioners
for executing the Acts. Those officials who failed to meet their quotas
were fined for neglect of duty.

The list of men raised for London survives. There a bounty of 25
guineas was offered to attract seamen, 15 guineas for landsmen; 5,704
men were raised including 1,371 seamen and 2,522 landsmen, to whom
were attached 440 pressed men.[58] The list of 47 men for Beaumaris in
Anglesey included 20 seafarers, 17 labourers, 6 miners, 2 shoe makers,
a miller and a gardener.[59] The navy seems to have been satisfied with
the result. Indeed in November 1796 the quota system was tried again.
Coastal counties were directed to find 6,124 men; the inland counties
and those in Wales, 6,525 men; Scotland, 2,108 men. However, at the
end of 1796, further men were difficult to find, and the quota system was
not tried again.[60]

The partial demobilisation of the navy in 1802–3 recreated the problem
of restocking ships for commissioning in 1803–5. The entry of Spain into
the war on the side of France late in 1804 necessitated a further effort.
Middleton, as Lord Barham, became First Lord of the Admiralty in May
1805 and in August consulted a committee of City businessmen involved
in shipping on the means they favoured for increasing the supply of men
to the navy.[61] They observed that coastal craft were navigated by men
either ineligible for impressment or by apprentices while outward-bound
merchant ships had hitherto been protected from impressment. But, in
their opinion, both could be stripped in the following proportions: one
man in six from coastal vessels; one in eight from those bound as far as the

---

[58] Lloyd, *British Seaman*, 181–2, citing TNA, ADM. 7/361.
[59] A. Eames, *Ships and Seamen of Anglesey 1558–1918* (NMM, Greenwich, 1981), 555–9.
[60] Emsley, *British Society and the French Wars 1793–1815*, 53.
[61] The committee included John Julius Angerstein (Chairman of Lloyd's Insurance Company), Thomas King and Robert Taylor.

Mediterranean or Baltic; one in ten from ships bound on distant voyages. 'The reason for the distinction in the proportion' was 'that vessels on short voyages can more easily replace the men than those on long'. They thought ships sailing with letters of marque should be prevented from taking men eligible for the navy, except officers and carpenters. Otherwise 'the only means' of raising 'stout able bodied men would be a more active and efficient impress in every part of the kingdom', including the inland counties where they suspected there were many persons who had been at sea. No other system would 'be efficient to the extent and with the dispatch required'.[62]

The impact of impressment

The committee's observations ring true, at least regarding former seamen living inland. From fear of the press gang, at the age of forty-seven the seaman John Nichol in 1803 fled Edinburgh to Cousland, 9 miles inland, where he lived for eleven years, and even then at times took refuge elsewhere.[63]

The committee also tacitly acknowledged the hardship of impressment for seamen as well as ship owners. Popular protest, rescues and riots were common throughout the country and indeed often attracted empathy, rather than opposition, from local authorities. Press officers could not expect unbiased trials if charged with injuries, damages or deaths during the course of their operations.[64] Members of Parliament were influenced by this local opinion and in 1787 the House of Commons called for an account of deaths during the American War after men were impressed and before they were put on board ships. The Navy Board had to report a total of 180; 10 at Liverpool, 18 at the Nore, 29 at Portsmouth, 12 at Plymouth and a remarkable 64 at Sheerness, presumably among men brought down from London.[65]

Apart from the human suffering, impressment created administrative complications. To permit naval stores and victuals to be shipped in barges, lighters, coasters, victuallers, transports and contractors' ships, the Navy Board had to issue 'protections' from the press: 23,242 between

---

[62] NMM, MID.1/5, Angerstein *et al.* to Barham, 3 Aug. 1805.
[63] *The Life and Adventures of John Nichol, Mariner*, ed. Tim Flannery (1822, republished Edinburgh, 2000), 185–9.
[64] N. A. M. Rodger, 'The mutiny in the *James & Thomas*'; T. Barrow, 'The Noble Ann Affair'; N. McCord, 'The impress service in the north-east of England during the Napoleonic War' – all in *Pressgangs and Privateers*, ed. T. Barrow (Whitley Bay, 1993), 5–11, 13–22, 23–37; D. Clammer, 'The impress service in Dorset, 1793–1805', *Maritime South-West* 21(2008), 101–13; Parkinson, *The Trade Winds*, 112, 245.
[65] NMM, ADM. BP/7, 18 Jan. 1787.

1776 and 1782. Parliamentary statutes and other departments of government protected about an equal quantity of men.[66] Yet on occasions of desperation, despite the difficulties for its subordinate boards, the Admiralty suspended 'protections' without warning and permitted the press to seize men in protected trades.[67]

Such suspensions broke trust. Royal dockyard workers regarded their freedom from the press as a right. In 1790 one lieutenant reported 'a large mob consisting principally of shipwrights belonging to Deptford Yard came to his rendezvous and with horrid imprecations, threatened him and his gang with death, and tore down his colours which they carried off in triumph'.[68] In 1801 Commissioner Coffin at Sheerness dockyard, who had had one of his boatmen impressed, was held for an hour against the rampart wall by artificers shouting 'throw him over, kill him' until he agreed the release of the man.[69]

Inhumane, socially divisive and administratively inconvenient, impressment was nevertheless regarded as indispensable. To what extent did it support British naval power? Accounts of recruits exist for the Seven Years' War, the American War of Independence and the Napoleonic War, but most pose problems of interpretation.

For the Seven Years' War, an account for the period 1755–7 states there were 70,566 recruits over the three years, of which 20,370 were volunteers and 16,953 were pressed men. Yet the figures appear to be based only on the returns of the Impress Service ashore. There remain 33,243 recruits who could have been either volunteers or pressed men.[70] Other figures for the Seven Years' War produced from a sample of muster books show that at least 55.6 per cent were volunteers, 15 per cent were definitely pressed, while another 25.9 per cent were men 'turned over' from one ship to another and whose method of first entry could only be clarified by research through muster books for their preceding ships.[71]

For the American War of Independence, a detailed account of men raised by the Impress Service survives. The account (see table 6.2) indicates that 113,925 men were raised by the Service and that 74,241 bounties were paid to volunteers. Those not paid bounty would appear to have been pressed. However, the Navy Board was at pains to point out the

---

[66] Ibid.; Gradish, *The Manning of the British Navy during the Seven Years War*, 66–7.
[67] For Navy Board objection to crews of transports being pressed, see NMM, ADM. BP/3, 29 July 1782.
[68] NMM, ADM. BP/10, 16 June 1790.   [69] TNA, ADM. 106/1844, 13 Apr. 1801.
[70] NMM, ADM. B/161, 10 Jan. 1759, a copy of which is in NMM, ELL/9, 10 Jan. 1759. Cited and tablified in Gradish, *The Manning of the British Navy in the Seven Years War*, 69, 212. These figures are examined closely in Rodger, *Wooden World*, 145–7.
[71] Rodger, *Wooden World*, 353.

number of men raised was that raised on shore, and did not include those raised at sea. By contrast, the number paid bounty for volunteering was not confined to those raised by the Impress Service on shore but included men who volunteered at sea or directly to particular ships from on shore.[72] The figures are thus incomparable.

For the Napoleonic War more reliable figures survive, produced by John Finlaison, the Admiralty Keeper of the Records between 1809 and 1816. He produced accounts of men recruited in 1810, 1811 and 1812 (see table 6.3).[73] They include and distinguish marines. The total number of recruits over those three years amounted to 72,603 men. In the three years, recruits at sea totalled 31,076, of which 18,604 (60 per cent) were volunteers and 12,472 (40 per cent) pressed men. Recruits on shore amounted to 41,473, of whom 24,594 (59 per cent) were volunteers and 16,933 (41 per cent) were pressed men. Elsewhere, pressed men in the Napoleonic War have been calculated at 47–50 per cent of the sailing complement.[74] Clearly, according to Finlaison's account, this was an over-estimate.

The efficiency of employment

Organisation and workload

Men recruited into the navy were delivered to receiving ships at the main ports from which they were sent to ships short of men. They reinforced an existing crew and supplemented men 'turned over' from other ships. Once on board a ship, new men were examined and rated within a few days. In the mid eighteenth century, one year at sea was considered sufficient to make a landsman into an ordinary seaman, and two years at sea to make an able seaman. However, sea time did not always make for qualified seamen.[75] Much care was thus taken to ensure men were rated according to their skill.

On board the *Warspite* in 1809 thirty-seven new men with pretentions to rating as seamen were 'minutely' examined by the master and first lieutenant. Captain Henry Blackwood reported only fifteen were 'passable men of war seamen' capable of doing the work of able seamen. They

---

[72] NMM, ADM. 7/3/2. The bounty became due only after volunteers had received three musters on board the ships where they were sent to serve; it was then paid by the Clerk of the Cheque at the first dockyard where ships happened to arrive.

[73] NMM, MLN. 153/3/41.

[74] M. Lewis, *The Navy of Britain*, 317–18, cited by Duffy, 'The foundations of British naval power', 71.

[75] Rodger, *Wooden World*, 26.

Table 6.3 *Men recruited to the British navy, 1810–1812*[a]

| | (Seamen + Marines = Total) | | |
	1810	1811	1812
At sea by the ships themselves			
Volunteers	7,774 + 68 = 7,842	5,362 + 51 = 5,413	5,293 + 56 = 5,349
Pressed Men	6,050 + 0 = 6,050	3,206 + 0 = 3,206	3,216 + 0 = 3,216
On shore by the recruiting parties and rendezvous			
Volunteers	4,534 + 2,948 = 7,482	4,352 + 3,381 = 7,733	5,406 + 3,973 = 9,379
Pressed Men	4,161 + 0 = 4,161	4,289 + 0 = 4,289	8,483 + 0 = 8,483
Total raised	22,519 + 3,016 = 25,535	20,641	26,427

[a] NMM, MLN.153/3/41

were 'willing and active young men' but inexperienced, so that 'unless they serve in a ship where there are many other seamen, they are not in themselves sufficiently so, to be depended on as good seamen'. Fourteen were designated ordinary seamen, being 'found extremely deficient on almost every point of seamanship and particularly as to the lead, tho' it is true, like all other men of war men, they go aloft and reef'. Six had 'no pretention whatsoever to a superior rating [other than landsman] except in their own assertion'. Appearances could be deceptive: one Swede looked like a seaman but knew no part of the duty. Another had maimed himself to avoid doing his duty: 'therefore in justice to good and willing men . . . ought never to rise higher'. Blackwood's nephew, though termed a midshipman, was paid as an ordinary seaman but, not having served three years at sea, was demoted to landsman.

In the eighteenth century, one-third of a ship's seamen were expected to be able, one third ordinary and the remainder landsmen. This was necessary not only for the safe handling of ships in working their sails, but for the crew to have the strength to work their guns. In 1809 the *Warspite* mounted eighty-four guns, the main deck carrying thirty 24-pounders. Yet she had the same complement of 640 men as ships which carried only 18-pounders on their main decks. In Blackwood's opinion, on account of the heavier guns, the ship 'ought not only to be strong but completely full as to number' of ordinary seamen and landsmen.[76]

Seamen, whether rated able, ordinary or landsmen, composed only 60 per cent of a crew. Sometimes over 20 per cent consisted of marines. Between 6 and 10 per cent were boys designated servants, by regulation at least eleven years old but sometimes, in contravention of the rules, as young as six. Commissioned and warrant officers comprised between 7 and 9 per cent, excluding petty officers who were skilled seamen appointed from the ratings of a ship.

The men on board a ship were also classed according to where they worked. Topmen handled the top-sails; forecastlemen handled the fore-sails; the afterguard heaved and hauled on the poop; the waisters did the same in the waist; idlers were specialists who did not have to stand watches. At sea, men worked seven days a week in four-hour watches, watch-on, watch-off. There were two two-hour dog watches between 4 and 8 each evening which, worked alternately, made the number of watches in a day an odd number and varied the pattern on alternate days.[77]

---

[76] NMM, MS88/090, Duckworth Papers, Blackwood to Duckworth, 4 Feb. 1809.
[77] Rodger, *Wooden World*, 18–29, 37–43, 348–51; Duffy, 'The foundations of British naval power', 71.

Compared to merchant ships, the complements of men in warships were heavy. This meant the ratio of ship-tons to each man in a warship was much smaller than in a merchant ship. In the mid eighteenth century, the full complements of warships created a ratio of 1 man to 2 tons. When the officers, servants and marines are excluded, the ratio of men to tonnage rose to about 1 man to 3 tons for first and second rates, 1 to 4 for third rates, and 1 to 6 or 7 tons for frigates. This may be compared to the ratio in merchant ships which ranged between 1 man to 10 tons to 1 man to 20. But they were under pressure to economise and, indeed, their mean ratio grew between 1751 and 1766 from about 13 tons to 15 tons per man.[78]

These ratios suggest that work was heavier in merchant ships and in smaller warships. Commissioned warships, moreover, spent on average more than half their time in port or moored in a secure harbour. This varied according to the size of ship and where they were stationed. According to Rodger, in home waters in 1756–63, first and second rate battleships spent 77 per cent of their time in port; the comparable figure for fifth and sixth rate frigates was 50 per cent; that for sloops, 51 per cent. Overseas, warships were more active. In 1756–63, warships in the West Indies spent 48 per cent of their time in port while in the Mediterranean they spent 43 per cent in port.[79] In the mid eighteenth century, seamen were thus at sea less in larger vessels than in smaller vessels, and less in home waters than overseas.

Deployment and turnover

By the end of the eighteenth century, with the long-term maintenance of the close blockade, warships were kept longer at sea. As the size of the commissioned fleet grew, there was a constant necessity to secure more men and to keep those available in constant use at sea. When vessels were taken out of commission, seamen were 'turned over' from ship to ship, recycling them for further use. Their intensive use was vital for large numbers of men were not employed at sea, were fictitious, incarcerated, sick or lost by discharge, death or desertion each year. Caution is thus needed in using the figures available for naval manpower. Great differences existed between the numbers of men for which Parliament voted wages, those 'borne' for wages and those mustered for victuals.[80]

The most accurate figures for the deployment of seamen were compiled by John Finlaison, the Admiralty Keeper of the Records, for the

---

[78] Rodger, *Wooden World*, 351; Davis, *Rise of the English Shipping Industry*, 71.
[79] Rodger, *Wooden World*, 352.   [80] Lloyd, *British Seaman*, 261–4.

Table 6.4 *State of the British navy with respect to men, 1811–1813*[a]

	Seamen + Marines = Total		
	1811	1812	1813
Total complement of ships in commission:	113,435 + 26,620 = 140,055	106,640 + 26,192 = 132,832	104,018 + 25,818 = 129,866
The number actually at sea in the whole service:			
Bound as part of complement:	98,811 + 24,341 = 123,152	96,136 + 24,082 = 120,218	97,156 + 24,450 = 121,606
As supernumeraries, lent men, etc.::	9,025 + 650 = 9,675	10,043 + 1,188 = 11,231	8,527 + 1,007 = 9,534
Total at sea:	107,836 + 24,991 = 132,827	106,179 + 25,270 = 131,449	105,683 + 25,457 = 131,140
In hospitals, at rendezvous or marine headquarters:	1,622 + 4,132 = 5,754	864 + 4,465 = 5,329	1,481 + 5,703 = 7,184
Total available force:	109,458 + 29,123 = 138,581	107,043 + 29,735 = 136,778	107,164 + 31,160 = 138,324
There were moreover in the enemy's prisons:	2,121	2,954	3,930
The fictitious men borne for wages as 'Widows' Men':	1,350	1,328	1,314
Grand total in pay:	142,052	141,069	143,568
Of the above were foreigners:	11,693 + 3,039 = 14,732	12,324	11,727 + 1,478 = 13,205
And of these foreigners were real or pretended Americans:	3,647 + 38 = 3,685	3,298	—

[a] NMM, MLN.153/3/41

period 1811–13 (see table 6.4). In 1811 142,052 seamen and marines were being paid. But of these only 123,152 or 87 per cent – 98,811 seamen and 24,341 marines – were part of the complements of specific ships. They were supplemented by 9,675 supernumeraries and loaned men including 650 marines. This provided 132,827 men actually at sea – 107,836 seamen and 24,991 marines. In hospitals, at the recruiting rendezvous on shore and at the various Marine headquarters were another 1,622 seamen and 4,132 marines. With these shore-based men, the total available force rose to 138,581 seamen and marines.

They naturally excluded 1,350 fictitious men and 2,121 prisoners of war. The former were often known as 'widows' men'.[81] In the eighteenth century, 2 were borne on a ship's books at the wage rate of able seamen to every 100 of a ship's complement. Their wages were assigned to the fund for paying officers' widows' pensions.[82] By 1811 1 was borne to every 100 men.

The small number of seamen listed as prisoners of war reflected a reluctance to return them on the part of Britain's enemies, who captured relatively few. During the French Revolutionary War, in 1798 after five years of war, there were around 4,000 British seamen in French hands and over 30,000 French prisoners of war in British camps. By 1801 there were over 70,000 French prisoners, nearly all of them seamen,[83] a number comparable to the 85,000 seamen France was said to possess at the beginning of the French Revolutionary War.[84]

A far greater number of British seamen were lost by discharge, death or desertion. In 1811 (see table 6.5) the number invalided or otherwise discharged from the service amounted to 11,071 seamen and 1,950 marines, a total of 13,021 men. Another 3,331 seamen and 934 marines – altogether 4,265 men – died or were lost at sea. Meanwhile, 8,155 seamen and 604 marines, in all 8,759 men, deserted. This brought the total loss for the year to 26,045 men. This was nearly 19 per cent of the total available force. Simply to maintain the status quo, that number had to be replaced each year. In fact only 20,641 men were pressed or volunteered for naval service in 1811. Losses thus exceeded additions to the naval workforce in that year.

These losses were substantial and were far greater than those from the regular army which ranged between 7 and 11 per cent during the Napoleonic War.[85] Why were they so great? An explanation may be derived from comparison of figures for 1755–7 in the Seven Years' War

[81] NMM, MLN.153/3/41.   [82] Rodger, *Naval Records for Genealogists*, 50.

[83] E. H. Jenkins, *A History of the French Navy* (London, 1973), 241; Clowes, *The Royal Navy*, IV, 185.

[84] M. Acerra and J. Meyer, *Marines et revolution* (Rennes, 1988), part 3.

[85] Holmes, *Redcoat*, 135.

Table 6.5 *Seamen and marines lost from the British navy, 1810–1812*[a]

The casualties in the navy in the years 1810, 1811 & 1812 were as follows:

	Seamen + Marines = Total		
	1810	1811	1812
Invalided or otherwise discharged from the service	$14,272 + 1,948 = 16,220$	$11,071 + 1,950 = 13,021$	$11,848 + 1,828 = 13,676$
Died or lost at sea in the same time	$4,092 + 1,093 = 5,185$	$3,331 + 934 = 4,265$	$3,397 + 814 = 4,211$
Deserted in the same time[b]	$11,597 + 896 = 12,493$	$8,155 + 604 = 8,759$	$7,551 + 599 = 8,150$
Total loss in the year	$29,961 + 3,937 = 33,898$	$22,557 + 3,488 = 26,045$	$26,037$

[a] NMM, MLN.153/3/41

[b] Note by J. Finlaison, Admiralty Keeper of the Records, 1809–16:

The Navy Board called the number of deserters in 1810 10,735 in an account which they compiled from the muster books. The reason of the excess here stated will be that many ships had not sent muster books and were not included in the Navy Board's return, neither were the marines on shore.

Memo. It will be observed that the total loss in the year 1810 was 33,898 men, whereas the total raised is only 25,535 making a deficiency of 8,363 men. We have no account however of the number of deserters apprehended in the year 1810, and it is therefore to be inferred that of the 12,493 deserters as above, nearly to the amount of 8,363 must have been reclaimed. That the number of men raised on shore is accurate is proved by its agreeing (within 119 men) with the abstract kept in the office of rendezvous returns.

Table 6.6 *Seamen lost from the navy in 1755–1757 and 1810–1812[a]* (*figures exclude marines*)

	1755	1756	1757	Total 1755–7	% 1755–7
Men 'mustered'	29,268	50,037	60,548	139,853	
Died	2,236	2,992	2,562	7,790	5.9
Deserted	4,310	3,339	4,647	12,296	8.8
Discharged	1,227	1,326	1,478	4,031	2.8
Total lost	7,773	7,657	8,687	24,117	17.2
% lost of men 'mustered'	27	15	14		
	1810	1811	1812	Total 1811–12	% 1811–12
Men 'at sea'	[b]	107,836	106,179	214,015	
Died	4,092	3,331	3,397	6,728	3.1
Deserted	11,597	8,155	7,551	15,706	7.3
Discharged	14,272	11,071	11,848	22,919	10.7
Total lost	29,961	22,557	22,796	45,353	21.2
% lost of men 'at sea'	–	20.9	21.5		

[a] Lloyd, *British Seaman*, 262–3; NMM, ADM.B/161, 10 Jan. 1759, in Gradish, *The Manning of the British Navy in the Seven Years War*, 212; NMM, MLN.153/3/41.
[b] Precise number unknown.

with those for 1810–12 (see table 6.6). The former exclude marines, so those for the period 1811–12 do the same. In the earlier period 17.2 per cent of the seamen mustered in ships were lost on average each year; in the later, over 21 per cent. Losses from death fell, from 5.9 per cent to 3.1. So did losses from desertion, from 8.8 per cent to 7.3. However, discharges of seamen as invalids rose from 2.8 per cent at the beginning of the Seven Years' War to 10.7 per cent towards the end of the Napoleonic War. This large proportion of discharges made for an efficient workforce. Yet it added considerably to the pressure to recruit and to the turnover of seamen.

Discharges and the death rate

The discharge of a far greater number of seamen as invalids before they died reduced the number of sick men who would die in service. The credit for a reduction of the death rate was claimed by the navy's medical profession. But they can also be credited with the higher proportion of seamen who were discharged as invalids. For their greater knowledge of the nature of diseases suffered by seamen permitted them not only to preserve life, but also to decide who were best discharged before they died. They were aided by the establishment of hospitals and by the rise of preventative medicine.

Hospital provision

Hospitals facilitated the concentration, classification and study of the sick. Until the mid eighteenth century, contractors were employed to look after men in rented accommodation around the coast of Britain and overseas. However, a naval hospital at Plymouth from 1689 was followed by others at Minorca and Jamaica by 1712. Gibraltar had a 1,000-bed hospital by 1740. Haslar Hospital with 2,000 beds was built at Gosport in 1746–62, Stonehouse Hospital with 1,000 beds at Plymouth in 1756–60.[86] Others were built in England after 1793 at Deal and Yarmouth, and overseas at Antigua, Bermuda, Barbados, Cape Town, Halifax, Kingston, Lisbon, Madras and Malta. By 1815 there were more than thirty.[87] Separate asylums were created for the mentally ill and insane while the great London hospitals – Guy's, St Bartholomew's and St Thomas's, and Bethlem – took naval patients.[88]

Staffing had to keep pace with this growth in infrastructure. Surgeons for ships as well as hospitals had to be recruited more widely. During the eighteenth century only a certificate from the Royal College of Surgeons of London would suffice as a qualification. But from 1797 certificates from the universities of Edinburgh and Dublin were accepted. Numbers of assistant surgeons and surgeons' mates also grew.[89] They did not keep pace with the growth of naval manpower after 1800 but their eventual release from service had a wider public benefit.[90]

In practice only a small proportion of the sick were ever referred to a hospital. Thomas Trotter, while Physician of the Channel Fleet, collated sick lists from ships and in 1794 distinguished those confined to bed and those who were 'objects for hospital'. Most of the sick did not fall into either of these categories: on 16 April 1794, of 725 men recorded sick, only 53 were confined to bed and only 20 were 'objects for hospital'.[91] This was probably typical. The seamen who actually reached a hospital

---

[86] K. Harland, 'The Royal naval hospital at Minorca, 1711: an example of an admiral's involvement in the expansion of naval medical care', *MM* 94(2008), 36–47.

[87] NMM, ADM. DP/38A, 6 Mar. 1811; ADM. DP/37B, 29 July 1817.

[88] E. H. Turner, 'Naval Medical Service, 1793–1815', *MM* 46(1960), 119–33.

[89] NMM, ADM. DP/40A, 15 Jan. 1820; ADM. DP/41A, 22 Jan. 1821.

[90] P. Mathias, 'Swords and ploughshares: the armed forces, medicine and public health in the late eighteenth century' in *War and Economic Development*, ed. J. M. Winter (Cambridge, 1975), 73–90.

[91] *The Health of Seamen: Selections from the works of Dr James Lind, Sir Gilbert Blane and Dr Thomas Trotter*, ed. C. Lloyd (NRS, 1965), 175, 198, 224, 232.

Table 6.7 *Admissions to Haslar Hospital in 1753–1757[a] and 1780[b]*

	1753–7	% of total cases	1780	% of total cases
Fevers	7,014	53	5,539	57
Scurvy	2,411	18	1,457	15
Rheumatism	430	3	327	3
Fluxes/Dysentery	286	2	240	2
Ulcers	277	2	979	10
Cutaneous disorders / Itch	273	2	165	17
Venereal disease	119	1	183	2
Consumption	114	1	218	2
Smallpox	103	1	42	0.5
Malaria / Intermittent fevers	71	1	33	0.5
Measles	40	0.5	28	0.5
Dropsy	28	0.5	24	0.25
Total cases	13,099		9,787	

[a] Gradish, *The Manning of the British Navy during the Seven Years War*, 202, who derived his information from TNA, ADM. 98/7, 117–20, 7 Feb. 1758 ('54 diseases, wounds and undiagnosed cases' producing 1,933 admissions are here omitted).
[b] *The Health of Seamen*, 200–1 (33 diseases, wounds and other cases are here omitted).

were thus a small proportion of the sick and, of this small proportion, the great majority suffered from a limited number of conditions.

Figures have survived for admissions to Haslar Hospital, the naval hospital at Portsmouth, for 1753–7 and 1780 (see table 6.7). Fevers, scurvy, rheumatism, dysentery or 'flux', ulcers and skin diseases were the most common problems. Stubbon ulcers and skin ailments arose from sores, sea water and continual friction. Venereal disease was widespread and only the most severe cases were admitted to hospital. With most men below the age of twenty-five,[92] and only 20–25 per cent married to actual wives, many engaged with prostitutes. The result was that venereal infections extended to 8–20 per cent of crews in 1756–63, and perhaps further because many men nursed infections and known numbers derive from those taking cures.[93] Indeed, because cures cost 15 shillings, many men took quack cures or ignored symptoms until they became chronic, prompting abolition of the charge in 1795.[94]

In 1753–7 seamen were admitted to Haslar Hospital suffering from one of sixty-six types of disorder; in 1780, forty-five types were listed. Not all caused death. Of 1,002 deaths in 1753–7, 65 per cent were caused

[92] 53 per cent of able seamen and 85 per cent of ordinary seamen, according to Rodger, *Wooden World*, 78.
[93] Rodger, *Wooden World*, 79–80, 360–8.   [94] *The Health of Seamen*, 228–9.

Table 6.8 *Numbers of seamen who died in the British navy, 1755–1757*[a]

Place or cause	1755	1756	1757	Total
On board ships	639	962	776	2,377
In hospitals	1,329	1,627	1,425	4,381
In sick quarters	194	256	169	619
Slain in combat	13	40	45	98
Died of wounds	3	17	17	37
Drowned	58	90	130	278
Total	2,236	2,992	2,562	7,790

[a] Based on NMM, ADM. B/161, 10 Jan. 1759, cited in Gradish, *The Manning of the British Navy during the Seven Years War*, 212.

by fevers, justifying fear of them in crowded ships, to which new recruits could easily introduce disorders like typhus. Meanwhile, 13 per cent of the deaths were caused by scurvy. Together, fevers and scurvy killed 78 per cent of the patients in Haslar Hospital in 1753–7.[95]

Although most patients were discharged, most seamen who died in the navy did so in hospital. In 1755–7 (see table 6.8), of 7,790 men who died, 56 per cent died in hospital, almost double the number who died on board ships. The way in which seamen were treated in hospital thus had the greatest potential to make an impact on the death rate.

Writing in 1815, Gilbert Blane was therefore on firm ground when he pointed out that, while a diminishing proportion of seamen were sent to hospital, the number of those who died in hospital had declined even more drastically. He referred to tables of sick sent to hospitals on shore or in ships on the home station between 1779 and 1813. Those sent comprised 1 man in every 3 voted for the navy in 1779–80, 1 in 4 in 1782–3, 1 in 8 in 1804, and 1 in 10 by 1813. But he also showed that deaths in hospital had declined from one in 42 of seamen voted to one in 143 in 1813.

Blane's ratios are supported by the figures derived from John Finlaison, Admiralty Keeper of the Records, who showed that only 3.1 per cent of seamen died in service in 1810–12 compared to 5.9 per cent in 1755–7. Blane argued that it was the reduction in the death rate among seamen that permitted the growth in the scale of naval manning. Comparing death rates in 1779 and 1813, Blane maintained that, by the latter date, about 6,674 men survived each year who would, under conditions in 1779, otherwise have died. He maintained that 'under such an annual waste of life, during the French Revolutionary and Napoleonic Wars the

[95] Gradish, *The Manning of the British Navy during the Seven Years War*, table 5, 202.

national stock of mariners must have been exhausted'.[96] But Blane made no reference to the number of men discharged as invalids, which was double the number who survived on account of better conditions and treatment.

Preventative medicine

Survival, possibly to be discharged as an invalid, was increased by the rise of preventative medicine. Miasmic theory, in which infections were thought to arise and spread on account of foul air, remained influential into the nineteenth century. Medical men therefore looked to the environment of seafarers for the sources of their ills. Efforts were made to keep ships clean, dry and ventilated.[97] Hence wind funnels and pumps were employed to channel air below decks which were regularly washed. Clothes too were kept clean.[98] In the Mediterranean fleet from 1798 and in the Channel fleet from 1800, sick bays in ships of the line were shifted from below decks to the space beneath the forecastle.[99]

From the mid eighteenth century increasing attention was paid to the diet of seamen. The harmful effect of excess alcohol was recognised in the 1740s, from which time the daily rum ration was watered and progressively reduced to a quarter of a pint in 1824. Their basic diet was not questioned, and survived until the nineteenth century.[100] The deleterious effect of insufficient fresh food was nevertheless recognised, compounded by the consequence of continuous employment at sea. Scurvy was the common ailment of seamen, perhaps the more insidious from being reclusive, emerging and retreating as food supplies dictated.[101]

The defeat of scurvy, following the shocking scale of mortality on Anson's circumnavigation of 1740–4, is now relatively well known. James Lind's investigation of extant literature, his experiment on board HMS *Salisbury* in 1747, and his publications on the subject are generally taken

---

[96] G. Blane, 'On the Comparative Health of the British Navy, from the Year 1779 to the year 1814, with proposals for its farther Improvement', first printed in *Transactions of the Medico-Chirurgical Society*, 6(1815), reprinted in *The Health of Seamen*, 175–201.

[97] *The Health of Seamen*, 177–85.   [98] Rodger, *Wooden World*, 106–7.

[99] The new location did not become universal: B. Lavery, *Nelson's Navy: The Ships, Men and Organisation 1793–1815* (London, 1989), 213–14; for a plan view of the sick bay of the *San Domingo, c.* 1812–14, see NMM, LOG.W/3.

[100] Gradish, *The Manning of the British Navy during the Seven Years War*, 141.

[101] For its regular occurrence, see *Surgeon's Mate: The diary of John Knyveton, surgeon in the British fleet during the Seven Years War 1756–1762*, ed. E. Gray (London, 1942). Although this diary was clearly embellished for readers, the regular references to the scorbutic cases seem authentic.

as the beginning of the campaign.[102] Anson's appointment of Lind in 1758 over the heads of professional colleagues to be senior physician at the newly opened Haslar Hospital until 1783 contributed to the environment of inquiry. But before 1793 the Sick and Hurt Board was composed of commissioners without medical qualifications and in 1753 they referred Lind's *Treatise* to land-based consultants, and in 1767 backed a treatment recommended by the surgeon David MacBride,[103] who proposed a 'wort, or infusion of malt, as a substitute for fresh vegetables' that was supported by Sir John Pringle, President of the Royal Society and consultant to the army.[104] The result was a period of continued uncertainty that encompassed Cook's trials of recommendations.[105]

Recently the medical men and naval officers who had experience at sea have been given the credit for the introduction of lemon juice.[106] At the time of the American War of Independence it had the good opinion of Gilbert Blane who became chief surgeon in Rodney's fleet in the West Indies. After the war Thomas Trotter demolished the antiscorbutic properties of a succession of treatments and maintained the case of fresh vegetables and 'those fruits abounding with an acid, such as the citric class'. They tapped experience that went back to the Seven Years' War.[107] Trotter became physician to the Channel fleet while Blane became a commissioner at the Sick and Hurt Board in 1795.[108] Yet already, from 1793, the board possessed two commissioners, Johnstone and Blair, who had been naval surgeons and advised the use of lemon juice.

In 1794–5, with the sanction and support of these commissioners, Commodore Peter Rainier used lemon juice to make a continuous voyage to Madras of nineteen weeks, with only the temporary appearance of scurvy. With this example, the Sick and Hurt Board successfully urged the introduction of lemon juice from 1795. 'Manufactured' and

---

[102] Lind's *Treatise of the Scurvy* (1753) and *An Essay on the Most Effectual Means of Preserving the Health of Seamen* (1757) are printed in *The Health of Seamen*, 7–110.

[103] D. MacBride, *An Historical Account of a New Method of Treating the Scurvy at Sea; containing ten cases which show that this destructive disease may be easily and effectually cured without the aid of fresh vegetable diet* (1764).

[104] NMM, ADM. FP/10, 1 July 1767, for which reference see J. Watt, 'Some consequences of nutritional disorders in eighteenth-century British circumnavigations' in *Starving Sailors: The Influence of Nutrition upon Naval and Maritime History*, eds. J. Watt, E. J. Freeman and W. F. Bynum (NMM, Greenwich, 1981), 51–71.

[105] J. C. Beaglehole, *The Life of Captain James Cook* (Hakluyt Society, London, 1974), 170.

[106] B. Vale, 'The conquest of scurvy in the Royal Navy 1793–1800: a challenge to current orthodoxy', *MM* 94(2008), 160–75.

[107] Gradish, *The Manning of the British Navy during the Seven Years War*, 162–70.

[108] C. Lloyd, 'The introduction of lemon juice as a cure for scurvy', *Bulletin of the History of Medicine* 35(1961), 123–32.

bottled under that board,[109] lemon juice was distributed by the Victualling Board from June 1796, to be issued daily as a preventative in
ships going on foreign stations. Owing to inadequate supplies, it was
only available as a cure to ships on home stations. Indeed Trotter maintained it was weakening if taken regularly. Scurvy thus still broke out
among men in the Channel fleet. By 1799 the situation had become
intolerable and it was Rear Admiral Berkeley who wrote a 'public letter'
to the Sick and Hurt Board emphasising the necessity for the juice as a
preventative, not simply a cure, on home stations.[110]

In 1800 Trotter was briskly replaced as chief adviser to the Channel
fleet by St Vincent, who had seen the benefits of lemon juice in the
Mediterranean where case-loads had been consumed since 1793. Equally
important, in England by 1799 bottled stocks of the juice were adequate
for a small dose to be issued daily to every man in the navy. This was
instituted in 1800.[111] Following the resumption of war in 1803, the lemon
juice was accompanied by sugar for a sweetener to make sherbert,[112] and
only seamen in ships employed on the English coast were not required
to take their daily dose.[113]

Preventative medicine made contributions to health in the British navy
in other respects. James Lind used Peruvian Bark or cinchona at Haslar
Hospital for the cure of malaria and recommended daily doses for men
sent on shore in the tropics to prevent attacks of the 'ague'. It was in
common use by the end of the eighteenth century.[114] The introduction
of inoculation against smallpox in 1799–1800, though voluntary, had
a similar preventative effect.[115] Prevention too entered thinking about
surgery, with regard to sepsis, trauma and pain, supported by the British

---

[109] For the management of its supply after the duties of the Sick and Hurt Board were
taken over by the Transport Board in 1806, see NMM, ADM. DP/32B, 14 Aug. 1812.

[110] Berkeley maintained he 'never knew an instance of a ships being out nine weeks that
the scurvy did not begin to shew itself, although kept under and certainly very much
lessened by the lime [sic] juice which is medicinally allowed to all ships. But this lime
juice is never made use of until a scorbutic patient discovers himself which is rarely or
ever until the disease has gained a considerable head, where as if it was mixed with his
drink from the time the beer was expended and that he was allowed sour krout with
his beef or to eat as a salad it might keep him free from the scurvy, or at least operate
upon him so as to keep the disorder from bursting forth in the violent manner which we
always see instances of at the period I have mentioned': *Channel Fleet and the Blockade
of Brest*, 396–7.

[111] *Ibid.*, 28–30, 77, 133–4, 142–5, 395–6, 457, 522–3, 543.

[112] In 1813 the 'exorbitant' cost of sugar contributed to suspension of issues of lemon juice
except to the sick: NMM, ADM. DP/33B, 8 Dec. 1813.

[113] NMM, ADM. DP/29, 27 June 1809.

[114] M. J. Cardwell, 'The Royal Navy and Malaria, 1756–1815', *Trafalgar Chronicle*
17(2007), 84–97.

[115] *Channel Fleet and the Blockade of Brest*, 546–7, 552, 559, 561.

passion for cleanliness and hygiene. By the time of Trafalgar, the patients of naval surgeons had a better survival rate than those undergoing surgery in civilian hospitals.[116]

Looking back in 1815, Gilbert Blane observed not only the long-term improvement in health at sea since the mid eighteenth century but 'another sudden decrease of sickness in the first years of this century'. He attributed the decrease in sickness to 'the ample and general supply of lemon juice, the superior attention to cleanness, dryness and ventilation, the improvement in victualling, vaccination and superior medical treatment'.[117] They all contributed to the prevention of disease. Meanwhile those who could not be cured were discharged.

Desertion and discontent

The British navy, like the army, continued to lose men by desertion. In view of the importance of morale and motivation in manpower, did this reflect a disinclination to serve the state or a more local cause of detachment? With the press supplying as many as 40 per cent of the new seamen each year, while those remaining in service were 'turned over' at regular intervals to new ships, did relationships become more impersonal? Rodger suggests a gradual decline in the paternalistic recruitment of crews by regional patrons may have contributed to the mutinies of the 1790s.[118] At the same time discontent within the navy gained a political complexion which posed a real management challenge.

The persistence of desertion

In 1755–7 desertion stood at 8.8 per cent of the seamen mustered at sea (see table 6.6). During the whole course of the Seven Years' War, 1756–63, 36–40,000 men were marked as 'Run', the naval term for desertion. That was an average of 4,500–5,000 a year from about 62,500 men, which was 7 per cent a year.[119] On the Leeward Islands station between 1784 and 1812, 7 per cent also ran, though losses varied between 1 per cent and 19 per cent of the complements of individual ships.[120] In the two years 1811–12, according to figures compiled by John Finlaison, an

---

[116] J. Watt, 'Surgery at Trafalgar', *MM* 91(2005), 266–83.

[117] *The Health of Seamen*, 181, 186.

[118] N. A. M. Rodger, 'The inner life of the navy, 1750–1800: change or decay?' in *Guerres et paix 1660–1815* (Vincennes, 1987), 171–9.

[119] Rodger, *Wooden World*, 203–4.

[120] J. Byrn, *Crime and Punishment in the Royal Navy: Discipline on the Leeward Islands Station 1784–1812* (Aldershot, 1989), 154, 221–8.

average of 7,853 seamen were marked as 'Run' from a mean 107,007 men at sea each year, or 7.3 per cent. Finlaison stated that the Navy Board in 1810 under-estimated numbers of deserters because many ships had not returned their muster books and were not therefore included in the Navy Board's returns; neither were the marines on shore.[121] If this also applied to 1811–12, the proportion of deserters may have been higher.

It certainly was during the American War of Independence. Between January 1776 and September 1780, 42,069 were recorded as 'Run'. That was an average of 8,400 a year from about 61,000 men, which was a remarkable 13.8 per cent a year.[122] Was this because of the opportunities for desertion and the shelter given to deserters by the colonists? Or, like some officers, did more seamen prefer to opt out of a conflict with which they disagreed?

For the period of the Seven Years' War, N. A. M. Rodger points out that figures have to be treated with care. Unauthorised absentees from ships in three successive musters were marked as 'Run' in the muster books. He points out that many seamen were detained on shore and thus marked as 'Run' though, in the language of the time, they were 'straggling'. Some also 'rambled' but did not intend to desert. Others temporarily absented themselves to secure alcohol.

He also suggests men were often pushed by the break-up of ships' companies and before they settled in to new ships. More than half ran in the first six months on board a ship. Virtually none ran after eighteen months. Small uncomfortable ships had more deserters than large ones. But so too did captains known for their brutality.[123]

Contemporaries attributed desertion to the character of seamen. George Cockburn, who served between 1786 and 1846, thought nothing could be done which 'would have the slightest effect towards checking the generality of our sailors from following the bent of their inclinations when seized by any whim or attracted by any present temptation, especially if holding out to them novelty and change'.[124] At present there is no evidence to suggest that volunteers were less likely to desert than impressed men. Some, once taken by the press, simply volunteered to get a large 'bounty' or more liberty while their ship remained in port.[125]

---

[121] NMM, MLN.153/3/41.   [122] Clowes, *The Royal Navy*, III, 339.

[123] Rodger, *Wooden World*, 189–200.

[124] LC, Cockburn papers, letters sent, private, container 13, Cockburn to Adam, 10 Aug. 1835.

[125] *The Nagle Journal: A diary of the life of Jacob Nagle, sailor, from the year 1775 to 1841*, ed. J. Dann (New York, 1988), 189; *Landsman Hay: The Memoirs of Robert Hay 1789–1847*, ed. M. D. Hay (London, 1958), 219.

And, once turned over, they were not distinguished in the ships' books from pressed men.[126]

Desertion was certainly discouraged. During the Seven Years' War, of 254 retaken, 13 were acquitted by courts martial, 9 found guilty on lesser charges, 176 sentenced to be flogged, and 53 to hang. The latter were reduced by pardoning to about a dozen.[127] On the Leeward Islands station between 1784 and 1812, 8 per cent of deserters were retaken; of these, 90 per cent were subject to court martial; 98 were convicted, 5 sentenced to death, 92 to receive between 50 and 500 lashes, and 1 to imprisonment for eighteen months.[128]

Severity of punishment tended to encourage resistance to recapture.[129] John Nichol was forced to 'skulk like a thief' to avoid the press gang. He only returned to Edinburgh after eleven years when, at the age of fifty-eight, he thought 'they would not have taken me had I wished to enter the service'.[130] The thoroughness with which ex-seamen were sought is attested by John Finlaison who remarked that, of 12,493 deserters in 1810, 'nearly to the amount of 8,363 must have been reclaimed'.[131]

Antidotes to desertion were practised. The system of withholding payment has been noticed and arrears owing were considered sources of discontent sufficient to trigger mutiny.[132] At the Nore mutiny, despite the preceding Spithead mutiny having won an increase in rates of pay, the mutineers appealed for the payment of 'all arrears of wages down to six months, according to the old rules', that is, as expressed in the 1728 and 1758 Acts of Parliament. Another tactic was to avoid sending men to hospital on shore whence men deserted or took a ramble.[133] It was promoted by preventative medicine and determined discipline. In November 1800, when the Channel fleet came into Torbay, only 16 men were sent to hospital out of a total fleet complement of 23,000 men.[134]

Crews could be isolated from shore. But in 1774 Captain Robert Tomlinson asserted that the indeterminate length of their service was more irksome to seamen than their low wages.[135] Hence in 1797 the

---

[126] Rodger, *Wooden World*, 163.   [127] *Ibid.*, 201–2.

[128] Byrn, *Crime and Punishment in the Royal Navy*, 164–5.

[129] Rodger, 'The mutiny in the *James & Thomas*'.

[130] *The Life and Adventures of John Nichol, Mariner*, 185–9.

[131] NMM, MLN. 153/3/41.   [132] NMM, ADM. BP/1, 8 May 1780.

[133] Rodger, *Wooden World*, 193–7; Byrn, *Crime and Punishment in the Royal Navy*, 156–64; NMM, MID.1/159, G. Rose to Barham, 6 Sept. 1807, who provides an account of the men who ran from Haslar Hospital, 1 Jan. 1775 to Aug. 1807.

[134] P. K. Crimmin, 'John Jervis, Earl of St Vincent 1735–1823' in *Precursors of Nelson*, ed. Le Fevre and Harding, 325–50.

[135] *Tomlinson Papers*, 120–1.

Spithead mutineers pleaded for freedom within a limited boundary that they might 'somewise have grant and opportunity to taste the sweets of liberty on shore, when in any harbour, and when we have completed the duty of our ship'. So too at the Nore they wanted 'liberty (a certain number at a time, so as not to injure the ship's duty) to go and see their friends and families; a convenient time to be allowed each man'.[136] Denial of shore leave consequently tended to aggravate relations, especially when others were seen to be treated differently.[137]

The politics of discontent

In December 1805 the crew of the *Royal Sovereign* reached England after the battle of Trafalgar and petitioned the Admiralty with a description of their usage since the action. On 21 October they had been served no provisions or liquor of any kind; at Gibraltar only 6 men were allowed to go on shore each day contrary to custom and an allowance of 100 men per day in other ships; the postman was denied admittance to the ship and seamen were denied permission to send their clothes on shore for washing. They wished 'only to receive the same libertys and usage as they rest of His Magestys ships'.[138]

Seamen were not immune or indifferent to the ideas embodied in Thomas Paine's *Rights of Man*. The new thinking did not easily penetrate military society but the state had to respond to changing expectations. There was thus a long-term dimension to the management of seamen. There were minor mutinies and acts of defiance throughout the eighteenth century but the principal collective demands of the state's seamen emerged before and during the Spithead and Nore mutinies in 1797.

The orderly Spithead mutiny has been regarded as a matter of 'labour relations'. Under the threat of invasion from France, government responded to the mutineers with remarkable alacrity. In addition to higher rates of payment, they achieved the abolition of the purser's eighth – previously deducted from victualling rations – and the promise of payment of wounded men until they recovered or were discharged on a pension. The more spontaneous and disorderly Nore mutiny grasped after desires of a more abstract nature which probably inhabited the head of every seaman, and expressed discontent at inequalities

---

[136] J. Duggan, *The Great Mutiny* (London, 1966), 103–4, 201–2.
[137] Cockburn to Nelson, 24 July 1797, BL, Add. MSS. 34,906, fo. 205.
[138] *British Naval Documents*, 552–4.

consistent with criticisms of the British state on shore. Among a range of demands, it challenged the distribution of prize money, the severity of punishments under the Articles of War, and the trial of seamen and marines at courts martial by naval officers instead of by social equals.[139]

Although in 1797 government was assured by magistrates sent from London to Portsmouth and Sheerness that the mutineers were not in direct touch with radicals on shore,[140] there can be no doubt from the Nore demands that political ideas infused the British fleet and that seamen were motivated by them. In 1798 there was a spate of mutinies which coincided with the rebellion in Ireland. Of 719 mutineers seized between 1793 and 1801, 302 or 28 per cent possessed Irish birth places; in 1797 22 per cent of those arrested were Irish; and in 1798 75 per cent of those charged with mutiny were Irish. Although these mutinies collapsed and were punished, not all failed. In March 1800 the *Danae* was sailed to Le Conquet by mutineers of whom five leaders were probably American.[141]

These mutinies enveloped the crews of whole ships. But most mutinous acts were those of individuals or small groups of seamen. During the French Revolutionary War those charged with offences that challenged authority, from mutinous expressions to sedition, were mostly young, without rank or local responsibility, and had only short-term experience of the navy. C. Doorne discovered that 73 per cent were under the age of thirty; 77 per cent were below the rank of warrant or petty officer; 85 per cent had served less than $2\frac{1}{2}$ years, while 59 per cent had served less than 1 year and a half.[142]

The response of the state to these challenges was to strengthen the guard placed on crews. The marines on ships acted as sea soldiers but they also served to enforce regulations and prevent desertion.[143] Their number increased significantly between the Seven Years' War and the Napoleonic War. In 1755 there were only about 5,000, in 1756 about 7,500, and in 1757 8,000. At sea in 1756–7, this was a ratio of 1 marine to 7 seamen. By 1811–13 the ratio had fallen to 1 to 4. Moreover, when the marines on shore were included, the ratio fell to 1 marine to 3.6 seamen.[144]

---

[139] Duggan, *The Great Mutiny*, 202, 252.
[140] TNA, ADM. 1/4172, 12 May 1797, 25 June 1797; Duggan, *The Great Mutiny*, 161–6, 331–2.
[141] Doorne, 'Mutiny and sedition in the Home commands of the Royal Navy'.
[142] *Ibid.*   [143] See Byrn, *Crime and Punishment in the Royal Navy*, 29.
[144] For the 1755–7 figures, thanks are due to Britt Zerbe; the 1811–13 figures are from MLN.153/3/41.

Management and motivation

Disaffection in the navy was also countered by the expansion of the officer corps, which included men from the lower deck not only in the hierarchies of warrant and petty officers but among the commissioned officers. It was moderated too by the regulation of informal punishments that occurred on board ships, and by the care with which formal punishment was adjudicated. However, opportunity for most sea officers was limited and their frustrations contributed to the development of a damning critique of the navy's discipline regime. Employing contemporary utilitarian ideas, criticisms gave expression to the radical aspirations of seamen as well as the disaffected officer.

Officers and opportunity

During the eighteenth century there was cumulative growth in the number of officers (see table 6.9). Officers rose to the rank of captain by patronage; thereafter they rose by seniority. Entry was by connection and training either on shore at the Portsmouth Naval Academy, renamed the Royal Naval College in 1806, or on a greater scale at sea under the eye of patron officers.[145] The terms applied to these volunteers altered in 1794 but there was no reduction in the numbers of boys entered by captains and no necessity for those entrants to possess Admiralty approval until after 1815.[146] The lieutenant's exam ensured competence, after which officers were appointed to vacancies in ships as they occurred on account of death, illness, resignation or advancement. Favoured lieutenants were taken into flagships where they were given commands or other vacancies. Commands demanded commissions,[147] and indicated suitability for advancement to larger vessels to which officers were appointed, or confirmed in post, by Admiralty authority. Even then, the personal knowledge of a commander-in-chief could influence the selection of the officers employed on his station.[148]

---

[145] H. W. Dickinson, *Educating the Royal Navy: Eighteenth- and Nineteenth-Century Education for Officers* (Abingdon, 2007), 10, 39.

[146] Clowes, *The Royal Navy*, III, 19; M. Lewis, *England's Sea Officers: The Story of the Naval Profession* (London, 1948), 84–99.

[147] See also the commission of James Saumarez, 27 May 1780, printed in *British Naval Documents*, 536–7.

[148] From off Toulon in October 1796, Sir John Jervis listed to the Admiralty Secretary twenty-one post captains and three commanders whom, he made clear, 'I prize very highly and nothing can gratify me more than having them with me wherever I go': NMM, NEP/7, Jervis to Nepean, 4 Oct. 1796.

Table 6.9 *The growth in the size of the officer corps of the Royal navy,*
1740–1815[a]

	1740	1780	1793	1815
lieutenants	315	1,349	1,417	3,211
commanders	36	184	163	762
captains	171	412	446	824
admirals	10	57	55	219

[a] A. S. Turberville, ed., *Johnson's England: An Account of the Life and Manners of his Age* (2 vols, Oxford, 1933), I, 63; W. James, *Naval History of Great Britain*, I, 53: V, 225.

Captains who transferred from one ship to another in service took with them 'followers'. Many were mature petty officers appointed by the captain[149] for their skills and management of men.[150] Patronage and paternalistic care for their interests helped to win the support of the remainder of the crew.[151] Warrant officers, appointed by board warrant to a ship and part of its accounting system, could not so easily move on.[152] Nevertheless, with a ship-board hierarchy of numerous lower posts, a skilled, able man could advance to warrant rank and even pass the lieutenant's exam.

For the Seven Years' War, Rodger found that 9 per cent of the men passing to become lieutenants were not gentlemen by birth.[153] For the French Revolutionary and Napoleonic Wars, Michael Lewis found 10.6 per cent of the officers in his sample came from business, commercial or working-class backgrounds; 50 per cent had professional parents, of whom half, that is about 25 per cent of the total, were from the naval profession. Meanwhile, nearly 40 per cent were from the peerage, baronetage or landed gentry (see table 6.10).

However, the commissioning of men from business or working-class origins was a wartime phenomenon. After 1815, there was virtually no demand for their services. The proportion with naval parentage also

---

[149] Petty officers included the captain's coxswain, mates to warrant officers, captains of the tops and of the waisters, forecastlemen and afterguard.

[150] Morriss, *Cockburn and the British Navy in Transition*, 37.

[151] *Ibid.*, 43–6. See also NMM, ADM. BP/11, 20 May 1791; ADM. BP/31A, 15 June 1811.

[152] Warrant officers included the master, gunner, carpenter, sailmaker, caulker, purser, surgeon, chaplain, cook, boatswain and master-at-arms.

[153] Rodger, *Wooden World*, 266–70.

Table 6.10 *The social background of the commissioned officer corps, 1793–1849*[a]

	1793–1814	and	1814–49
sample size	1,800		834
	%		%
parents in peerage	7.3		10.7
baronetage	4.7		7.1
landed gentry	27.4		27.6
professions	50.0		54.2
business and commerce	3.9		0.4
working-class	6.7		0.0

[a] Lewis, *The Navy in Transition*, 22.

declined to a third of the professional group. Meanwhile, the proportion of officers originating in the peerage and baronetage increased.[154]

Even during wartime, cultural background placed a limit on the distance lower-deck men could rise in the naval hierarchy. Calculations for 1793–1815 indicate a lower-deck man stood a 1 in 2,500 chance of obtaining an officer's commission and that seven out of ten of those who were commissioned remained a lieutenant. Those who did gain a commission had to distinguish themselves, an opportunity that sometimes only came late in life. James Bowen was forty-three on being made a commander, from being master of the *Queen Charlotte* at the battle of the First of June 1794; Phillip Lovell and William Millett were forty-one and thirty-eight on being made lieutenants.[155]

For men of working-class or business origins, reward for industry was more easily achieved through the ranks of the warrant officers. During the Seven Years' War, the demand for skilled seamen to act as ships' masters induced many experienced seamen to enter the navy from the merchant service. Perhaps the best-known example was James Cook who entered the navy in 1755, having been a master in the east coast coal trade. Rodger claims that, between 1745 and 1757, 21 per cent of those passing for lieutenant had served in merchant ships before or after joining the navy.[156]

[154] M. Lewis, *A Social History of the Navy 1793–1815* (London, 1960), 44–5; and *The Navy in Transition 1814–1864: A Social History* (London, 1965), 19–26.

[155] Lewis, *A Social History of the Navy 1793–1815*, 48.

[156] Rodger, *Wooden World*, 266–70.

After the Seven Years' War, many masters left the navy to join merchant ships. In 1773 the Admiralty saw fit to increase the number retained during peace on half-pay and in 1779 to increase its amount. However, during the American War of Independence there was a 'great want of masters', many 'quitting that line to be made lieutenants'. So many masters were passing for lieutenant that in 1779 the Navy Board had to stop examining them. The examination of masters resumed in 1788, but in 1793 there were only 297 remaining and, in lieu, the Admiralty had to appoint 'extra' lieutenants to perform the master's duties on some ships.[157] The shortage continued through the French Revolutionary War and extended to other warrant officers. As a result, an order in council of January 1803 advised all of them to resist looking forward to promotion beyond their grade.[158]

Despite patronage and meritocracy, there were therefore limits to the opportunity offered by the navy. Did this dull ambition and make some careless of their private opinions? Lord St Vincent, Commander-in-Chief of the Channel fleet in 1800, condemned 'the licentious conversation of the wardroom officers' which 'occasioned infinite mischief' by soon diffusing through a ship. In February 1801 he was still not satisfied, observing that the officers in general 'with very few exceptions are so licentious, malingering and abominable that their conduct must bring about another mutiny'.[159] His friend, Captain Troubridge, had occasion to complain too of 'the improper, growling conversations held in the wardrooms' that 'caused the seamen to murmur'.[160]

Punishment and discipline

St Vincent was the product of an earlier age and he complained too of 'the much lamented change in the character of seamen, which I find far worse in this [Channel] squadron than I did in the Mediterranean. Here they are a drunken, slovenly, lazy and filthy race and although the discipline and subordination is somewhat improved I despair of seeing it perfectly restored during the present war.'[161] Having risen on his merits with little money or influence,[162] he had gained a reputation as a

---

[157] NMM, ADM. BP/4, 18 Feb. 1783; ADM. BP/8, 30 June, 11 Nov. 1788; ADM. BP/10, 19 Apr. 1790.

[158] The Parliamentary Debates... from 1803, ed. T. C. Hansard, 2nd series, vol. X, 173–4, 16 Feb. 1824. See also the memorial of masters in the navy reviewing the terms of their half-pay and widows' pension, NMM, ADM. BP/36A, 13 Feb. 1816.

[159] Letters of 10 June 1800 and 8 February, BL, Add. MSS. 31,167.

[160] Troubridge to Nelson, 20 Sept. 1801, cited in Knight, The Pursuit of Victory, 416.

[161] Letter of 7 Sept. 1800, BL, Add. MSS. 31,163.

[162] R. F. Mackay, 'Lord St Vincent's early years (1735–55)', MM 76(1990), 51–65.

stern disciplinarian, whose tough stance was approved by officers like Nelson.[163]

Punishment appeared an appropriate response to rebellion, dissidence and unsatisfactory shipboard conduct. Yet excesses raised retaliation. The *Hermione* mutiny in 1797 revealed how bloody could be the revenge of a violent crew on a hated captain.[164] Most captains, by good judgement, managed the difficult balancing act between lenience and ruthlessness. Yet the degree of punishment tended to take an upward spiral.

Summary punishment by flogging remained the most common informal sanction. Distant from other ships, where courts martial could not be convened, such punishment was expedient. Moreover, summary sentences were invariably more humane than sentences for the same offence imposed by court martial. They were thus accepted by seamen who anyway preferred instant punishment to the ordeal of awaiting trial. Captains accordingly summarily punished many of the offences listed in the Articles of War, which included desertion, disobedience, making a disturbance, uncleanness, immorality, insolence or contempt, mutiny or sedition, neglect of duty, theft and violence. Many of these were committed under the influence of alcohol, but this only added to the weight of punishment.[165]

Summary punishment was inflicted on 7 per cent of men in the 1740s, and on 9 per cent in the Leeward Island station between 1784 and 1812. Admiralty regulations limited captains to one dozen lashes but many captains awarded two, three or more dozen lashes for repeated offences. On the Leeward Islands station before 1806, 33 per cent of all floggings were more than a dozen lashings; after 1806, 53 per cent were more than a dozen.[166]

During the Napoleonic War, the Admiralty realised this increase in flogging represented an increase in shipboard violence. Seamen were sometimes expected to participate in hitting offenders like thieves who 'ran the gauntlet'. Seamen were themselves hit with a stick or rope's end by a bosun's mate on the orders of an officer, in what was termed 'starting'. Informal and unrecorded, starting could inflict painful and temporarily crippling bruising. In 1806 the Admiralty prohibited the practice of running the gauntlet and in 1809 prohibited the use of starting. From 1811 captains had to make returns of summary punishments by flogging.

---

[163] Knight, *The Pursuit of Victory*, 239–40.
[164] J. D. Spinney, 'The Hermione Mutiny', *MM* 41(1955), 123.
[165] Rodger, *Wooden World*, 220; Byrn, *Crime and Punishment in the Royal Navy*, 120–9.
[166] P. Kemp, *The British Seaman: A Social History of the Lower Deck* (London, 1970), 90; Byrn, *Crime and Punishment in the Royal Navy*, 68–88, 108.

Complaints from seamen of excessive punishment by officers were generally investigated. The demand for the removal of tyrannical officers during the Spithead mutiny was in many cases respected by the Admiralty. During the Napoleonic War captains who punished informally with unwarranted severity were likely to be subject to court martial.[167] The most notorious was Captain Robert Corbett who became a captain in 1806, suffered a mutiny against 'inhuman cruelty' in 1808, was censured by the Admiralty after court martial in 1809, suffered a near-mutiny in 1810, and died from wounds received in action in 1811. Rumour maintained his wounds had been inflicted by his own crew, who had also failed to load and aim their guns properly. After the action some surviving seamen swam to another British warship.[168]

The necessity for just and fair punishment gained expression in the carefully conducted and penetrating cross-examinations of courts martial. Officers, seamen and marines were all subject to the Articles of War.[169] While captains awarding summary punishments were comparable to local magistrates on shore, courts martial were comparable to county assizes that awarded punishments for 'capital' offences. Officers sat as both judge and jury at courts martial, but the witnesses could be cross-examined by any of the officers who formed the court as well as by the defendant who was able to receive advice. Cross-examinations were often exhaustive; proceedings were conducted in a spirit of respect for the rule of law; and minutes were monitored by the legal officers of the Admiralty.[170]

The system of courts martial remained the backbone of British naval discipline. Yet in 1813 it was attacked by Thomas Hodgskin in *An essay on Naval Discipline*. Hodgskin, a naval lieutenant from 1806 until 1812, has been regarded as a malcontent. But he had long experience of the navy and had educated himself in contemporary philosophical thinking. The son of the Deptford dockyard storekeeper, with a religious education, he had entered the navy in 1800 at the age of thirteen, and by 1812 had served in seven ships. Between 1809 and 1811 he served under Captain William Ferris who in February 1810 was charged at court martial of treating his crew with severity. The crew appealed to the

---

[167] Byrn, *Crime and Punishment in the Royal Navy*, 19, 116; *Shipboard Life and Organisation, 1731–1815*, ed. B. Lavery (NRS, 1998), 374–5, 403–8.

[168] *DNB*: James, *Naval History of Great Britain*, v, 181–3; *Shipboard Life and Organisation*, 373–4, 401–8.

[169] N. A. M. Rodger, ed., *Articles of War: The Statutes which Governed our Fighting Navies 1661, 1749 and 1886* (Havant, 1982).

[170] Byrn, *Crime and Punishment in the Royal Navy*, 121–2, 185. For these examinations, see *Naval Courts Martial, 1793–1815*, ed. J. Byrn (NRS, 2009), 1–146.

Commander-in-Chief on the Baltic station partly on account of the death of a boy, aged sixteen. Hodgskin was also tried because, as first lieutenant, he had placed the boy in the care of the sergeant of marines to have him cleaned at the time he was hurt and died.[171] Acquitted in 1810, Hodgskin was tried again in 1812 for neglect of duty when an impressed man deserted from his care.[172] The court martial went badly for him, possibly because, shortly before, he protested at another lieutenant being brought into the ship over his head and being overlooked in succession to the senior vacancy.[173]

Although the product of a disappointed men, Hodgskin's views on naval discipline are instructive as they indicate the new ideological environment in which the British armed forces had to operate. Hodgskin had read Locke, Paley, Malthus and Bentham as well as naval writers like Patten. A disciple of Jeremy Bentham, he thought in 'utilitarian' terms, applying the test of the greatest happiness of the greatest number to the Articles of War.

He believed the Articles created a legal environment which gave the seaman a bad character which was then used to justify application of the penal code. Repeated summary punishments simply became the norm, were no disgrace and more a source of sympathy than remorse. Sentences in courts martial depended on the disposition of the officers to the opinions and reasoning of the accused. Yet officers and men were conditioned by their early entry, their imitation of superiors, and the constant company of those 'accustomed to set religion at defiance'. Officers neglected their education and simply fell in with existing practice. Meanwhile, crowded together, seamen were stripped of individuality and of 'all principles of morality and religion'. He claimed seamen were regarded as little better than animals, and that their degree of discontent was taken as a measure of their subjection.[174]

Hodgskin claimed the Articles of War were 'a dreadful system of laws'. He wanted complete reform. Like Nore mutineers fifteen years earlier, he wanted trial by equals. He believed small punishments were more efficacious in preventing crime than severe ones and that every crime should have an appropriate punishment attached to it according to a

---

[171] TNA, ADM. 1/5402, 5–7 Feb. 1810, 7 Feb. 1810.
[172] TNA, ADM. 1/2934, lieutenants' letters H239, memorial of Thomas Hodgskin, 1 Aug. 1812.
[173] TNA, ADM. 1/5425, 25 Apr. 1812.
[174] D. Stack, *Nature and Artifice: The Life and Thoughts of Thomas Hodgskin, 1787–1869* (London, 1998), 40–5; T. Hodgskin, *An essay on Naval Discipline shewing part of its evil effects on the minds of officers, on the minds of men, and on the community, with an amended system by which pressing may be immediately abolished* (London, 1812), 30, 58–9, 69, 116–17, 133.

schedule: 'There should be nothing left to be punished at the caprice or discretion of members of courts martial, according to any customs they might ever have seen practiced.' It was the legislature which should prescribe the penalties attached to naval crimes, not the officers, who maintained what Hodgskin termed a 'coercive system'. Without that, in his opinion, impressment would have become unnecessary.[175]

Hodgskin had an axe to grind, but his criticisms and proposals should not be dismissed as simply those of a malcontent and radical. His views were drawn from the ideas of the time. His writings suggest he knew what the people of the West Country thought of the navy. He knew the language of Americans and appreciated their frequent use of words expressing their constitutional liberty. He had observed the manner in which American ships were treated by British warships. He had a perspective that placed the laws governing the navy and the treatment of seamen in their context. As Hodgskin revealed, the ideological environment of the navy was changing and raising new standards by which its procedures were judged.

This desire for codification of naval law, for new weighting in the award of punishments and equality of class in the attitudes of those who judged offenders echoed utilitarianism on shore. The navy could not avoid changing political opinion, which affected the attitudes of seamen as well as officers. However, with the persistence of a high rate of desertion and an increase in the rate of discharges, recruitment too had to be maintained at a high level. This demanded the impressment of about 40 per cent of recruits in 1810–12. In view of the relatively low rates of naval pay compared to the merchant service, poverty must have aided in the recruitment of volunteers. In the circumstances, the state must be credited with filling the navy's ships with seamen, which was a bureaucratic achievement aided by the maritime economy. It was nevertheless an achievement that ran counter to developments in philosophical thought and opinion, to which the Admiralty responded only by the prohibition of a few customary punishments.

[175] Hodgskin, *An essay on Naval Discipline*, 95–6, 120–1, 127, 135–6.

7   Foodstuffs and victualling

However they were recruited, the state's relationship with the members of its armed forces depended on its ability to feed them. The supply of food was fundamental. Its supply was also a logistical challenge which reflected well on the capabilities of the responsible bureaucracy. Most of the state's food originally came from within Britain. However, as domestic population grew, resources within England became inadequate and in the mid eighteenth century imports from Ireland, and later from the continent, became necessary. The Treasury organised the supply of provisions to the army while the Victualling Board supplied those to the navy, but in 1793 the latter also became responsible for supplying the army overseas. Foodstuffs were purchased centrally by contract or on commission, locally by agents at the yards, or by officials of the forces overseas. All stocks, however, were monitored meticulously in London, from where orders for new deliveries were made. At the point of distribution, food became the responsibility of the naval purser or army Commissariat. The role of the latter is examined later. That of the purser is examined here as a vital point of contact between the state and its employees.

The sources of foodstuffs

The agricultural revolution in England facilitated not just population, industrial and urban growth but the supply of Britain's armed forces in their operations overseas. E. A. Wrigley states that in England between the sixteenth and nineteenth centuries gross cereal yields per acre (the quantities harvested) doubled, while net yields (the amount available for consumption) rose by 135 per cent. He suggests English agricultural productivity was double that of France.[1] Nevertheless, by the mid eighteenth century domestic supplies of foodstuffs became insufficient to meet all demands and the supply of Britain's armed forces owed much to the

---

[1] Wrigley, 'Society and economy in the eighteenth century'.

geographical enlargement of the sources of supply available to the state's contractors.

Wet provisions

Until the mid eighteenth century London was the central market of agricultural produce in Britain and the navy was able to tap that supply. London itself, with just over half a million people around 1700 and nearly a million by 1801, drew supplies by drove road, river, canal and coastal shipping from north, east and west. Regions specialised and were tapped by middlemen and wholesalers. Cattle came from as far afield as Scotland and Wales; much wheat and barley came from East Anglia; butter from York, Hull, Scarborough, Stockton and Newcastle; cheese from Cheshire, Lancashire, Warwickshire, Suffolk and Somerset. Cattle as well as hogs were fattened in counties local to London before being herded in for sale.[2]

The London market was sufficiently large to accommodate naval purchases that fed 48,000 seamen at the height of the Spanish Succession War in 1706, and 51,000 at the end of the War of Austrian Succession in 1747. The Victualling Board set a minimum weight for oxen of 6 hundredweight; most on average weighed 7 hundredweight. Each ox was estimated to yield 160 4-pound pieces, which permitted the entire navy to be fed on 12,000 head of cattle a year. Likewise hogs: the average animal purchased for the navy supplied about 100 pounds of meat; 40,000 hogs were enough to meet the navy's annual needs. This may be compared to the scale of sales at London's Smithfield market: over 73,000 head of cattle in 1725 and over 187,000 hogs in 1731.

But the navy purchased much of its supplies by contract, so they came directly from their sources of supply. Contracts for beef required the delivery of cattle for slaughter between late September and early May. Quotas were delivered progressively, by certain dates. The dealers who won contracts for the supply of Portsmouth and Plymouth generally supplied those yards from counties in the south and west. Their bids were competitive because they saved on transport or droving costs. In the 1740s prices for fresh beef at Portsmouth tended to exceed those for London, but were lower for delivery at Plymouth. Likewise Kent could supply Dover cheaper than it did London.[3]

During the Seven Years' War, this pattern of supply began to change. English national resources gave out. Poor harvests in England in 1755–6

---

[2] F. J. Fisher, 'The development of the London food market, 1540–1640', *EconHR* 5(1935), 46–64, repr. in *Essays in Economic History*, ed. E. M. Carus-Wilson (London, 1954), 135–51.

[3] Baugh, *British Naval Administration in the Age of Walpole*, 407–10.

and 1761–2 reduced the availability of forage grain and livestock. Food riots occurred throughout the country in 1757. Foot and mouth disease also diminished cattle stocks until the disease waned in 1759. Thus, instead of livestock production responding to the demands of the Seven Years' War, prices rose. Indeed, the price situation of the previous war was completely reversed. By 1758, beef prices at Dover and Portsmouth were as high as in London, while the Plymouth price was higher than at any other point of delivery. Pork remained cheaper in London than at Portsmouth or Plymouth but was more expensive for supply at Dover.[4]

Aggravating this situation was a marked growth in the size of the armed forces. The number of men required at sea exceeded previous musters from the outset. By 1757 over 63,000 were borne on ships' books. By 1762 nearly 85,000 were borne, with nearly 82,000 being mustered.[5] There were also armies serving overseas and a number of expeditions. By later standards, the number of soldiers outside Britain that had to be provisioned was small; in 1759 there were 6,000 British and 5,000 American soldiers available for the assault on the French in North America. Moreover, for the campaigns in Canada and the Caribbean, supplies of food were obtainable in North America. Nevertheless, shipments to the allied army in Germany added strain to food supplies in Britain.[6] To meet demand, supplies were increased from Ireland. Hitherto, the import into England of Irish animal products had been prohibited by the Cattle Acts of 1666–7. During the 1740s the laws had occasionally been circumvented but in March 1758, when prices of pork and beef began to soar, the laws were suspended for six months, a suspension repeatedly renewed until 1762 when the prohibition was lifted altogether.[7]

From 1758, therefore, Irish pork, beef and butter supplemented English sources, the prices in Ireland tending to moderate those in England. It was a situation of which the Treasury and Victualling Board took full advantage, keeping down costs for the army and navy and assuring them of a reliable supply.[8] The expedient of importing from Ireland was maintained even during the following peace. In 1770, when prices of pork rose even higher than during the Seven Years' War, the Victualling Board proposed an import of 4,000 hogs at a price one-third cheaper than those already acquired in England.[9] During the second half of the

---

[4] Gradish, *The Manning of the British Navy during the Seven Years War*, 150.
[5] Lloyd, *British Seaman*, 262.   [6] Barnett, *Britain and Her Army*, 206.
[7] Acts 31 Geo. II c. 28 (1758); 2 Geo. III, c. 6 (1761).
[8] Gradish, *The Manning of the British Navy during the Seven Years War*, 149–52; for the board's purchases of pork in 1758–9, see fn. 1, p. 151.
[9] NMM, ADM. DP/102, 16 Nov. 1770.

century, Ireland became Britain's stock yard. The value of imports from Ireland into England, in large part 'wet' animal products, quadrupled. Between 1760 and 1800 exports to Britain of beef rose by a factor of 5, pork by a factor of 7.5, and butter by 6. After 1778, Ireland also increased its exports of grain, flour and oatmeal.[10] But, for the armed forces, these never rivalled in quantity the 'dry' provisions – flour, peas and oatmeal – obtained in England.

During the American War of Independence, the inability of the Treasury Commissariat to purchase sufficient food supplies for the army in North America added the burden of feeding a larger army overseas to the demands of a larger navy. In 1778 the Treasury had to supply rations to 40,000 soldiers and to make contracts for 7,280,000 pounds of pork, 1,820,000 pounds of beef and 780,000 pounds of butter. By 1781, when it had to feed 86,000 soldiers overseas, these quantities more than doubled.[11] Meanwhile the Victualling Board in 1781 had to feed over 95,000 seamen. Hence the boards together were obliged to secure and despatch food supplies for at least 181,000 men.

With Ireland now the main source of the army's pork, beef and butter, and a principal source of these for the navy, supply did not fail. Beef and pork were purchased 'by the means of merchants of respectability in the Irish provision trade'. They were reimbursed their actual *bona fide* costs and charges, and received a commission for their trouble of 3 shillings for each half puncheon or tierce, and 2 shillings for each barrel of beef and pork.[12] Irish provisions for the army were mainly delivered at Cork. Here export merchants placed orders with wholesale butter buyers and with butchers who travelled the hinterland to contract for the herds of graziers. The butchers were key middlemen who employed salters and coopers and whose business connections extended to France, Spain and North America. In 1793 an Irish Victualling Office opened at Dublin to secure Irish provisions for the army and navy.[13]

Dry provisions

During the French Revolutionary War, it was the supply of 'dry' provisions – flour, peas, oatmeal – from England that ran short. Shortages

---

[10] L. M. Cullen, *Anglo-Irish Trade 1660–1800* (Manchester, 1968), 45–9, 67–73.

[11] In 1778 the Treasury also had to supply 1,040,000 pounds of oatmeal, 97,500 bushels of pease and 14,560,000 pounds of flour: Baker, *Government and Contractors*, 22–7.

[12] 32nd Report of the Select Committee on Finance, 1798, *Reports of Committees of the House of Commons 1797–1803*, 1st series, XIII, 3.

[13] D. Dickson, *Old World Colony: Cork and South Munster 1630–1830* (Cork, 2005), 137–46, 366–9.

were localised, especially in the south and west, and were exacerbated by contracts to feed the navy. Certainly some contemporaries associated food shortages with supply to the navy. After the poor harvest of 1794, a gentleman living in the south-west explained the high food prices by

the long continuance of the outward-bound fleets on the coasts near Plymouth, from which market and that of Dock 20,000 inhabitants more than usual were fed for several months; a number nearly equal to all the stationary inhabitants of both those places. During their continuance here, after the vicinity had been much drained of oxen, sheep, corn and potatoes, supplies were drawn from the several counties of Somerset, Dorset, Gloucester and Worcester.[14]

Food shortages were to recur in 1800 and 1801, by which time the number of men borne on the books of ships amounted to 126,000. Then the prices of foodstuffs received at Deptford reflected the shortages of 'dry' provisions of England rather than the 'wet' obtained in Ireland. In 1800 'wet' prices were on average 55% higher than they were in 1793: butter up by 28%, salt pork by 50%, and salt beef by 86%. Cheese, although obtained from various places in the British Isles, rose in price by 58%. However, the prices of 'dry' commodities rose on average 114%: flour by 161%, peas by 92%, and oatmeal by 89%. The poor harvests between 1793 and 1800 seriously affected wheat and biscuit meal prices which rose 120% and 149%, respectively.[15]

By 1801, after the cost of beef, biscuit and biscuit meal were the second most expensive foodstuffs consumed by the navy. Wheat and flour, from which biscuit was made, were the third most expensive.[16] It was thus natural that the prices of these commodities should be of particular concern during the Napoleonic War. A scarcity was much dreaded. In case of a failure of supply, in 1809 the Victualling Board tried flours made of other materials. One combined potatoes with molasses and biscuit meal; another supplemented wheat meal with barley and molasses. The board believed either flour would be acceptable in the event of a wheat failure.[17]

In the event, failure was averted by the import of wheat from the continent of Europe despite Napoleon's continental system. In April–May 1810, 200,000 quarters of foreign corn were sold in a 'few weeks' on the London market. Imports that year were nearly six times their size in 1808–9. This was an unprecedented quantity and put down to the temporary lifting of Napoleon's prohibition against Britain, a window of opportunity that would not remain open for long. Indeed, there was

[14] PRO, HO.42/34, item 99, W. Elford to the Duke of Portland, 6 Apr. 1795, quoted in Emsley, *British Society and the French Wars 1793–1815*, 42.
[15] NMM, ADM. DP/23, 9 Aug. 1803.   [16] NMM, ADM. DP/26, 31 Dec. 1806.
[17] NMM, ADM. DP/29, 4 Sept. 1809.

also a scarcity of grain on the continent and later the countries supplying wheat sought to purchase grain in Britain. The demand ensured that prices on the London market in 1810 reached 106 shillings a quarter, twice the price paid in 1794, and 150 per cent that of 1802.[18]

Yet there was no further influx of imports. Continental exporters had to pay highly for licences and freight, especially after Napoleon got to know of the British scarcity, and British importers were deterred by the risks involved in shipments from the continent. With another poor harvest in prospect, the navy's domestic suppliers looked unlikely to meet their commitments and the Admiralty was advised the state itself might have to purchase grain on the continent in the winter of 1810–11.[19]

Further poor harvests drove up the domestic price of wheat to 126 shillings a quarter in 1812.[20] To that shortage was added a scarcity of vegetables. In November 1812 the Agent for the victualling yard at Chatham reported prices in Kent enhanced by the 'enormous expenditure of vegetables for the [merchant] shipping'. The Agent foresaw even higher prices that coming winter and recommended the allowance of vegetables to seamen on naval vessels be cut by one-third on sea-going ships, and by one half on stationary vessels. The Victualling Board referred the matter to the commanders-in-chief on the home station. They canvassed the opinions of naval captains at the Nore and in the Medway who differed, some thinking the vegetable allowance insufficient, others that it could be cut by to two-thirds.[21]

Shortages of 'dry provisions' thus affected the military services. But in Britain they were less affected than the civilian population because the state generally paid the high prices, while overseas they were fed from elsewhere. Sicily, for example, supplied the Mediterranean fleet and the army in the Spanish Peninsula, which also benefited from Iberian imports from the United States. Yet the domestic shortages did have the effect of stimulating imports from the continent and, incidentally, from Ireland which, after the turn of the century, became a major exporter of grain, flour and oatmeal as well as of pork, beef and butter.

Imports

By contrast with the prices of domestic foodstuffs, those of imported commodities rose very little. Sugar, molasses, rice and raisins remained

[18] Mitchell and Deane, *Abstract of British Historical Statistics*, 95, 488.
[19] NMM, ADM. BP/30A, 18 May 1810.
[20] Emsley, *British Society and the French Wars 1793–1815*, 153.
[21] NMM, ADM. DP/32B, 27 Nov. 1812.

stable in price, the cost of the first two tending even to decline. Despite distant origins, alcohol remained plentiful and steady in price.

Rum and wine were important for the navy on account of the inability of ships to stow more than about six weeks' supply of beer. Rum was normally purchased for the navy in quantities ranging from 5,000 to 92,000 gallons. But at the beginning of the American War of Independence, a four-month stock of rum for 10,000 men was established at Portsmouth and Plymouth and 100,000 gallons of West India rum were purchased in London from a variety of dealers. The price remained relatively stable: it rose early in 1776 but settled down in the summer to little more than its pre-war price. That year the Victualling Board went on to purchase a total of 691,500 gallons, with another 470,000 gallons in 1777.[22]

Wine served as a supplement to rum and was purchased in Lisbon and Oporto. The price increased in 1769 with mobilisation purchases but thereafter remained steady through the American War, only falling a few pence in 1784.[23] In 1781 the navy purchased nearly 16,000 gallons in Lisbon and 493,000 gallons in Oporto. Occasionally French red wine was also purchased. In 1781 the navy was offered up to 100 tuns of red wine then at Guernsey, and the Victualling Board favoured the purchase as the price was 10 shillings a tun cheaper than wine at Oporto.[24] For the same reason, later in 1781, the board accepted an offer of 800 puncheons of 100,000 gallons of French brandy, to be imported in neutral ships from Guernsey or France. The board had sampled it, considered it 'very good', and the price more reasonable than that of Spanish brandy.[25]

The Victualling Board and its yards

The naval Victualling Board was the principal customer of merchants in the provision trade, even more so after 1793 when it received responsibility for supplying the army overseas.[26] The army's provisions were still paid for by the Treasury, which before that had arranged supplies by contract, but their management by the naval board aimed at economy and efficiency by the application of greater expertise. Overseas, the army's supplies were supplemented by the local purchases of the Commissariat, while those of the navy were supplemented with local purchases by fleet

---

[22] NMM, ADM. DP/110, 12 May 1778.   [23] NMM, ADM. DP/7, 11 May 1787.
[24] NMM, ADM. DP/1, 5 Mar. 1781.   [25] NMM, ADM. DP/1, 7 Sept. 1781.
[26] A general account of the work of the board may be found in J. MacDonald, *Feeding Nelson's Navy: The True Story of Food at Sea in the Georgian Era* (London, 2004), 45–60.

agents and ships' pursers. The organisation of supply had developed on a pragmatic basis and by the end of the eighteenth century was firmly founded on work in the board's yards, and on systems of purchase, quantification and distribution.

The Board's functions and reputation

The Victualling Board was created by the appointment of commissioners in 1683. It met in offices on Tower Hill and, from early in the War of Spanish Succession, was composed of seven commissioners, three of whom formed a quorum that could take decisions. Initially, when there was much business, the board met daily, including Saturdays and Sundays. At this time there was no fixed order to business. However, by 1704, specific days were allotted to particular concerns[27] and, by the time of the American War of Independence, the commissioners met as a full board only about three times a week.[28] To streamline business, from 1800 the commissioners were divided into two specialist committees: one for general business, the other for cash and store accounts.[29] It was a necessity, for weekly board business had increased from 52 items in 1702, to 130 in 1793, and would peak at 228 in 1809.[30] But the number of commissioners remained at seven,[31] and the business of the full Board, measured by its letters, remained at little more than twice the amount it conducted over a century earlier.[32]

The Victualling Board performed four main functions. It calculated costs: of feeding the number of men for the coming year (see table 7.1), to which was added the payment of staff and the amount of interest on victualling bills. It made contracts: for deliveries of livestock and raw foodstuffs for processing, for their packing and preservation, and for their delivery to ships and depots. It supervised its yards and depots, commissioners often visiting to investigate proceedings, victuals and alleged abuses. It checked the accounts of agents, storekeepers and pursers,

---

[27] P. K. Watson, 'The Commission for victualling', 70.

[28] 8th Report of the Commissioners on fees, 36.

[29] The Commissioners on fees recommended a committee for correspondence and cash, and a committee for stores. See *ibid.*, 23.

[30] TNA, ADM. 111/1, 126, 191, items for 30 Nov. – 5 Dec. 1702, 1–7 Feb. 1793, 1–7 Apr. 1809.

[31] NMM, ADM. DP/8, n.d. [1788]; ADM. DP/29, 26 Apr. 1809; ADM. DP/34A, 19 Feb. 1814.

[32] In November–December 1683 letterbooks record that the board sent about fourteen letters each week; by January–February 1812 they still only record about thirty-two a week: TNA, ADM. 110/1, 64.

Table 7.1 *Estimate for victualling 110,000 men in 1798[a]*

Item	Quantity	Rate	Amount
Bread	27,500 cwt	@ £1 0s 0d per cwt	£27,500 0s 0d
Spirits	192,500 galls	@ £0 4s 6d per gall	£43,312 10s 0d
Beef	220,000 pces[b]	@ £8 0s 0d per 38 pces	£46,315 15 9s ¼
Pork	220,000 pces[c]	@ £9 0s 0d per 80 pces	£24,750 0s 0d
Pease	13,750 bshls	@ £2 0s 0d per quarter	£3,437 10s 0d
Oatmeal	20,625 bshls	@ £15 0s 0d per ton	£5,800 15 7s ½d
Butter	165,000 lbs	@ £3 15s 0d per ton	£5,524 11 0s ¾d
Cheese	330,000 lbs	@ £2 16s 0d per cwt	£8,250 0s 0d
Vinegar	27,500 galls	@ £12 12s 0d per ton	£1,443 15s 0d
Necessary money and contingencies		@ £0 0s 1d per man/day	£12,833 6s 8d
Casks, hoops and bags		@ £0 0s 1d per man/day	£12,833 6s 8d
			£192,001 10s 9 ½d
One ninth[d] on those articles which pursers are allowed for waste, cost at the credit prices			£5,550 18s 6d
One ninth on the same articles calculated at the present prices, to cover losses, condemnations and waste in the several stores			£11,872 0s 1d
Expense of extra articles furnished by this Office such as sugar for lemon juice, wine for the sick, vegetables etc., estimated @ 6d per man/month			£2,750 0s 0d
Amount for 13 lunar months			£212,174 9s 4 ½d
			£2,758,268 1s 10 ½d

[a] NMM, ADM. DP/17, 19 Oct. 1797.
[b] Pieces of 8 lbs.
[c] Pieces of 4 lbs.
[d] See p. 318 for the transition from the eighth to the ninth allowance to pursers in May 1797.

corresponding with sea officers regarding issues, and rendered its own accounts to the Navy Board.[33]

Until the end of the eighteenth century, the Victualling Board laboured under a reputation for inefficiency. This had arisen during the War of Spanish Succession and gave rise to the division of business to provide for its complete supervision. Business practice was also tightened, the commissioners being directed 'to sign all certificates and other papers at your office, and not at taverns and coffee houses as hath been heretofore practiced'.[34] The responsibilities established in 1710 were retained largely unchanged.[35] Board commissioners had either accountant responsibilities for cash or for stores or supervisory responsibilities at the main victualling yard for the operations of the hoytaker,[36] the brewhouse, the cutting house, the bakehouse or the cooperage.

Board commissioners were appointed by Letters Patent under the Great Seal, but their choice was subject to political patronage. During the War of Spanish Succession less than half the eighteen commissioners were appointed for their experience or knowledge of the navy.[37] By mid-century inefficient appointments were opposed.[38] In 1751 Anson wrote to the Duke of Newcastle of a nominee for a position as a commissioner that his patron 'might as well have asked for him to be made a captain of a man of war... This gives me an opportunity of observing to your Grace that, instead of adding to the useless people that are allowed in that office, (if we should have a war with France) more people of business must be brought into it'.[39]

Finding suitable people, however, was a challenge. Former contractors, or those closely connected with dealers, could corrupt proceedings.[40] But business experience was an asset. Jonas Hanway, the most famous of

[33] Watson, 'The Commission for victualling', 71–2; Ehrman, *Navy in the War of William III*, 159; NMM, ADM. DP/1–3.

[34] PRO, ADM. 7/648, quoted in Baugh, *British Naval Administration in the Age of Walpole*, 54–7.

[35] *Naval Administration, 1715–1750*, ed. Baugh, 401.

[36] Victualling Office employee who hired shipping in the River Thames.

[37] Five had served either in the Victualling department or in other branches of government such as the dockyards, Greenwich Hospital, Post Office and Trinity House; three had been in the navy, two later going on to higher-status seats at the Navy Board. However, another ten were Members of Parliament appointed to the board for political purposes; two were from families with considerable electoral influence: Watson, 'The Commission for victualling', 297–314.

[38] Rodger, *Wooden World*, 333.

[39] Baugh, *British Naval Administration in the Age of Walpole*, 61, quoting S. W. C. Pack, *Admiral Lord Anson* (1960), 185.

[40] After the appointment of Sir John Houblon, Governor of the Bank of England and City merchant, business men who had formerly acted as contractors were no longer appointed: Watson, 'The Commission for victualling', 297–314.

the victualling commissioners, was appointed in 1759. For the previous thirty years he had worked in the English 'factories' at Lisbon and St Petersburg and recommended himself to the Admiralty from 1756 by his foundation and management of the Marine Society – a charity to raise, clothe and equip men and boy volunteers for the navy.[41] He was made a commissioner with superintendence of the bakehouse and mills and was to serve mainly in that capacity until 1783.

Based in London, the commissioners' superintendence of yard functions was criticised in 1788 as 'rather nominal than real'. It also required 'practical skill and knowledge . . . which few, if any, gentlemen in their habits of life can be expected to possess'. There was a series of abuses during the War of American Independence which, in 1781 and 1782, gave rise to two complete post reshuffles, but after that the commissioners again settled into their respective roles. The dual service of the Accountant for Stores, then from 1784 the Accountant for Cash, as board chairman also prevented the 'strict or proper attention' to their respective departments.[42]

Nevertheless, by the end of the century the Victualling department had shaken off its reputation for inefficiency. In 1788 the Commissioners on fees noticed that the department had 'for some time past' been the focus of public attention, but acknowledged 'the zeal and integrity' which had motivated its 'many able servants of the Crown', and observed 'the ability, practical knowledge and character' of those in post.[43] And indeed, by the 1790s, possessed of unrivalled knowledge of food supply, the board did not fail to direct senior naval officers how best to provide for their ships.[44] It was this authority which the Treasury harnessed in 1793 when the department was given responsibility for supplying the army overseas.

The victualling yards

The victualling yards were the backbone of the state's food supply system. In 1808 they processed and despatched about 68 per cent of the value of food supplied to the navy and 63 per cent of that supplied to the army

---

[41] Hanway also came from a family of naval officers and administrators: his father had worked in the Navy Office from 1691 before becoming agent victualler at Portsmouth in 1711; William Hanway was a clerk in the Navy Office between 1731 and 1757; Thomas Hanway was a naval captain who became the yard commissioner at Chatham dockyard in 1761 and a controller of victualling accounts at the Navy Office in 1771. See J. S. Taylor, 'Jonas Hanway: Christian mercantilist', and A. W. H. Pearsall, 'Jonas Hanway and naval victualling', in *Journal of the Royal Society for the Encouragement of Arts, Manufactures and Commerce* 134(1986), 641–5, 657–9.

[42] 8th Report of the Commissioners on fees, 15.   [43] *Ibid.*, 13, 16.

[44] NMM, ADM. DP/12, 31 May 1792.

overseas. The remainder was supplied to depots and ships on station by contractors or purchased by agents, pursers and the Commissariat from local sources. This was both convenient and necessary because the growth of the yards did not match that of the navy and army overseas during the eighteenth century. At mid century the main Deptford, Portsmouth and Plymouth yards were still dispersed collections of buildings, some of which dated from the reign of Elizabeth I.[45] This was because facilities had evolved piecemeal, largely after 1683, as the Victualling Board gradually made the navy independent of contractors.[46] Their consolidation by the 1790s reflected recognition of the primary importance of food to the armed forces.

At the beginning of the eighteenth century, the main victualling yard was on Tower Hill in London.[47] The yard had a slaughterhouse, which could deal with up to 200 oxen or hogs at a time, a cutting house, hanging sheds and pickle yard; a baking house for bread, known as biscuit; and a cooperage. The Hartshorn brewery in east Smithfield, near the Tower, was purchased in 1701 to supplement a rented brewery at St Catherine's wharf. But mills for grinding grain were at Rotherhithe and storage was short: during the 1701–13 war, five ships were moored in the Thames to serve as extra storage.[48] These facilities remained much as they were at the outbreak of the 1739–48 war. More storehouses were rented near Tower Hill but they were inconveniently scattered. Tower wharf was also proving too small.

To expand, the Victualling department had to move from central London. One option was to move to Rotherhithe, farther down the Thames, where since 1698 the board had leased water mills.[49] However, space at Rotherhithe was limited and the better option was to expand farther down the Thames at Deptford, adjacent to the dockyard. As early as 1742 the purchase of Sir John Evelyn's Redhouse estate was discussed. This would provide 11 acres of space, an 800-foot timber wharf and some storehouses. In the event, the property was leased and the board envisaged creation of an 'entire victualling office'. Fires at Rotherhithe and at

---

[45] J. Coad, *Historic Architecture of the Royal Navy: An Introduction* (London, 1983), 118–19.

[46] M. Oppenheim, *A History of the Administration of the Royal Navy and of merchant shipping in relation to the Navy from 1509 to 1660 with an introduction treating of the preceding period* (London, 1896, repr. Aldershot, 1988), 140–4, 325; Coad, *Historic Architecture of the Royal Navy*, 115.

[47] TNA, ADM. 49/59, instructions 1704–96.

[48] Watson, 'The Commission for victualling', 76–7, 146.

[49] The mills were still retained in the 1790s but needed renovation and modernisation. They were tide mills and the board considered installing Boulton and Watt steam engines: TNA, ADM. 114/41, 25 June 1698, 30 Nov. 1780, 3, 6 Dec. 1790, 4 July 1791.

Deptford stimulated new building and by 1756 there were two new flour mills, a kiln to dry pease and another mill to grind oats.[50] By 1780, with a range of new storehouses, space at Deptford was already running short but building still went on. By 1785 most of the London work had been transferred to Deptford.[51] In 1793 a distinction was drawn between the London and Deptford establishments when that at St Catherine's in London was dedicated to feeding the army overseas.

Chatham possessed few facilities. Distant from the sea in terms of sailing time, ships fitting for sea usually took in their provisions at the mouth of the Medway where they were supplied from London.[52] By 1821 the Crown owned only one old storehouse in Rochester and kept provisions on board ships.[53] Dover was better equipped with a small slaughterhouse, bakehouse, cooperage and brewery dating from the early eighteenth century, a flour mill and slaughterhouse built in 1755–6. The main south coast sources of provisions were Portsmouth and Plymouth. However, in both places facilities were scattered until the mid eighteenth century.

At Portsmouth the original yard in the old town was little more than a small slaughterhouse; ships were used for storage.[54] During the Spanish Succession War, a large bakery was established in King Street, Portsmouth, and negotiations opened for a bakehouse at Weevil, Gosport, which was rented from 1714 and purchased in 1753. In 1744 a new mill was added close by to ensure the supply of good meal and flour. There was also a good supply of fresh spring water at Gosport, and a brewery was built there during the 1702–13 war. This was producing 40 tuns of beer a day by 1782, by which time a second larger brewhouse had been added, capable of producing 50 tuns a day,[55] production that was further extended during the Napoleonic War. A cooperage was created in 1711 which in 1766 was centralised at Gosport around a purpose-built yard. This establishment became known as Weevil and was only renamed the Royal Clarence Victualling yard in 1831.[56]

[50] *Victoria County History of Kent* (3 vols., London, 1908–32), 372–3; Gradish, *The Manning of the British Navy during the Seven Years War*, 148; for instructions given to the victualling officer at Deptford in 1756, see TNA, ADM. 49/59, 5 May 1756.

[51] Baugh, *British Naval Administration in the Age of Walpole*, 433–5; Gradish, *The Manning of the British Navy in the Seven Years War*, 148; NMM, ADM. DP/103, 8 Feb., 1 Apr. 1771; ADM. DP/112, 18 Aug., 1 Nov. 1780; ADM. DP/3, 17 Feb. 1783.

[52] TNA, ADM. 7/658, quoted in Coad, *The Royal Dockyards, 1690–1850*, 274.

[53] See Coad, *Historic Architecture of the Royal Navy*, 117, 119.

[54] For Portsmouth victualling yard pay books 1721–8, 1736–44, see RNM, 321/73 and 322/73; for the officers' weekly returns 1718–20, 1767–70, see RNM, 1973/320 and 1984/373.

[55] NMM, ADM. DP/2, 6 Mar. 1782.

[56] Coad, *Historic Architecture of the Royal Navy*, 119, 122–3; Coad, *The Royal Dockyards, 1690–1850*, 274–81; see also Coad's account of the creation of the Royal Clarence

Plymouth yard was equally divided. In 1707 the victualling agent was busy establishing a yard at 'Entry Comb' or Lambhay on the north-east side of the Citadel, stretching along the waterfront to the Citadel. The site was neither spacious nor easily accessible, vessels over 30 tons having to load 'after half flood'. Yet in 1745 two bakehouses, a slaughterhouse, a new wharf and further storehouses were built there, buildings supplemented until 1831 by flour mills leased from Plymouth Corporation.[57] However, between 1729 and 1733 a brewhouse and cooperage had also been established on the Cornish side of Hamoaze at South Down, near the mouth of Millbrook Creek. The brewery could produce 80 tuns a week. There was also a wharf convenient at high tide for loading, which, in 1749, the board of Admiralty observed, was well adapted to the service of the fleet.[58] Subsequently, therefore, further development took place.

In southern Ireland, Kinsale served as a temporary depot for naval stores and victuals during every war between 1689 and 1748 but was not revived during the Seven Years' War.[59] During the French Revolutionary and Napoleonic Wars Kinsale again proved useful, but a victualling agent was also maintained nearby at Cork and in 1811 the Kinsale establishment was transferred to premises on Hawlbowline Island, near Cork.[60]

As well as these home yards, by 1814 overseas victualling establishments existed at Gibraltar, Malta, Minorca, Lisbon, Rio de Janeiro and the Cape of Good Hope.[61] These were primarily depots, supplied from England or by contract. In Rosia Bay at Gibraltar work began on a huge underground reservoir in 1799 and a large storehouse in 1807. This became the largest storehouse for victuals hitherto constructed for the

---

Victualling yard in his 'Historic architecture of H.M. Naval Base Portsmouth, 1700–1850', *MM* 67(1981), 3–59.

[57] They were only superseded after the Royal William Yard was built between 1825 and 1835 on filled ground beside Cremill peninsula near Stonehouse: A. D. Lambert, 'Preparing for the long peace: the reconstruction of the Royal Navy 1815–1830', *MM* 82(1996), 41–54; Coad, *Historic Architecture of the Royal Navy*, 120–2; Coad, *Royal Dockyards, 1690–1850*, 283–90.

[58] TNA, ADM. 7/653, 10 Aug. 1749, quoted in J. Coad, 'The development and organisation of Plymouth Dockyard, 1689–1815' in *The New Maritime History of Devon*, ed. M. Duffy *et al.* (2 vols., London, 1992, 1994), I, 192–200; and in J. Coad, 'Historic Architecture of H.M. Naval Base Devonport 1689–1850', *MM* 69(1883), 341–92.

[59] Except where references are made to other sources, the above information is derived from Watson, 'The Commission for victualling', 76–94, 141–7, 370–3; Baugh, *British Naval Administration in the Age of Walpole*, 436–40; Gradish, *The Manning of the British Navy during the Seven Year War*, 19, 148–9.

[60] For the quantities of provisions, casks, hoops and staves issued by the agent at Cork at the height of the Napoleonic War, see NMM, ADM. DP/33B, 31 Aug. 1813. In 1903 the facilities near Cork were named the Royal Alexandra Yard.

[61] NMM, ADM. DP/34A, 19 Feb. 1814.

navy and reflected the importance of the Mediterranean station.[62] At Malta in 1800 the navy inherited a number of purpose-built buildings including a bakery and storehouses at the Marina in Valletta. The bakery here owed its position to great underground corn stores in this part of the city. Most of the other victualling stores, houses and offices were at Vittoriosa alongside Dockyard Creek.[63] But together they could store large quantities of food and were fundamental to the support of the Mediterranean fleet after 1800.

Systems of purchase

For 1798, when the Victualling Board was catering for 110,000 men in the navy, it had to supply, every twenty-eight days, 1,375 tons of bread, 786 tons of beef, 393 tons of pork and 147 tons of cheese. At that rate, each lunar year the board had to provide 17,875 tons of bread, 10,218 tons of beef, 5,109 tons of pork and 1,911 tons of cheese.[64] During the subsequent war, these quantities increased. They excluded the food necessary to feed the army overseas. How were such quantities obtained? The Victualling Board used three methods of purchasing foodstuffs. Most were obtained by deliveries under contracts for which merchants competed by tendering offers, from which the board selected the most reasonable. Occasionally individual merchants were commissioned to make purchases on the board's behalf. In addition, captains, pursers and fleet agents were authorised to make purchases for ships on station.

Purchases by tender

To warn potential contractors of impending demands, the Victualling Board issued a 'declaration of victuals' that would be required to meet the needs of the fleet during the following year. In the mid eighteenth century that declaration was made in late September. This was because in previous centuries most bulk buying was done in winter to supply ships in the spring for a summer campaign. This particularly applied to the purchase of meat which was best preserved when packed under cool conditions. The declaration was based on an estimate of the number of

---

[62] Coad, *Royal Dockyards, 1690–1850*, 323–4; Coad, *Historic Architecture of the Royal Navy*, 126.

[63] Between 1841 and 1845, farther along Dockyard Creek near the San Lorenzo landing place, a new mill and bakery were built on a palatial scale, thereby concentrating the victualling establishment within one area: Coad, *Royal Dockyards, 1690–1850*, 349–50; Coad, *Historic Architecture of the Royal Navy*, 127–8.

[64] NMM, ADM. DP/17, 19 Oct. 1797.

men that would be needed in the fleet the following year and was secured by the Admiralty from the Privy Council. By the mid eighteenth century it was a formality. It nevertheless served the purpose of setting a figure to which all involved could work.[65]

The Victualling Board gave 'public notice' when it required provisions 'for all persons inclinable to serve the same to come to our office and make tenders'.[66] A day was specified for the delivery of sealed written tenders. Public notice was given by advertisement. In the early eighteenth century, advertisements were placed in the *London Gazette* and on the door of the London Exchange.[67] By the early nineteenth century, the Victualling Board routinely placed advertisements[68] and notices in five London newspapers,[69] two Kent papers for supplies to Chatham and Dover,[70] three Hampshire papers for supplies to Portsmouth,[71] and in two West Country papers for supplies to Plymouth.[72] For supplies to Cork the board placed adverts and notices in twelve Irish newspapers, though some were used only occasionally.[73]

In London a copy of the intended contract was made available at the Secretary's office for inspection by any interested dealer. To receive tenders, a box was located outside the door of the Victualling Office board room. In 1813 the box had two locks and keys, the latter held by the chairman and secretary. Having received tenders in the morning of a specified day, the box was carried into the board room at one o'clock precisely.[74] The board's procedure was then to call 'in each person to know if they will not abate [the price] and then contract with such as come lowest'.

Sometimes, to obtain the quantities needed, the board was forced to accept a number of tenders at a variety of rates. Sometimes, too, when the board sensed that dealers were 'under combination or the prices be too high we put off the same until means can be found to disappoint them'. The board was sensitive to deviations from market trends as food

---

[65] Watson, 'The Commission for victualling', 119; Baugh, *British Naval Administration in the Age of Walpole*, 386–90; NMM, ADM. DP/32B, 11 Aug. 1812.

[66] TNA, ADM. 110/4, 100, Dec. 1708, quoted in Beveridge, *Prices and Wages in England*, I, 515.

[67] Watson, 'The Commission for victualling', 99.

[68] In 1816 the cost of advertising for supplies to the different yards was recorded as £594 2s 0d for London, £118 11s 6d for Deptford, £37 17s 0d for Plymouth, £73 19s 6d for Portsmouth, £30 1s 0d for Chatham, £10 9s 0d for Dover and £141 18s 6d for Cork: NMM, ADM. DP/41A, 20 July 1821.

[69] *The Times, Morning Herald, Morning Chronicle, Courier* and *Sun*.

[70] The *Kentish Courier* and *Maidstone Gazette*.

[71] The *Salisbury and Winchester Journal, Hampshire Telegraph* and *Sussex Chronicle*.

[72] The *Sherbourne and Yeovil Mercury* and an Exeter newspaper.

[73] NMM, ADM. DP/41A, 20 July 1821.   [74] NMM, ADM. DP/33B, 30 July 1813.

prices reacted relatively freely to market conditions. Moreover, as communication became more rapid, the board had available to it the prices of agricultural products throughout the country. By the third quarter of the century the board, like dealers, was informed by printed guides of the average weekly corn prices around the country, classified for England by inland counties and those upon the coast.[75]

The system of competition was instituted in 1683. Then, most agreements were for definite quantities of provisions to be delivered within a few weeks. Nevertheless, some arrangements for commodities such as butter and cheese were supplied under annual contracts, while from 1713 quarterly contracts were usually made for meat. Some of these contracts were for the total quantity needed by the navy. For these commodities the navy became dependent on some large-scale contractors. However, though those contractors might supply the navy continuously or alternately over a number of years, their agreements were always arranged in competition with others.

In this process, experience, organisation and economies of scale ensured that some traders emerged to remain contractors for long periods. Large and small contractors can be distinguished in the contract books kept by the Victualling Office, which are still extant and detail every delivery and every payment.[76] Butter and cheese supply was usually in the hands of one man. This was the case in 1703–9, and again between 1745 and 1780 when it was regularly in the hands of a contractor named Barneveld. Except in the period 1781–4, one dealer supplied all five victualling ports. One family usually supplied all the beef too. Between 1760 and 1829, Peter, Samuel and William Mellish supplied beef initially to London and Chatham, also later to Dover and Portsmouth.

Yet, while the supply of butter, cheese and beef was usually granted to a single or very few contractors, the supply of wheat, malt, peas, oatmeal, hops, flour, biscuit, coal, rum, raisins, rice and sugar was invariably divided between a large number of merchants, sometimes as many as twenty to a single port! And, among these smaller-scale merchants, some were active in contracting over long periods, while others came and went from the books of the Victualling Board over very short periods. During wars there was a tendency for contracts to be concentrated in fewer

---

[75] The abstract covered North and South Wales and Scotland as well. For an example, see TNA, ADM. 114/70, average prices for the standard Winchester bushel of 8 gallons. A note by the printer asserted that the abstract was collected and published by the authority of Parliament.

[76] For details of contracts made by the board 1776–1826, see ledgers TNA, ADM. 112/162–212. For abstracts of contracts by place of delivery 1784–1807, see TNA, ADM. 112/140–3.

hands – a trend apparent from 1778 – but when peace resumed there was always a return to greater competition.

Until 1760, except in the supply of butter and cheese, local dealers dominated the supply of the smaller ports.[77] After 1760 the London dealers in beef and salt and occasionally in other commodities muscled in on the supply of these ports. Competition was obviously intense, for at intervals local suppliers regained the port contracts by undercutting the London men. The latter were not inviolable for, despite their economies of scale, in some commodities they had to pay higher transport costs. This was not the case where the London men used local agents to obtain their supplies – which they did in the case of beef. However, they did tend to dominate the supply of commodities which were bought most cheaply in bulk in London: for example, vinegar, rum, raisins and sugar.[78]

The London dealers supplied the victualling yard at Tower Hill, and that which developed at Deptford after 1742. Generally, the London contractors were large-scale dealers. This was an advantage in times of scarcity for they had the connections and financial means to obtain scarce supplies. At times when naval credit was weak they were also able to accept payment in bills that carried a high discount rate. At these times the large London dealers could, separately or together, charge the navy higher prices. Even so, in the eighteenth century these were often prices worth paying. For, if there were complaints, they could the more easily be placed at the doorstep of the dealers and precautions taken against their reoccurrence.[79] These factors tended to drive up the cost of naval food prices: indeed between 1769 and 1778 the prices of all commodities purchased by the Victualling Board rose, some substantially. The only price which fell was that of iron hoops for casks.[80]

The employment of a small number of contractors could have given rise to political influence. However, there is no evidence to suggest victualling contracts were used for political purposes. Until the late eighteenth century, the Admiralty had more influence through the dockyards, and then through employees rather than contractors.[81] The Victualling Board was thus able to adhere to the principle of competition, even though it admitted deviating from that principle in two circumstances.

---

[77] For the contractors who supplied Portsmouth yard in 1721, see the Clerk of the Cheque's account of provisions issued to HM ship *Windsor*: RNM, 2003/77.

[78] Beveridge, *Prices and Wages in England*, 515–17; Watson, 'The Commission for victualling', 99–104.

[79] Watson, 'The Commission for victualling', 117–18.

[80] Prices paid for provisions 1769–78: RNM, Admiralty Library MS collection, MS 35.

[81] Watson, 'The Commission for victualling', 110–11; Morriss, *Naval Power and British Culture*, 40–1.

In 1798 it acknowledged making contracts 'not always' by means of public advertisement but by 'receiving private tenders coming spontaneously from merchants, dealers and others; which they are often induced to accept, not only by the reasonableness of the prices . . . but also by the conviction that if they should themselves invite proposals from different persons, they might alarm the market and thus enhance the price'. The practice of receiving spontaneous tenders was justified too as a means of deterring any 'combination among the merchants or dealers to the detriment of the public, as used sometimes to be the case formerly'.[82] The board also acknowledged arranging for commodities to be purchased by individual merchants on commission. Experience dictated, however, that this brought hazards as well as benefits for both the merchant and the state.

Purchases on commission

During wartime after 1778 the Victualling Board obtained large quantities of provisions by commissioning a single dealer to act as agent for the board and to buy all or part of the amount needed. The dealer received a percentage fee in reward. The Victualling Board did this when the number of seamen to be victualled was increasing and prices were rising. Because rising prices tended to suggest a relative shortage, the board argued the system of public advertisement and tender fuelled fears of a scarcity and caused further price rises. In these circumstances, the board preferred to write directly to merchants asking for their lowest terms for supply, to consider spontaneous offers from dealers or to commission a single agent.[83]

However, there were dangers in the employment of a single agent. That of Christopher Atkinson during the War of American Independence taught the board a harsh lesson. Between 1775 and 1778 Atkinson supplied malt to the London victualling yard. It was a supply that was so good his contract was extended to the naval ports as well as London. Indeed in February 1779 the board decided to purchase all its wheat, malt, flour, pease, pot barley, oats and oatmeal for the ports and London by commissioning Atkinson alone to make the purchases.[84]

The decision gave Atkinson enormous purchasing power and excited the envy of those prevented from tendering and obtaining contracts. He

---

[82] 32nd Report of the Select Committee on Finance, in *Reports of Committees of House of Commons*, XIII, 508, 533, referred to in Beveridge, *Prices and Wages in England*, I, 515.

[83] Beveridge, *Prices and Wages in England*, I, 518.

[84] D. Syrett, 'Christopher Atkinson and the Victualling Board, 1775–82', *HR* 69(1996), 129–42.

also invited hostility by his own conduct. When taking over the malt contract, Atkinson revealed 'a gross imposition' by the previous contractor, William Bennett. In retaliation, Bennett proceeded to attack Atkinson's character as a contractor, placing twelve letters or paragraphs in the *General Advertiser and Morning Intelligence* between October 1779 and April 1780. Bennett was clever enough to avoid using words that would expose him to a legal prosecution. But on 31 January 1781 he printed a letter signed by himself alleging that Atkinson had defrauded the Victualling Board by purchasing malt at a cheaper price than that which he claimed to have purchased it on commission.

Bennett distributed his letter to members of the House of Commons, of which Atkinson was a member. Atkinson answered him by making a deposition in the Court of Kings' Bench on 7 February 1781. Bennett had numerous supporters and they began a public 'clamour' which embarrassed the Victualling Board. The latter claimed later it had already come to suspect the quality of flour supplied by Atkinson, and to feel 'their confidence [had] outrun their discretion'. They introduced vouchers to attest the quality of his supplies but, they claimed, 'the board had not from time to time such vouchers delivered to them as were found upon experience to be necessary to justify themselves'. According to Atkinson, the quality conditions set by the board were 'totally impracticable from the nature of the corn market'. He attempted to satisfy the board in 1781 but the Victualling Board felt obliged to discontinue his services.[85]

The board claimed it did this because Atkinson, acting both as corn factor and as agent for the navy, combined private and public roles which were open to criticism outside the Victualling Office. As it told Atkinson:

Your situation exposed you to suspicion because you did business for yourself as for the Crown; and the power you possessed of commanding the market, in consequence of an employment of such a magnitude, gave you too much influence; it created envy and discontent in several quarters, and from various motives. Thus it appeared . . . our duty to drop a connection attended with such inconvenience. The plan we had adopted, however pleasing with respect to the quality of the grain you supplied, the execution could not remain in the hands of one man without exposing ourselves to the imputation of partiality.[86]

Rashly, Atkinson appealed to the Treasury for an investigation of his conduct, offering also to reveal flagrant practices, injurious to the health

---

[85] Montague Burgoyne, Victualling Board commissioner 1781–5, claimed particular opposition to Atkinson's operations and seems to have been instrumental in the exposure of his fraud. See Appendix 7 to the 8th Report of the Commission on fees, which records his evidence to the commission, 9 and 19 July 1787.

[86] WLC, Shelburne Papers, vol. 151, fo. 96.

of seamen. The Treasury invited a submission of evidence from Atkinson, which, early in 1782, it placed with a committee of the House of Commons instructed to investigate the Victualling Board's methods of purchasing provisions. The investigation did not go well for Atkinson. Evidence came to light which permitted the Treasury to prosecute him, even though he was 'so rooted at the Victualling Office that the most decisive proofs of his guilt could hardly obtain his removal from an employment of near £4,000 per annum'. In 1784 Atkinson was convicted of defrauding the public and expelled from Parliament for perjury.[87]

During the peace following the War of American Independence, the Victualling Board had no need to purchase on commission. In 1788 the Commissioners on fees were relieved, believing it held out 'temptations to men of a fraudulent disposition'.[88] But, when war resumed, so did the practice. Rising prices after 1790 encouraged the board between 1793 and 1799 to obtain all its extra biscuit for the ports by this means; between 1794 and 1804 another agent supplied pease to the London yards; between 1795 and 1803 all the yards received their wheat and malt the same way; and between 1795 and 1808 an agent was employed to purchase beef, both fresh and for salting.[89]

Extraordinary was the mission of William Eton in 1803 to purchase victuals in southern Russia for the supply of the navy and the army at Malta. He was directed to purchase quantities of beef, pork, wheat and pease but also to purchase naval stores and open trade from the ports of the Black Sea to Malta so as to render that island independent of supplies from Sicily and North Africa. His instructions directed him to find out what British manufactures and colonial products might be sold there and whether countries in Europe might be supplied by road through Poland. He was to make observations of navigation by rivers, the state of the roads and the time that would be involved in conveying goods into the interior. There were fears that France might exploit supplies from the region, so Eton was to see how far Britain might engross the trade.[90]

Eton made purchases to the value of almost £10,000. The supply of wheat and pease proved good, but that of pork and beef did not. Pork originating in Poland arrived at Malta smelling 'nauseous'. Except with regard to the supply of wheat and pease, his mission was deemed by the Victualling Board to offer no 'advantage to this department'.

---

[87] NMM, ADM. DP/4, 15 Nov. 1784; J. Norris, *Shelburne and Reform* (London, 1963), 108.

[88] 8th Report of the Commissioners on fees, 16.

[89] Beveridge, *Prices and Wages in England*, 517–18.

[90] NMM, ADM. DP/24, 25 June 1804; draft instructions enclosed in ADM. DP/26, 8 Aug. 1806.

Eton himself had difficulties in obtaining adequate documentation of his purchases and, although he closed his accounts in 1806, he was still trying to obtain settlement in April 1810.[91]

The experiences of Atkinson and Eton proved the danger of an individual taking a commission too great in scale or risk for one person. However, the persistence of the Victualling Board in the practice of making purchases on commission suggests a continuing public benefit. It was not usually used to supply stations overseas which demanded contracts of particular intricacy and variation, but for deliveries to the yards in England where commodities could be inspected.

Purchases on station

Ships not supplied from the King's victualling yards or by contractors could, when necessary, purchase their own provisions. Three methods of purchase were possible.

Firstly, captains and pursers of ships far from England or contractors' regular supply points could make purchases themselves directly from local tradesmen.[92] This was done in the 1740s, for example, off the coasts of Cape Breton Island and the thirteen American colonies. Captains paid with their own money, or the pursers drew bills of exchange on the Victualling Board. Meanwhile, the purser kept accounts and vouchers of prices paid, often recorded in foreign as well as the English currency. The practice was convenient and permitted the acquisition of fresh local food. But the Victualling Board distrusted local arrangements which could work to the financial benefit of the officers involved. Purchases were thus made only on the orders of the captain,[93] who was supposed to act as a check upon his purser, and each element of a transaction was well documented.[94]

Secondly, in more frequented places, merchants or others were appointed to serve as 'correspondents'. They could be appointed by the Admiralty, the Victualling Board or the local Commander-in-Chief and they made purchases on behalf of ships, drawing bills of exchange on the Victualling Board.[95] In the Mediterranean in the 1740s, at Genoa and Leghorn, the correspondents were the British consuls. In the 1760s

---

[91] NMM, ADM. DP/30A, 26 Apr. 1810.

[92] 8th Report of the Commissioners on fees, 13–14.

[93] NMM, ADM. DP/3, 15 Apr. 1783.

[94] 8th Report of the Commissioners on fees, 13; 9th Report of the Commssioners on fees, 7; Baugh, *British Naval Administration in the Age of Walpole*, 391–7; P. K. Crimmin, 'Letters and documents relating to the service of Nelson's ships, 1780–1805: a critical report', *HR*, 70(1997), 52–69.

[95] 8th Report of the Commissioners on fees, 14.

the consul at Malta became the means of supply of two ships, for which the consul's servant drew three bills of exchange upon the Victualling Board. In 1811 the consul at Gothenburg arranged supplies of live oxen, fresh beef and vegetables to the squadron under Vice-Admiral Sir James Saumarez. At Buenos Aires in 1817 the consul supplied ships on the South American station, including a commission of $2\frac{1}{2}$ per cent upon the prices 'by way of remuneration for his trouble and services in the transaction'.[96]

The bills drawn on the Victualling Board would not be honoured, even if drawn by consuls, unless they were accompanied by certificates, from the captains and two principal resident merchants, stating that the provisions supplied were purchased at the current wholesale market price of the place where bought, and by an affidavit from the consul that the prices of the provisions 'were actually paid by him without any profit or advantage to himself, or to any person or persons on his account'.[97] Nevertheless, such commissions could become self-serving. Purchases could become anticipatory, a regular resort of local commanders, and a regular business sideline, even of consuls. Although the Admiralty condoned the service as helpful, the Victualling Board thought the system expensive and liable to abuse.[98]

On this account, the Victualling Board preferred a third method of arranging purchases: through agents formally appointed either by itself or by the local commanders-in-chief. These officials had much in common with those senior officials at the victualling yards in England who were also known as agents. In 1718 an agent victualler was appointed to serve British ships putting into Lisbon.[99] By 1739 there were agents at Port Mahon in Minorca, and Gibraltar; in 1740 two joint agents were appointed at Jamaica; and in 1745 an agent was appointed at Kinsale in southern Ireland. The two agents at Jamaica proved dishonest, obtaining for themselves a kick-back on local contracts they arranged: 10 per cent on the value of meat contracts, 5 or 6 per cent on the value of rum contracts. By 1746 they had made £50,157 on their meat contracts alone. Discovery by the Victualling Board resulted in the supply at Jamaica being placed with a contractor appointed in London.[100] Nevertheless,

[96] NMM, ADM. DP/102, 24 Jan. 1770; ADM. DP/31B, 9 Dec. 1811; ADM. DP/37B, 15 Sept. 1817.

[97] NMM, ADM. DP/37B, 15 Sept. 1817.

[98] Baugh, *British Naval Administration in the Age of Walpole*, 397–8.

[99] A copybook of instructions, forms and precedents exists for John Sargent, appointed to Lisbon, 22 May 1718; see J. D. Alsop, 'Royal Navy Victualling for the eighteenth century Lisbon station', *MM* 81(1995), 468–9.

[100] Baugh, *British Naval Administration in the Age of Walpole*, 403–5.

agents at depots abroad continued to be necessary and operated under comprehensive instructions drawn up by the Victualling Board.[101]

Agents appointed by the Victualling Board were also carried afloat with squadrons or fleets.[102] From 1700 agents afloat were routinely appointed to squadrons destined for operations in the Baltic.[103] From 1739, during wartime, the East Indies squadron also routinely carried an agent who purchased food supplies, drawing bills on the Victualling Board.[104] In 1770 an agent was appointed by the Commander-in-Chief in the East Indies,[105] setting 'on foot a correspondence with proper persons in all the principal ports in India' even before he had received instructions from the board in London.[106] Likewise, the agent employed on the Mediterranean station during the French Revolutionary War, while receiving and distributing supplies from the yards in England, Port Mahon and Gibraltar, also made a range of contracts with merchants all round the western Mediterranean and bought a good deal of fresh produce in local markets.[107]

Quantification and quality

Since the early eighteenth century, the British seaman was permitted each week the basic food ration shown in table 7.2. The rations provided for soldiers were similar, though in slightly different proportions, to those for seamen. For seamen, the quantity of butter was increased – to 8 ounces a week – by the end of the century, while other commodities were added to the diet.[108] But the basic rations did not alter until 1847,[109] and the quantities formed the basis of all calculations of requirements to feed the navy. Before 1793 the Treasury, and after 1793 the Victualling

---

[101] For a departure from instructions by the Malta agent, see NMM, ADM. DP/30B, 6 Oct. 1810.

[102] For instructions to Mr Richard Rosewall, agent to ships under Curtis Barnett, see TNA, ADM. 49/59, 20 Apr. 1744.

[103] Sir George Rooke's expedition in 1700 was the first to the Baltic but had not gone beyond The Sound. Between 1715 and 1727 there were nine more: D. D. Aldridge, 'The victualling of the British naval expeditions to the Baltic Sea between 1715 and 1727', *Scandinavian Economic History Review* 12(1964), 1–25.

[104] NMM, ADM. DP/112, 19 Jan. 1780.

[105] In 1795 the Victualling Board admitted it was too far removed to make arrangements for food supply in the Indian Ocean, which it had to leave to 'the discretion of the commander-in-chief upon the station': NMM, ADM. DP/15, 11 Feb. 1795.

[106] NMM, ADM. DP/102, 19 Apr. 1770.

[107] Crimmin, 'Letters and documents relating to the service of Nelson's ships'.

[108] C. Lloyd, 'Victualling of the fleet in the eighteenth and nineteenth centuries' in *Starving Sailors: The Influence of Nutrition upon Naval and Maritime History*, ed. J. Watt, E. J. Freeman and W. F. Bynum (NMM, Greenwich, 1981), 9–15.

[109] MacDonald, *Feeding Nelson's Navy*, 10.

Table 7.2 *Basic weekly food rations for seamen and soldiers, 1755–1847*

	Seamen[a]	Soldiers[b]
Biscuit, bread or flour	7 pounds	7 pounds
Beef	4 pounds	7 pounds
		or
Pork	2 pounds	4 pounds
Butter	6 ounces	6 ounces
		or
Cheese	12 ounces	8 ounces
Pease	2 pints	3 pints
Oatmeal	3 pints	$\frac{1}{2}$ pound
Beer	7 gallons	

[a] Watson, 'The Commission for victualling', 112, quoting TNA, ADM. 110/5, 20; MacDonald, *Feeding Nelson's Navy*, 10, quoting the 1733 edition of the printed *Regulations and Instructions*; Baugh, *British Naval Administration in the Age of Walpole*, 375, quoting the 1743 edition; Gradish, *The Manning of the British Navy in the Seven Years War*, 141, quoting TNA, ADM. 110/22, 316.
[b] Baker, *Government and Contractors*, 22.

Board, used the quantities of the soldiers' basic rations to calculate the requirements of the army. The methods by which total requirements were calculated, and by which defects and deficiencies were avoided, are given below for the navy.

Stocks

As numbers of seamen to be fed increased during the eighteenth century, and as more were supplied on foreign stations, so the necessity grew for the Victualling Board to know precisely what provisions were located on these stations and would be available to meet the needs of the armed forces deployed there. Accounts maintained by the Victualling Office acted as the medium of control and gave the board absolute quantities of the provisions on each station. For the period of the American War of Independence, separate statements of the victualling situation on the different stations survive in the Admiralty Secretary's in-letters.[110] However, from 1808 the Victualling Board prepared periodic global statements 'of the quantities of provisions remaining in the several stores at home and

[110] NMM, ADM. DP/107, 28 Sept. 1775; ADM. DP/108, 11 May 1776; ADM. DP/109, 20 Jan. 1777; ADM. DP/111, 9 Aug. 1779; ADM. DP/1, 23 Feb. 1781; ADM. DP/3, 30 July, 4, 13 Aug. 1783.

abroad shewing the number of days the same will serve the men at the respective stations'.[111]

One of the most remarkable is that for 1 February 1809, when the number of ships in the Royal Navy reached a peak and the number of men 'borne for wages' on the books of ships reached almost 142,000. The storehouses for the home station comprised those at the main ports, depots at Yarmouth, Leith, Falmouth and Cork and depot ships at the Nore and Downs anchorages. Storehouses on foreign stations comprised 'sundry victuallers' in the Baltic, depots at Malta, Gibraltar, Lisbon, the Cape of Good Hope, 'the Brazils', and warehouses belonging to contractors in the Leeward Islands, Jamaica, Bermuda and Halifax. The only station for which recent statistics were lacking was the East Indies, where a contractor supplied all the provisions except the pork, beef and suet. Figures for the latter were included, derived from returns from the East Indies dated in May and June 1808, six months earlier.

The men to be served by the stocks were given for each station as follows:

Main yards and depot ships	80,000 men
Yarmouth, Leith, Falmouth	4,000 men
Cork	6,000 men
Baltic	11,000 men
Malta, Gibraltar	29,000 men
Lisbon	4,000 men
Leeward Islands and Jamaica	20,000 men
Bermuda and Halifax	5,000 men
Cape of Good Hope	2,300 men
East Indies	7,000 men
Brazils	5,000 men

The storehouses had enough provisions available to serve 173,300 men, about 30,000 more than actually borne on ships' books and 40,000 more than voted by Parliament.

The stocks available were prodigious (see table 7.3). For each of the locations they were detailed in precise quantities – pounds, gallons, hundredweights, pieces or bushels – and according to the number of days they would last the men on the station. To maintain them, supplements were delivered periodically by contractors or naval victuallers. For loading purposes, the tonnages of each commodity were calculated with precision as indicated by the quantities needed to feed 20,000 men for one month serving in the Channel fleet (see table 7.4).

---

[111] See statements in NMM, ADM. DP/29, 2 Jan., 1 Feb. 1809; ADM. DP/37B, 7 July 1817.

Table 7.3 *Stocks of food on the home and Mediterranean stations on 1 February 1809*[a]

AT THE MAIN YARDS AND DEPOT SHIPS, for 80,000 men.[b]

77,462 hundred weight of bread, which would last	108 days
421,682 gallons of beer	5 days
507,447 gallons of spirits	101 days
778,250 gallons of wine	77 days
1,278,624 eight pound pieces of beef	223 days
2,681,126 pounds of flour	78 days
248,302 pounds of suet	86 days
3,204,057 four pound pieces of pork	560 days
50,975 bushels of pease	142 days
41,352 bushels of oatmeal	154 days
1,532,670 pounds of sugar	357 days
696,137 pounds of butter	162 days
1,663,713 pounds of cheese	194 days
481,969 pounds of rice	56 days
55,727 gallons of vinegar	78 days
254,429 pounds of tobacco	89 days

AT GIBRALTAR AND MALTA, for 29,000 men.

26,752 hundred weight of bread, which would last	103 days
76,218 gallons of spirits	42 days
236,442 gallons of wine	65 days
58,108 eight pound pieces of beef	28 days
703,627 pounds of flour	56 days
127,781 pounds of raisins	61 days
5,952 pounds of suet	5 days
263,468 four pound pieces of pork	127 days
12,892 bushels of pease	99 days
10,959 bushels of oatmeal	112 days
46,960 pounds of sugar	30 days
343,918 pounds of cheese	110 days
234,426 pounds of rice	75 days
7,882 gallons of vinegar	30 days
59,055 pounds of tobacco	57 days

[a] NMM, ADM. DP/29, 1 Feb. 1809.
[b] Excludes stocks at Yarmouth, Leith, Falmouth and Cork.

Supplements

The calculation of supplements to stocks was straightforward. From knowledge of the number of men on a particular station and of the number of days which their stock of victuals would last, the Victualling Board estimated the quantity of foodstuffs remaining on any particular day, how long those stocks would last thereafter, and the quantities that

Table 7.4 *The tonnage of provisions for 20,000 men for one month*[a]

'A tonnage of the species for one month for 20,000 men for Channel Service supposing 3 weeks beer and 1 week spirits, and half the beef in flour, suet and raisins with the tonnage.'

		Tons	cwt	lb
Bread	5,000 bags	416	0	2
Beer	1,575 tons	1,575	0	0
Spirits	8,750 galls	34	3	0
Beef	20,000 double[c]	75	3	0
Pork	40,000 double	83	0	1
Flour	120,000 pounds	66	1	1
Suet	10,000 pounds	7	0	0
Raisins	20,000 pounds	11	$\frac{1}{2}$	0
Pease	2,500 bushels	78	$\frac{1}{2}$	0
Oatmeal	3,750 bushels	78	$\frac{1}{2}$	0
Butter	30,000 pounds	13	1	$\frac{1}{2}$
Cheese	4,000 pounds	30	3	0
Vinegar	5,000 galls	20	2	1
		2,486	14 $\frac{1}{2}$	5 $\frac{1}{2}$[b]

[a] RNM, Admiralty Library MS 82, f. 50, Estimate book, 1790–99.
[b] The weights of the butter and the final total have been altered in the original, and the total is unclear.
[c] Pieces.

would be needed for restocking. As the number of men on any station was liable to change, so rates of consumption of stocks could increase or decrease. But the Victualling Office was alerted to such changes by the receipt of regular returns from home yard and overseas agents, fleet and squadron commanders.

Weekly, monthly and quarterly returns from the yards and squadrons were registered at the Victualling Office. Rough estimates of remainders were formed, copies of which survive in the Admiralty Library. These were remainders in store from 1790 to 1799 and were of two types: of all commodities at particular places or of a single commodity. Returns from new stations appeared as Britain extended her naval presence throughout the world: for example, from the Leeward Islands in November 1794, the West Indies in November 1795, New South Wales in August 1796, and the Cape of Good Hope in June 1798.[112]

[112] RNM, Admiralty Library MS82.

This system of calculating cumulative supplements had long been evolving. It was honed to a fine art during the American War of Independence. Then, estimates took into account shipments on passage as well as deliveries by contractors. Hence 'A state of the Victualling Admiral Graves Squadron' of 28 September 1775 was premised on the fact that 'The Admiral in his letter of 19th of August 1775 informs the Commissioners of the Victualling that the squadron was complete to three months on average for the ships complements.' These complements totalled approximately 6,000 men.

The 'state' thus proceeded:

Say therefore that the squadron had provisions of all species on the 19th of August for .... 84 days

Sent out a supply by order of the Right Honourable the Lords Commissioners of the Admiralty of the 9th February 1775, for 4 months for 4275 men, which is for 6000 men .... 79 $\frac{2}{3}$ days

Sent out another supply, by order from their Lordships of the 16th June 1775, for 4 months for 5000 men – except rum which Mr Grant has contracted to deliver in the month of October next, which for 6000 men, is ...... 93 $\frac{1}{3}$ days

Sent out another supply by their Lordships order of the 5th July 1775 for 4 months for 6000 men, except rum which Mr Grant has contracted to deliver in the month of January next, and pork, which he has engaged to deliver to answer this supply, which say therefore is for 6000 men .... 112 days

_____

369 days

Abate, taken from the Trident victualler which carried part of the first supply to complete the squadron to 3 months on 19th of August .... 16 days

_____

for 6000 men – Remain   353 days

_____

353 days is 50 weeks 3 days – or from the 19th August 1775 to the 6th of August 1776.[113]

By mid 1776 the number of seamen on the North American coast had risen to 13,000 men and accounts had become more comprehensive and wide ranging in terms of both their time span and their geographical scale. Thus in May 1776, when the Victualling Board produced its 'State of the Victualling of H.M. Ships in North America', it took into account supplements carried in nine victuallers that had arrived at Boston

---

[113] NMM, ADM. DP/107, 28 Sept. 1775.

between August and December 1775, four that had arrived in Antigua in December 1775, two that had been ordered to Quebec and Halifax, and eight that had arrived in Spithead and were awaiting passage.

During the American War of Independence, from necessity, clerical calculations became comprehensive, in sequence linking consumption to contract on a projection that extended at least a year ahead. Necessarily, of course, write-offs were included – from decay, capture or foundering.[114] In time the contents of missing victuallers came to form separate and distinct accounts.[115] Regular reports of the state of the victualling on each station started from August 1779.[116] Agent victuallers appointed to particular fleets kept the Victualling Board informed of local distribution problems and of desirable priorities in the despatch of provisions. By 1780 tablifications had become standardised to indicate their flow: those victuallers arrived off New York, those on passage under convoy, those under sailing orders at Spithead, those complete in their loadings and on their way to Spithead, those taking in cargoes at Deptford.[117] At the subsequent peace, the tables contracted but the accounting procedure persisted.[118]

Deficiencies

During the War of Spanish Succession, shortages of victuals were blamed for hold-ups to expeditions, sickness, necessity to discharge men, and deaths among seamen and among the land forces embarked on board ships. In a wind-dependent communication and supply system, these shortages are easy to understand. Victuallers could take up to five months just getting from London to Portsmouth.[119] As voyages to distant stations increased in the eighteenth century, so shortages were anticipated and became an integral part of the supply process.

Ships bound on distant voyages could not always stow all the provisions they needed. Nor could victuallers always carry sufficient quantities to restock ships fully. There was, anyway, no point in despatching provisions that would decay before they could be consumed.[120] Indeed, the life of provisions determined whether ships overseas would be better receiving food from home or from contractors abroad. Victualling decisions thus had a critical impact on the survival of the state's forces distant from Britain.

[114] NMM, ADM. DP/108, 11 May 1776 enclosure.
[115] NMM, ADM. DP/109, 20 Jan. 1777.   [116] NMM, ADM. DP/111, 9 Aug. 1779.
[117] NMM, ADM. DP/1, 23 Feb. 1781.
[118] NMM, ADM. DP/3, 30 July, 4, 13 Aug. 1783.
[119] Watson, 'The Commission for victualling', 113–17, 132–40.   [120] Ibid., 74, 113.

Experience gradually diminished the sources of deficiency. For example, regulations were developed early against inadequate packing. The Victualling Board worked to ensure every container contained the correct or recorded contents. All casks were marked with their contents 'by some person that can, if there shall be occasion, testify on oath that there is in the said cask the quantity marked by him on the head thereof'. All beer casks were gauged, that is their capacity calculated from their dimensions.[121] Records of contents and capacity were preserved, for deficiencies on opening were always referred to the Victualling Board, both to clear the purser of any misdemeanour and to prevent any short measure on the part of packers.

When provisions ran short, seamen were put on short allowance: two-thirds the amount of normal rations.[122] Seamen were put on short allowance as a matter of course in the sixteenth and seventeenth centuries. In the eighteenth century money was paid in lieu of food. The amount of short-allowance money paid has been taken as a measure of efficiency of supply in the early eighteenth century. Daniel Baugh points out that the money paid in 1702–7 was twelve times the amount paid during the first five years of the war between 1739 and 1748, even though during the latter period a large portion of the British fleet was deployed in the Mediterranean and the West Indies. The decline in payments led him to suggest that the Victualling Board had managed to minimise deficiencies by the mid eighteenth century.[123]

When shortages did occur after that the officers were required to reduce their consumption accordingly. The 1787 *Regulations and Instructions* 'strictly charged' the purser not to supply any officer his whole allowance while the rest of the company went short.[124] In the Mediterranean in 1797, when bullocks were slaughtered for consumption, standing orders upon the *Alexander* required officers to take their share of the offal as well as of heads and tongues. The Commander-in-Chief took 'a head at every slaughtering by way of example'.[125]

But after the War of American Independence, shortages on foreign stations were unusual. On the contrary, abundance was more of a concern

---

[121] To gauge a cask, 'take the diameter at the head and also at the bung, add them together and take the mean, which multiply by three for the (vulgar) circumference, then take half the mean diameter and multiply half the circumference which gives the square inches, then take the length and multiply the square inches and divide the product by 221 (the square inches in a gallon) and that gives the contents in a gallon of wine or oil measure': BL, Add. MS 36,784, fo. 17, quoted in Watson, 'The Commission for victualling', 73.

[122] Gradish, *The Manning of the British Navy during the Seven Years War*, 141.

[123] Baugh, *British Naval Administration in the Age of Walpole*, 374.

[124] *Regulations and Instructions*, 1787, 61.

[125] Order book of HMS *Alexander* in National Library of La Valette, Malta, LIBR 1194, quoted by R. Monaque, 'On Board HMS *Alexander* (1796–9)', *MM* 89(2003), 207–12.

than deficiency. A new regulation of 1796 directed that the Victualling Board purchase back from seamen savings they made from their standard rations. While this saved food from being wasted, the procedure smoothly converted the rations of seamen into money that was added to the wages. It was a transaction that probably became all the more common following the concession won at Spithead for seamen always to receive their full allowance of provisions.[126]

Defects

During the first half of the eighteenth century, contemporaries delighted in shocking stories of poor victuals. The stories often attributed blame to the administrators ashore,[127] hinting at suppression within the military hierarchy.[128] Such stories undermined the reputation of the Victualling department of the navy. Complaints continued into the second half of the eighteenth century, sometimes with due reason. But, especially from about 1780, the Victualling Board waged an unceasing campaign to sustain a reputation for providing sound food. As late as 1811, rejecting a complaint through the War Office about cheese on board a transport, the Victualling Board enlisted the aid the Admiralty, maintaining that such complaints brought the King's service into disrepute.[129]

The board had most control over the quality of food issued by its yards. It required staff to observe every necessary precaution in curing, packing and preserving foodstuffs. Stowage on board victuallers and in warships was also calculated to avoid deterioration.[130] Provisions issued from a victualling yard for consumption north of latitude 27 degrees were supposed to hold good for six months. Those issued for consumption south of that latitude were supposed to last for twelve months.[131] Provisions thus had to be consumed according to their age. Much here depended

---

[126] NMM, ADM. DP/19, 2 July 1799.

[127] For example, one former Victualling yard meat cutter claimed that in 1701–2 'he cut several hogs full of matter and stinking corruption for sea service, which flew in his face when he chopped them, so as to make him spew and his nose to bleed. He saw those hogs salted and packed up and when he told several of the clerks of it, the teller of the meat returned it to the scale again and it was afterwards salted and packed up amongst the other meat.' This former employee alleged that much of the best meat was pilfered on the order of the chief clerk. Cited in Watson, 'The Commission for victualling', 125–6.

[128] During the 1740s Major Jonathan Lewis, transported in a naval vessel, wrote privately to John Russell, about his complaint 'to Admiral Balchen that 13 butts of beer out of 20 we received on board . . . stunk; that the man that opened the bung of the cask had like to a dropt down . . . I first sent the adjutant to the colonel who sent for answer I might complain to the admiral, and the two admirals Balchen and Stewart being together they put me right: . . . ' M. E. Matcham, *A Forgotten John Russell* (London, 1905), 144.

[129] ADM. DP/31A, 7 Mar. 1811.   [130] NMM, ADM. DP/2, 6 Mar. 1782.

[131] Watson, 'The Commission for victualling', 73.

on the purser. Only in 1811 did the Victualling Board recommend a plan, put forward by Captain John Yule, for dividing the bread room in warships into compartments to permit access to the oldest biscuit first.[132]

Naturally, neither the Victualling Board nor its suppliers took responsibility when provisions were opened in the wrong order or at the wrong time. This was underlined in 1791 when the captain of the *Andromeda* storeship returned into store at Falmouth beer that had just been delivered to his ship. He complained that the beer was unacceptable. However, the Victualling Board supported the contractor's agent who maintained that it was new beer, just brewed, and would have been acceptable had it been left to settle for two or three days and not immediately broached.[133]

Victuals condemned as defective upon survey were returned or destroyed. A ship's captain or commanding officer could order a survey. Surveys were usually performed by three officers, usually masters, drawn from three separate ships – if practicable, none from the ship within which the victuals had spoiled and, if possible, including a naval officer from on shore. No provisions were to be thrown over board except cheese, all else being returned to a victualling agent unless this was totally impractical. If condemned for causes occurring prior to receipt, provisions were returned if possible to the victualling officer from whom they came.[134]

If the defects were not attributable to the pursers, the latter were credited in their accounts with the quantity of the victuals condemned and issued with more.[135] The cost of condemnations was borne by the public unless their cause was traced to contractors. It was thus incumbent upon the Victualling Board to investigate their cause and to place the expense where the responsibility lay.

The board thus took close interest in all condemnations and demanded the severe punishment of those who condemned food without good cause.[136] This may have deterred condemnations. At any rate, the proportions of victuals condemned to the totals issued were small in the mid eighteenth century. Stephen Gradish observed, of an account of quantities issued and quantities condemned between 1749 and 1757 (see table 7.5a), that the figures indicated a marked improvement in victualling since Stuart times.[137] N. A. M. Rodger used the same figures,[138] remarking that 'with the exception of stockfish, there was no item of which as

[132] NMM, ADM. DP/31B, 11 Dec. 1811.   [133] NMM, ADM. DP/11, 10 Nov. 1791.
[134] *Regulations and Instructions*, 1787, 66–7, 118–19.
[135] Watson, 'The Commission for victualling', 73.
[136] NMM, ADM. DP/103, 2 Dec. 1771.
[137] Gradish, *The Manning of the British Navy during the Seven Years War*, 145.
[138] His percentages are given in table 7.5a with two differences, those for beer and stockfish which he gave as 0.9 per cent and 7.9 per cent, respectively.

Table 7.5 Condemnations of food, 1749–1757, 1778–1781, 1776–1780

(a) CONDEMNATIONS

| | Species quantity issued | 1749–57[a] | |
		Quantity condemned	Ratio
Bread	54,642,437 lb	185,761 lb	1 lb in 294 (0.3%)
Beer	110,049 tuns	1,062 tuns	1 tun in 103 (1.0%)
Brandy	351,692 galls	none	
Beef	4,498,486 lbs	2,659 lbs	1 lb in 1,691 (0.06%)
Pork	6,734,261 lbs	2,217 lbs	1 lb in 3,037 (0.03%)
Peas	203,385 bush	1,201 bush	1 bush in 169 (0.6%)
Flour	6,264,879 lbs	20,858 lbs	1 lb in 300 (0.03%)
Suet	809,419 lbs	895 lbs	1 lb in 904 (0.1%)
Raisins	705,784 lbs	702 lbs	1 lb in 1,005 (0.1%)
Oatmeal	138,504 bush	1,214 bush	1 bush in 114 (0.9%)
Vinegar	390,863 galls	none	
Stockfish	166,943 lbs	13,043 lbs	1 lb in 12 ¾ (7.8%)
Oil	71,668 galls	322 galls	1 gall in 222 (0.4%)

Table 7.5 *(cont.)*

(b) CONDEMNATIONS   1778–81[b]
(1) provisions condemned and thrown over board,
(2) those condemned and returned into store,
(3) those condemned in the stores and storeships.

Year	Condemned	Bread lbs	Flour lbs	Pease bush galls	Oatmeal bush galls	Rice lbs
1778	(1)	128,651	3,844	101–4	55–1	2,938
	(2)	84,523	2,504	160–4	49–5	–
	(3)	54,472	10,967	186–5	–	3,814
Total		267,646	17,315	448–3	104–6	6,752
1779	(1)	291,559	8,422	92–4	119–4	686
	(2)	171,946	1,312	53–6	16–6	–
	(3)	130,452	–	–	–	–
Total		593,957	9,734	146–2	136–2	686
1780	(1)	209,076	1,855	16–4	8 –	–
	(2)	187,731	20,076	76–4	596–6	–
	(3)	223,720	15,745	–	8 –	–
Total		620,527	37,676	93 –	612–6	–

(cont.)

Table 7.5 (cont.)

(c) BEER CONDEMNED

When brewed	Where brewed	Quantity brewed in tuns	Quantity condemned in tuns	Proportion of condemnations to quantity brewed
1776	London	5,343	150	1 tun in 35
1777		5,148	60	1 tun in 85
1778		9,505	778	1 tun in 12
1779		9,974	570	1 tun in 17
1780		9,582	565	1 tun in 17
1781		9,492	228	1 tun in 41
1776	Portsmouth	7,471	110	1 tun in 67
1777		13,777	894	1 tun in 15
1778		14,639	426	1 tun in 34
1779		18,512	510	1 tun in 36
1780		16,407	489	1 tun in 33
1781		15,510	270	1 tun in 57

1776–80[c]

[a] An account of the provisions issued from the Victualling Stores in England and the amount condemned by survey, 1 January 1749 (O.S.) to 31 December 1757: Victualling Board to Admiralty, 15 Feb. 1758, copy in BL, Stowe MSS, vol. 152, fo. 130. Printed in Gradish, *The Manning of the British Navy during the Seven Years War*, 144.

[b] NMM, ADM. DP/1, 8 June 1781.

[c] NMM, ADM. DP/2, 18 Feb. 1782.

much as 1 per cent was condemned, an astonishing fact considering the limitations of technology and the hazards to which the full casks were exposed after issue'.[139]

However, these figures may be incomplete.[140] They may be compared to a more comprehensive account of condemnations for January 1778 to March 1781 relating to bread and dry stores – flour, pease, oatmeal and rice (see table 7.5b) – produced by the Victualling Board for the House of Commons in June 1781. It classifies condemnations according to whether the food was thrown overboard, returned into store, or condemned in the stores and storeships.[141] The amounts condemned may be related to the quantities issued according to the number of men borne on the books of ships.[142] Taking bread alone, compared to the quantities issued in 1778, 1 pound in 85 was condemned (1.2%); in 1779, 1 pound in 49 (2.0%); in 1780, 1 pound in 53 (1.9%). With an average of 1.7%, the bread condemned in 1778–80 was a higher proportion of the total issued than the 0.3% of 1749–57.

Figures also exist for beer brewed and condemned between 1776 and 1781 (see Table 7.5c). These again are higher than the 1 tun in 103 (1%) condemned in 1749–57.

Condemnations may have been higher in the American War of Independence than in the period of peace 1749–57 because food was kept for longer or was subject to more damaging conditions of storage and transportation. The condemnations of some foods were comparable between the two periods. For example, in 1778 1 gallon of pease in 210 was condemned, or 0.48%,[143] which bears comparison to 0.6% in 1749–57. But condemnations did not, as one might assume, automatically take a downward path. Those of beer, for example (table 7.5c), increased sharply for

---

[139] Gradish, *The Manning of the British Fleet in the Seven Years War*, 144–5; Rodger, *Wooden World*, 85.

[140] The numbers of men borne on the books of ships for victuals each year between 1749 and 1757 together amount to 217,310. This suggests that, at the rate of 1 pound of bread a day per man, the amount of bread issued should have been: 28 days × 13 lunar months × 217,310 men = 79,100,840 lbs.

[141] NMM, ADM. DP/1, 8 June 1781.

[142] In 1778 there were 62,719 borne on the books of ships for victuals; in 1779, 80,275; and in 1780, 91,566. That means, at the rate of 1 pound of bread a day per man, the amount of bread issued should have been:
  1778 – 28 days × 13 lunar months × 62,719 men = 22,829,716 lbs
  1779 – 28 days × 13 lunar months × 80,275 men = 29,220,100 lbs
  1780 – 28 days × 13 lunar months × 91,566 men = 33,330,024 lbs.

[143] Men received half a pint of pease four times every seven days, which was 96 pints or 1.5 bushels of pease a year. In 1778, when 62,719 men were borne for victuals on the books of ships, about 752,628 gallons (94,078 bushels 4 gallons) were issued, of which 3,587 gallons (448 bushels 3 gallons) were condemned.

Table 7.6 *The total cost of food for victualling the navy and the army in 1808, distinguishing the value of deliveries to ships, garrisons and storehouses*[a]

Navy	
Supplies to ships	£ 717,095 13s 5d
Deliveries into store	£4,214,970 16s 8½d
Total	£4,932,066 10s 1½d
Army	
Supplies to garrisons	£ 294,869 19s 5d
Deliveries into store	£ 502,546 7s 3d
Total	£ 797,416 6s 8d
Total, army and navy	£5,729,428 16s 9½d

[a] TNA, ADM. 112/194, Victualling Office contract ledger, 1808.

Portsmouth in 1777 and for London in 1778 as the breweries at those ports stepped up their output.

Distribution and delivery

In 1808 (see table 7.6), of £5.7 million spent on feeding the navy and the army overseas by the Victualling Board, 86 per cent was spent on the navy, and 14 per cent on the army. Of that delivered by contractors for the army, 37 per cent was delivered directly to garrisons overseas, and 63 per cent into store at St Catherine's, London, for processing and despatch by the board. Of that spent on the navy, about 68 per cent was for food delivered into storehouses at the main ports (see table 7.7), 14.5 per cent for deliveries by contractors to ships on station, and 17.5 per cent for purchases probably by captains and pursers on station and paid for by bills of exchange drawn on the Victualling Board.[144]

Ships generally carried provisions, issued from storehouses at the ports, for at most six months. Operations, voyages or the maintenance of a naval presence on a station thus demanded the collection, delivery or purchase of additions to the original supply. Collections could be made from store-houses at the main ports or at depots around the coasts of Britain and overseas. Deliveries were made by contractors or by victuallers sailing from the main yards in England. The system of storehouses and deliver-ies gave rise to a far-reaching and complex network of supply.

[144] TNA, ADM. 112/194.

Table 7.7 *The value of food delivered into navy storehouses in 1808, with other naval victualling expenses[a]*

	£
Artificers bills	28,959
Balance of Pursers' Accounts	126,491
Barley (Pot)	6,461
Bavins	9,478
Beef Suet	41,774
Beef Irish	26,007
Bills of Exchange[b]	863,888
Bisket	419,288
Bisket Bags	51,559
Bisket Meal	125,245
Butter	82,957
Callavances	589
Candles	8,346
Cask	25,596
Cheese	113,749
Coals	14,413
Cocoa	15,164
Flour	86,532
Freight	29,041
General Average	4,045
Hogs	27,278
Hoops, iron	30,984
Hoops, wood & coopers' stores	7,977
Hops	9,656
Imprest	171,544
Malt	94,404
Necessary Money	55,154
Oatmeal	58,535
Pease	136,949
Pork Irish	155,801
Promiscous payments	11,127
Raisins	45,294
Rice	27,937
Rum	353,302
Rye	729
Salaries, pensions, extra rent, taxes	39,609
Salt	47,609
Staves	76,172
Sugar	108,134
Tobacco	16,325
Vegetables	8,173
Victualling men out of private provisions	2,689
Vinegar	17,337
Wheat	70,370
Wine	191,808
Total	4,214,971

[a] TNA, ADM. 112/194.
[b] Probably from victualling agents and from captains for purchases by pursers.

By contractor

By the time of the War of American Independence there were twenty-nine locations around Britain's coasts where contractors supplied ships.[145] In 1808 there were thirty-one, and in 1812, thirty.[146] (See table 7.8.) Contractors also supplied depots and ships overseas. In 1808 they supplied fourteen places, and in 1812, fifteen. Although they supplied fewer locations overseas than around the coast of Britain, the value of these overseas deliveries was 55, rising to 57, per cent of the total.

Contractors were employed because they simplified the work of shipment. Delivery by the local agent of a contractor could also be cheaper than shipment using state resources. What is more, the sources of their provisions were often near their place of delivery, which permitted them to supply provisions that were fresher and had a greater shelf life. Should a contractor fail to meet his agreement, or the number of men on station suddenly increase, the Victualling Board was usually able to step in and boost supplies from their own yards, if only temporarily.[147]

Contractors usually agreed to supply a range of 'sea provisions'. At the Victualling Office, the business of advertising, tendering and agreeing contracts for 'sea provisions' went on much like that for the supply of the yards in England. Summaries of tenders from the mid eighteenth century survive for victualling HM ships at Jamaica, Nova Scotia, New England, Philadelphia, the Bahamas, Grenada, Tobago, Barbados and the Leeward Islands. The first four, submitted between 1765 and 1778, show three competitors for the Jamaica and Nova Scotia contracts; five for the New England and Philadelphia contracts; six for Barbados and the Leeward Islands.

Each tender was for the proportion of provisions necessary for the supply of 100 men for seven days. Those for the three northerly stations – Nova Scotia, New England and Philadelphia – were for the supply of bread, beer, rum, beef, pork, pease, oatmeal, butter, Cheshire cheese, vinegar and necessary money. Those for southern stations – Jamaica, Barbados and the Leeward Islands – omitted beer and enlarged the portion of butter in lieu of smaller quantities of butter and cheese.[148]

Resources were clearly important to the acquisition of contracts. They attracted some of the biggest dealers (see also chapter 3). In 1778

---

[145] NMM, ADM. DP/3, encl. 11 Aug. 1783.   [146] TNA, ADM. 112/194, 198.
[147] NMM, ADM. DP/111, 14 July 1779; ADM. DP/37A, 4 Feb. 1817; ADM. DP/36A, 8 July 1816.
[148] NMM, ADM. DP/110, 29 Apr. 1778.

Table 7.8 *The locations and value of contract food supplies to ships in 1808 and 1812*

Overseas supplies in **bold**.

Contract ledgers	1808[a] Total £	1812 Bills on course[b] £	1812 Ready money £	1812 Total £
Antigua	**56,138**	**40,049**	**2,121**	**42,170**
Barbados	**110,418**	**49,168**	**391**	**49,559**
Belfast	2,961	2,069	–	2,069
Bermuda	**25,667**	**42,833**	**3,001**	**45,834**
Bristol	11,121	2,316	–	2,316
Cork, Kinsale, etc.	7,650	3,310	–	3,310
Dartmouth	119	152	–	152
Demerara	**3,163**	**695**	**–**	**695**
Dover	1,021	2,155	–	2,155
Dublin	3,566	3,545	–	3,545
Exmouth	58	16,815	–	16,815
Falmouth	56,191	–	–	–
Greenock	2,128	3,491	–	3,491
Guernsey & Jersey	21,240	22,948	–	22,948
Harwich	5,572	12,221	478	12,699
Hull	1,390	3,460	–	3,460
Jamaica	**111,822**	**41,809**	**10,507**	**52,316**
Isle of Wight	4,272	77	4	81
Leith	45,932	43,816	72	43,888
Limerick	1,283	494	–	494
Liverpool	14,616	6,754	–	–
Londonderry & Loughswilly	2,024	12,631	6	12,637
Madeira	**8,043**	**4,458**	**2,066**	**6,524**
Margate	566	449	–	449
Martinique	**–**	**2,905**	**1,810**	**4,715**
Milford Haven	1,263	1,613	–	1,613
Newfoundland	**–**	**–**	**3,009**	**3,009**
New Providence	**10,595**	**5,115**	**217**	**5,332**
Newry	12	–	–	–
New Romney	1,604	3,179	–	3,179
Norfolk, Virginia	**1,768**	**631**	**354**	**985**
Nova Scotia	**55,785**	**58,806**	**2,274**	**61,080**
Penzance	347	–	–	–
Promiscous Supplies	19,407	1,756	28,359	30,115
Quebec	**2,035**	**3,509**	**–**	**3,509**
Scilly	151	450	–	450
Seaford	1,322	1,821	–	1,821
Surinam	**4,036**	**2,179**	**–**	**2,179**
Swansea	–	882	–	882

(cont.)

Table 7.8 *(cont.)*

Contract ledgers	1808[a] Total £	1812 Bills on course[b] £	1812 Ready money £	1812 Total £
Thames	8,697	–	–	–
Torbay	12,993	1,573	–	1,573
Tortola	**1,551**	**–**	**–**	**–**
Trinidad	**6,570**	**3,677**	**–**	**3,677**
Tynemouth Haven	2,123	1,904	–	1,904
Waterford	3,133	163	–	163
Weymouth	1,630	1,806	–	1,806
Windward & Leeward Isles	**511**	**241**	**30,492**	**30,733**
Yarmouth	84,598	58,825	35	58,860
Total	717,096	466,750	85,199	551,949

[a] ADM. 112/194. Payments rounded to nearest pound.
[b] ADM. 112/198. Payments distinguished according to whether paid for by victualling bill or by cash.

contracts for five stations drew tenders from fifteen merchants.[149] Some tendered for several contracts. Robert Grant tendered to supply ships on five stations in the west Atlantic – at Jamaica, Nova Scotia, New England and Philadelphia, Barbados and the Leeward Islands – and was successful in providing the lowest tenders for the two northern stations – Nova Scotia and New England. Perhaps his main source of supply was in North America and thus his transport costs for those two stations were less than those of his competiters.[150]

The tender which was adopted by the Victualling Board resulted in a contract much like that made with John Blackburn on 7 February 1776 for supply of all His Majesty's ships and vessels that put into Barbados in want of provisions. Such contracts were repeatedly re-written to fill loopholes. Terms were remarkably demanding and specific, set to serve the needs of the navy not those of the contractor, who in many respects was undertaking a challenging task. To ensure the contractor understood what he was taking on, the contract was written in non-legal language, in the first person – 'I' – and had to be signed by him in the presence of two Victualling Office witnesses.

[149] They included Jonathan Blackburn, Thomas Barke, Jonathan McConnell, Alexander Dover, John Thomson, Isaac Levy, Samuel Smith, Thomas Fludyer, William Shelby, James McCord, Messrs Eden and Court, and Robert Biggin.
[150] NMM, ADM. DP/110, 29 Apr. 1778.

In 1776 Blackburn was required to supply all vessels that wanted provisions 'with good and wholesome sea victuals fit in all respects for the service of His Majesty's navy within forty eight hours' after they were demanded, winds and weather permitting. The food that he was to supply was quantified and priced in the smallest practical units: bread by the hundredweight; rum and vinegar by the gallon; beef and pork by the piece; pease and oatmeal by the bushel; butter by the pound. The prices included new casks of all kinds that might be wanted for packaging, the setting up and brimming of casks, cartage, labour, freight, boat hire and all other charges involved in delivering the victuals. For the due performance of every part of this contract, the contractor bound himself to pay a penalty of £5,000 in case he should fail in any part. Moreover, he was obliged to find two other persons, able, sufficient and approved by the board, to be bound jointly with the contractor in a bond to the amount of £5,000. While running initially for twelve months, thereafter the contract could only be terminated by either side with eight months' warning. Occasionally standard contracts, like that with John Blackburn, had exceptions or additions added as memoranda, much like codicils to a will.[151]

After 1793, when the Victualling Board became responsible for the supply of army expeditions and garrisons overseas, the board commissioners used their contracts for the supply of sea provisions as a model to draw up those for the supply of the army. Ships' officers were replaced by commissioned army officers; the purser by the senior officer of the Commissariat. Certificates of receipt had to be returned via the Treasury, one copy to be forwarded to the Victualling Board, to whom other vouchers, affidavits and documents were to be sent under sealed cover.[152] Although in 1808 the value of foodstuffs supplied to garrisons was only 41 per cent that of sea provisions supplied to ships, by then much was purchased for the army overseas by its own Commissariat.

By victualler

Ships operating in home waters were usually assigned provisions for less than six months, often three or four, especially when they were expected to return to the coast of England for replenishment. In 1800 food was shipped from Plymouth to Torbay for ships sheltering there, where a supply of water was established. Small Brixham vessels were hired, some to

[151] NMM, ADM. DP/110, 29 Apr. 1778.
[152] NMM, ADM. DP/36B, 11 Apr. 1816. Specimen contract for the supply of rum to troops.

convey fresh beef, others to carry water out to ships loading in Torbay.[153]
Some warships carried provisions back to other ships which remained
on station, for example blockading Brest. In 1800 the *London* and *Ajax*
each carried back 10 tons of vegetables; in 1801 the *Mars* carried twenty
bullocks and 6 tons of vegetables.[154]

Otherwise, ships on station received their provisions by victualler
loaded at a main yard in England. The Victualling Board had long experience of using these vessels. During the War of Austrian Succession,
victuallers had carried provisions to the Mediterranean and West Indies.
By the end of the American War of Independence, the board had over
100 victuallers supplying ships in North American waters.[155] In April
1782, 39 were listed as at New York, Charlestown, St Lucia or ordered
for the Leeward Islands and Jamaica (see table 7.9). For the supply of the
ships blockading Brest in 1800, the Transport Board initially provided
7 vessels, all between 113 and 163 tons; by June 1801 18 vessels were
so employed. In May 1801 they conveyed out to the squadron off Brest
3,000 tons of water. In May, June and July 1801 they carried 10,062
pounds of cabbages, 27,883 pounds of potatoes, 919 pounds of turnips,
and 8,571 pounds of onions. Convoys of victuallers were despatched
from Plymouth. One vessel of 170 tons was fitted to carry twenty-five
head of cattle.[156]

This scale of supply to ships on station was an achievement. But it
did subject the provisions to considerable risk of damage. The victuallers
were convoyed out to Ushant and back by sloops and cutters, very occasionally by a ship of the line travelling in the same direction. But the
victuallers were generally not 'weatherly' vessels, and tended to straggle
and suffer damage in bad weather, some even sinking in strong gales.
The weather was especially important during the process of transfer of
provisions from victualler to ships on station. In 1800 St Vincent ordered
that fresh provisions, including bullocks, were to be delivered to the inshore squadron at least once a month. Damage generally resulted in some
returns, especially of bread and beer. The victuallers returned with the
hides, tallow and tongues of the animals that had been slaughtered.

Despite the use of victuallers, the ships blockading Brest between 1800
and 1805 suffered regular shortages.[157] In 1811 it was proposed that
these shortages be reduced by the monthly despatch from Plymouth to
the fleet off Brest of two coppered transports (or a converted warship),

---

[153] Steer, 'The blockade of Brest'.   [154] *Ibid.*
[155] Syrett, 'The Victualling Board charters shipping, 1739–1748'; and 'The Victualling
Board charters shipping, 1775–82', *HR* 68(1995), 212–24.
[156] NMM, ADM. DP/42, 11 June 1800, encl. 6 June 1800.
[157] Steer, 'The blockade of Brest', 314–15.

Table 7.9 *Navy victuallers in North America and the West Indies in April 1782*[a]

Ships' names	Masters' names	Tonnage	Station
			New York
Burstwick	Alex. Anderson	215 $\frac{3}{4}$	
Argo	John Marshall	343 $^{11}/_{12}$	
Fortune	Danl. Dale	400 $\frac{1}{2}$	
Myttle	Thos. Dawson	406 $\frac{1}{6}$	
Polly	Richd. Pickering	428 $\frac{3}{4}$	
Thornton	Thomas Boaz	416 $\frac{1}{2}$	
Essex	George Marshall	368 $\frac{1}{2}$	
Diana	Jos. Walrond	400 $\frac{1}{6}$	
Union	Danl. Thomason	358 $^{5}/_{12}$	
Centurion	John Disting	453 $\frac{1}{3}$	
Charlotte	Thos. Pearson	396 $^{1}/_{12}$	
Eagle	Andw. Harper	407 $^{11}/_{12}$	
Mentor	John Samson	302	
Ann	Moses Cadenhead	404 $\frac{1}{4}$	
Prince Willm. Henry	Walter Carr	436 $\frac{1}{4}$	
Etherington	Thos. Bogg	538	
London	Thos. Brodrick	446 $\frac{1}{2}$	
Liberty	John Davies	107 $\frac{1}{3}$	
			Charlestown
British Queen	Thomas Potter	380	
Ceres	Jon. Stevenson	248 $\frac{3}{4}$	
Stody	John Davison	224	
Prior Blessing	Christ. March	184 $^{5}/_{12}$	
Friendship	N.W. Seager	179 $\frac{1}{6}$	
Love	T. Bloomfield	168 $^{11}/_{12}$	
Minerva	Thos. Gleed	244 $\frac{1}{4}$	
Henley	Jas. Hinton	242 $\frac{1}{3}$	
Three Friends	George Wrodl	329	
Thorn	Joseph Leonard	241 $\frac{2}{3}$	
Freedom	Jas. Turner	421 $\frac{5}{8}$	
Benjamin	Jas. Wishart	307 $\frac{5}{8}$	
Portsmouth	E. Nicholson	512 $\frac{1}{3}$	
Fern	T. Tinmouth	427 $\frac{1}{2}$	
William	J. Randall	285 $\frac{1}{8}$	
			West Indies
Two Sisters	John Hull	537 $\frac{7}{8}$	St Lucia
Fanny	J. Cheap	271 $\frac{5}{6}$	Supposed ditto
Devonshire	Henry Furneaux	595 $^{5}/_{12}$	Ordered for
Earl Cornwallis	J. Mills	633 $\frac{5}{6}$	Leeward islands ...
Henry	Isaac Amory	455 $\frac{5}{6}$	Ordered for
Atterington	Cuthbert Parks	509 $\frac{1}{3}$	Jamaica ...

[a] NMM, ADM. BP/3, 13 Apr. 1782.

capable of carrying about 400 tons or one month's provisions of all types for 6,000 men. Two coppered transports capable of taking 336 tons of water (the amount consumed by 6,000 men every two weeks) would also be despatched fortnightly. As these vessels would return with empty casks, ballast was considered unnecessary. The arrangement was adopted and victuallers were despatched even through the winters of 1811 and 1812. A similar plan was implemented for the supply of the squadrons off L'Orient, Lisbon and the north coast of Spain.[158]

Victuallers also went north to the Baltic between 1808 and 1812 to supply the fleet under Vice-Admiral Sir James Saumarez. In May 1809 he possessed fifteen ships of the line and requested provisions for 12,000 men for two months. The following year the Victualling Board was instructed to communicate directly with Saumarez on the subject of the regular despatch of provisions for 15,000 men. In 1812 the rendezvous of the victuallers with the Baltic fleet was Vinga Sand off Gothenburg.[159]

During the Napoleonic War, the regular despatch of naval victuallers to supply fleets and squadrons in European waters became part of the greater system of global supply to Britain's armed forces, noticed in the second chapter of this book. Food was delivered by the Victualling Board to army garrisons much in the same way as it was delivered to ships, only officers of the Commissariat took delivery rather than a fleet agent and ships' pursers.

When victuallers arrived at a ship, their masters were responsible for delivering to the ship's captain a bill of lading by which he was expected to check the whole consignment brought on board. The victuallers' masters were responsible for seeing that the provisions were 'put into the slings or tackles of the ship . . . by careful men belonging to the ship'. But there the responsibility of the masters ended. Damage or loss by slippage from the slings, by malice or carelessness, was an offence for which the seamen were liable. At that same point of unloading, the provisions became the responsibility of the purser under the orders of the captain.[160]

By the purser

The purser was entrusted with 'keeping and distributing the provisions out to the ship's company'. Sea provisions were supplied for a specific

---

[158] NMM, ADM. DP/31B, 3 Aug., 30 Sept. 1811; ADM. DP/32B, 30 July, 30 Sept., 28 Oct. 1812.

[159] A Ryan, 'An Ambassador afloat: Vice-Admiral Sir James Saumarez and the Swedish court, 1808–1812' in *The British Navy and the Use of Naval Power in the Eighteenth Century*, ed. Black and Woodfine, 237–58; NMM, ADM. DP/30A, 29 Mar. 1810; ADM. DP/32B, 25 Sept. 1812.

[160] *Regulations and Instructions*, 1787, 66–7.

number of men for a specified number of days. In addition, the purser was responsible for receiving water, casks, slops and 'necessaries'. The slops included jackets, waistcoats, drawers, shirts, frocks, trousers, stockings, hats, shoes and blankets – all delivered by the bale. Before 1758 they were received from a clothing contractor; after 1758, from the Navy Board.[161] The 'necessaries' comprised coals, firewood, hammocks, bedding, turnery – wooden plates and bowls used by the crew, candles, lanthorns, to purchase which he was allowed 14 pence per man a month, or 17 pence if the crew was less than sixty men.[162] All had to be inspected for quality and quantity and signed for in triplicate by the captain, master and boatswain.[163]

The purser was responsible for the 'good order, stowage and preservation of the provisions'.[164] Those which were oldest had to be most accessible and issued first. To encourage him in this economical order of issue, he was warned that provisions condemned by survey after expiry of the period for which they were expected to serve would receive no allowance in his accounts. In other words, he would be charged for wasting them. Leakage from beer casks might be allowed if the explanation for the loss on survey was reasonable, but there would be no allowance for leakage from casks of wine, oil, brandy, rum or arrack. For them, the purser was totally responsible and was answerable for their mishandling or misuse.

The purser ensured these sea rations were issued daily in the quantities to which each man was entitled. By the regulations of 1787, not one man could be entered on the purser's books, nor any provisions issued to him, without a note or voucher from a Clerk of the Cheque on shore or the commanding officer on board. But, thereafter, the receipts of every authorised man were recorded in the purser's books.[165] Most provisions, most of the time, were issued to six-man messes. The issues made to each mess were recorded in the purser's account or mess-book. Although these records were kept by the month, they attempted to capture precisely what provisions were issued to each man, for the purser had to account for all his disbursements down to the single ration. It called for accounting of extraordinary detail for, though officers and men were each entitled to the standard ration, they did not have to receive and consume it. On the

[161] *Shipboard Life and Organisation*, 566–92; *British Naval Documents*, 469–70; Rodger, *Wooden World*, 88.

[162] In addition, the purser was also allowed 2 shillings a month loading money, 10 groats a month adz money, and 'four pence a tun for drawage of beer'.

[163] Watson, 'The Commission for victualling', 328.

[164] The storage and management of foodstuffs on board ships is discussed in MacDonald, *Feeding Nelson's Navy*, 71–96.

[165] *Regulations and Instructions*, 1787, 115–21.

contrary, they could eat all, part or none of the ration, and were entitled to credit for what they did not consume.[166]

To assist him in keeping these detailed records, the purser was assisted by a steward, sometimes also a steward's mate. To ensure casks were sound, repaired and 'shaken' (dismantled) as necessary, he was assisted by a cooper, sometimes also a cooper's mate. Usually the purser found these assistants and paid them a sum on top of their wages.[167] Assistance would have been essential. In ships the size of sloops, the commander acted as the purser. In larger ships, ranging from 200 to 1,000 men, just the physical work involved in feeding the crew required delegation. Seamen assisted as well, for bags and casks had to be brought up from the hold and the contents of those containing meat counted. Logs of ships conventionally contain references to the number of pieces of beef or pork contained in the latest cask. At least one lieutenant probably witnessed proceedings. For any cask that fell short of the number of pieces marked on its head by the packer, the purser was instructed to apply to the captain for a warrant for the master, and for one or more of the mates, to survey the contents. Thereafter the purser was required to keep an exact account of the pieces of meat in every cask broached so that a 'true balance' might be delivered to the Victualling Board.[168]

Despite these responsibilities, the purser was not trusted. To cover losses from wastage and spillage, he kept back 2 ounces in every pound of provisions he issued. Admiralty investigation in 1773 revealed that the 14-ounce pound had no official foundation and that it had evolved from custom and practice.[169] Seamen believed the purser deprived them of 2 ounces in every pound of food which was rightly theirs.[170] In May 1797 the seamen in the Spithead mutiny attacked his allowance of an eighth of all bread, beer, butter and cheese issued, and, to appease the mutineers, the eighth was formally abolished. After 1797 seamen had to be issued with provisions at the rate of 16 ounces in the pound. Indeed, thereafter, because they did not consume all their provisions, greater quantities were purchased back for money by the calculation of the purser.[171] However, the Admiralty acknowledged that an allowance for wastage was still necessary. It was thus officially reinstituted on 18 May 1797 by Admiralty order at the rate of one-ninth, calculated at the passing of the purser's accounts.[172]

---

[166] Rodger, *Wooden World*, 89–90.   [167] *Ibid.*, 91.
[168] *Regulations and Instructions*, 1787, 121–2.   [169] NMM, ADM. DP/105, 23 July 1773.
[170] J. Masefield, *Sea Life in Nelson's Time* (London, 1905, repr. 1972), 46–8.
[171] NMM, ADM. DP/19, 2 July 1799.
[172] NMM, ADM. DP/17, 18 May 1797. In 1820, on the model of the purser's allowance, storekeepers abroad who issued provisions to seamen were allowed a sixteenth of their issues: NMM, ADM. DP/40A, 4 Aug. 1820.

At the same time, the purser was also suspected of selling provisions preserved or saved. He was permitted an allowance in his accounts for savings in provisions issued but, to limit profit from any sale of savings, the value of this allowance was limited to less than half the real value of the food. Partly as a result pursers fell into debt; in 1788 they owed the state £78,000. Yet their debt was alleged to be an added inducement for pursers to sell provisions.[173] The Commissioner on Fees recommended the allowance for savings be raised, and in 1807 this became one of a number of improvements in the funding of the purser.[174]

Especially until then, the purser had to avoid any mistake or suggestion of fraud. As the ship's banker, shop-keeper and accountant for wages as well as food – he kept 'an exact muster book', recording every attendance and allowance to the seamen – the integrity of the purser was critical to his employment. He had to be punctilious in keeping records and was advised in 1767: 'One or two days' neglect of all or any of these will not only prove burthemsome to the memory, but in the end turn to a heap of confusion, and the least omission will cause an objection in your certificates which without very good reasons will not be removed.'[175] For this reason, pursers were exacting in their attention to detail. In 1812, on joining *La Volontaire*, Thomas Peckston 'found every book and paper not only complete to my hands, but a pattern of neatness... On examining the same, I find them equally correct.'[176]

As the representative of the state, the minutiae of his concerns balanced the vast scale of supply upon which the Victualling Board conducted its operations. Without his attention to detail, there would have been no economy or control. More than that, the purser's integrity would have been impugned and, with it, that of the state. As the representative of the state, indeed, he was not only responsible for the smooth operation of each ship's internal economy, to which food was central, but projected daily a political message crucial to social cohesion at a time of ideological change.

Food, and its efficient provision, thus performed a role fundamental to the service of the state. The same may be said of the food supplied to the army by the Commissariat as of that to the navy by the purser. Indeed, the victualling service epitomised the achievements of the state's bureaucracy. The pains taken to achieve safe delivery, adequate stocks and fitness for consumption reflected the value put on the state's

---

[173] 8th Report of the Commissioners on fees, 17, 26.
[174] TNA, ADM. 106/3085.
[175] W. Mountaine, *The Seaman's Vade-Mecum* (London, 1767), 233, quoted in Rodger, *Wooden World*, 90.
[176] RNM, Peckston papers, 17 July 1812.

employees. That the naval Victualling Board in 1793 should become responsible for the supply of food to the army overseas was more than just a sensible rationalisation of organisation, for the board had well over a century of service and, as the state enlarged its endeavours and responsibilities overseas, so the organs of the state had to respond. The board's experience and efficiency was accordingly extended to both the armed forces, making it the centre of food procurement and distribution throughout Britain's maritime empire.

8   Shipping and transportation

Food, manpower, guns and materials for the maintenance of Britain's armed forces throughout the world could not have been delivered without sea transport. The growth of the shipping industry was fundamental to the expansion of Britain's maritime economy; it was also fundamental to the projection of the British state. A state transport service had taken shape under William III. However, the dissolution of that service in 1717 required the different departments of government that used shipping to hire and manage their own. Added work, competition for ships, differences in terms of hire, weaknesses of control all thereafter complicated the use of shipping by the boards involved. Acute problems were experienced during the American War of Independence and a state transport service was re-instituted at the beginning of the French Revolutionary War. From 1794 the new Transport Board played a central role in Britain's overseas operations, achieving economies and efficiency on a large scale. It chartered shipping according to commercial practices and used it flexibly to suit the requirements of different departments. Key operational roles were performed by agents for transports at the main ports, who prepared and despatched vessels, and by agents who accompanied fleets of transports overseas. Britain's overseas achievements reflected this marriage between a key maritime resource and the state's bureaucracy.

The early transport service

After the Restoration the responsibility for hiring transports rested with the Treasury, whose commissioners delegated this task. The first Transport Board was established in 1689 to provide ships and 'necessaries' for the transportation of William III's army to Ireland, then to transport the army operating in Flanders. A copy book of orders from the Transport commissioners to the masters of transports survives for the period 1697–1700.[1] The Board consisted of eight commissioners who took orders

---

[1] Watson, 'The Commission for victualling', 277–9; RNM, Admiralty Library MS 51.

from the Secretary of State through the Treasury. Officially it was wound up in 1702, but the War of Spanish Succession demanded new commissioners. There were two until 1705, then three, who formed an unofficial board which received a patent in 1710 and survived until 1717. They received their first parliamentary vote of funds in 1704, from which time they were able to make payments in course, like other departments of government, for ships hired.[2]

The duties of these early commissioners and their agents are examined here because they indicate the work of the Transport Board established in 1794 was not new. The also indicate the range of services which were thrown upon other boards between the abolition of the first board in 1717 and the establishment of the new one in 1794.

The work of the commissioners

At the beginning of the eighteenth century, the Transport commissioners usually hired high-capacity 'fly boats' or square stern merchant ships to convey troops overseas. They paid a rate per ton per month, for example 14 shillings in 1702, 12 shillings in 1704. Before being hired, the ships were surveyed and valued in case they were lost at sea or captured, in which case the state had to pay compensation to the owners. They were hired for a minimum period of four or six months, although they were often retained in service much longer. To earn their fee, the owners had to maintain the ships, and hire, pay and victual the crews, who had to be supplied at the rate of one man to 10 tons of shipping.

The troops were put on board at the rate of 1 soldier to every 1.25 tons. The Transport Board hired carpenters to fit the ships with cabins and cradles, and in 1702, when ships were fitted to convey 5,000 soldiers to Cadiz, the board had 2,000 cabins and cradles made for the higher ranks, while the soldiers slept in hammocks. All the men were supplied with bedding, rugs and pillows, the board employing four women to mend and fill old beds and pillows.[3] In 1702 soldiers sent to Spain and the Mediterranean were embarked from Portsmouth, those for Flanders and Holland from Harwich.

Experience soon developed a mode of procedure which seems to have been relatively efficient. The secrets of success lay in three factors: a long run-in time for planning and preparation, attendance at the point of embarkation of a Transport commissioner to ensure ships were fitted and

[2] Rodger, *The Command of the Ocean*, 196.
[3] Watson, 'The Commission for victualling', 263.

victualled as ordered, and consideration for the needs of those embarked, both soldiers and horses.

In 1706, when 5,200 soldiers were sent with a fleet under Shovell to attack Toulon, orders to prepare ships for transports were received a year ahead of the planned attack. Then, in July 1706, Thomas Colby was sent to Spithead to oversee preparations and report proceedings to the Secretary of State. Colby checked that the owners of the ships had fitted their ships in conformity with their charter parties, and that the work of the carpenters employed by the Transport Board to make ship-board stabling for horses was satisfactory to the dragoon officers. After examining the transports in Portsmouth harbour and at Spithead, Colby visited the regiments to ensure they could embark easily and quickly. Some were on the Isle of Wight, the rest around Portsmouth. Colby came to the conclusion that embarkation could be completed within two or three days.

To embark the foot soldiers the local naval Commander-in-Chief supplied boats and tenders. The loading of horses posed a particular challenge. Colby considered the jetty heads within Portsmouth's naval dockyard were the best places from which to embark the animals safely and with least difficulty. However, as the dragoons were in the Isle of Wight, he considered their horses should be shipped off to the trans-ports by hoy from Cowes Road which would save the fatigue of a double embarkation, five days of trans-shipment, and a cost of 18 pence per horse.

Meanwhile, he had to consider the victualling of the soldiers and horses on board ships that came from Ostend. For short voyages the Transport commissioners bought and supplied food to the soldiers. However, for voyages to Portugal, Spain and the West Indies the food supplies were requested by the Lord High Admiral or the Secretary of State from the Victualling Board. For these longer voyages, the soldiers were given the same kind of food as the seamen but at two-thirds the quantity. The horses on these longer voyages were allowed 15 pounds of hay and 1 peck of oats a day. For a journey of three months, each horse was provided with two and a half tuns of water cask, of which half was iron-bound and half wood-bound.[4]

The shipment of horses posed a range of problems. They needed 7 to 8 tons of shipping to a horse, which in turn required large ships that could only be hired in England, usually in the River Thames. Those of 100–140 tons hired in 1703, though for only a short voyage to Ireland, proved too small as there was insufficient space for hay and water. Leaving horses behind was not an option, as their respective troopers usually had

[4] *Ibid.*, 264.

a financial interest in them. Later, in 1756, twenty-three ships were hired to carry horses, with an average size of 270 tons. Then the largest was 345 tons, which carried forty-four horses; the smallest 200 tons, which carried twenty-six.[5]

Everything possible was done to preserve the lives of horses embarked on board transports. In June 1702, when the wind turned against ships carrying 1,500 horses down the River Thames for Holland, the transports turned back to Woolwich to re-victual. There, the army officers persuaded a committee of the Lords of the Council to permit the animals to disembark, having 'several horses already dead and others so sick that they find it very difficult to keep them upon their legs'. They maintained that, were the horses to be kept on board, they would not 'be able to take the field this summer'. As a result all were disembarked, and re-shipped a week later when the wind became favourable.[6] Generally, horses did not ship well. Even on the short voyage from London to Holland between 27 February and 1 March 1702/3, 93 horses died from a total of 965 that were shipped.[7]

The work of an agent afloat

In 1708, when a fleet with troop transports was sent to the Mediterranean, an agent for transports accompanied it and was issued with a set of instructions which indicate how the transports were to be managed. The fleet sailed from Portsmouth and the agent was ordered to take up quarters in one of the transports, where he was to remain during the course of the voyage. Before sailing, he was to ensure that each transport had on board the necessary stores and number of seamen which, if short, were to be made up to complement by the master. Although the Victualling Board was responsible for putting the provisions on board and laying down the rate at which they were to be issued, the agent for transports was to 'take notice' of their rate of consumption and ensure by 'reports to the commanding officers by land or sea of the sufficiency of the stores for the intended voyages'.

On these voyages, the agent was to note any master who failed to respond to orders or signals to sail, or any other neglect, so that the board could formally protest against his proceedings and stop payment for the ship. While abroad, the master of any ship requiring repairs or stores was required to draw upon the account of the owners. Only when

---

[5] NMM, ADM. BP/109, n.d. [1758].
[6] Watson, 'The Commission for victualling', 265.
[7] Death rate figure from TNA, T.1/79/35, fo. 134, cited in *ibid.*, 266.

this was not possible was the master to be permitted to draw on the board's agent, for example at Lisbon, who would ensure the charge was forwarded to the board and placed to the account of the owners. Should a ship suffer severe damage and require considerable work, the agent was to provide for her speedy repair and, if handicapped by neglect or lack of cooperation on the part of the master, he was 'to protest against her and discharge her from her Majesty's service'.

Such ships as were ordered home were to sail with the first convoy. Vacant hold space in ships coming home was to be let out for the carriage of cargo. For replacement or extra ships hired by the agent, he was 'to have a proper order in writing for so doing' and 'to agree at as cheap a rate as you can'. He was to obey orders received from either of the Secretaries of State, the Secretary at War, the General or Commander-in-Chief of the land forces, the Admiral or Commander-in-Chief of the fleet, her Majesty's ambassador or envoy in Portugal, as well as the Transport Board.

Naturally, the agent was to listen to all complaints against masters who failed to provide soldiers with their rations or who managed the provisions badly. These cases were to be reported to the Transport Board. In all other matters, he was to cooperate with other agents, for example at Lisbon, but not to interfere in the conduct of their duty. He was to remain with the transports wherever they were ordered, and was to return with them to England when they were so ordered, ensuring they returned without delay.[8]

Treasury, Navy, Victualling and Ordnance Board problems

In 1717 the Transport Board was wound up. Thereafter, for the greater part of the eighteenth century, the Treasury Board retained the responsibility for hiring ships to transport the army and for supplying victuals to the army overseas, but delegated the tasks in wartime to the Admiralty, Navy and Victualling Boards. Meanwhile the Navy, Victualling and Ordnance Boards also hired shipping for their own purposes.

In 1740, when troops were sent to Carthagena, South America, and in 1742, when troops were sent to Flanders, the task of hiring the transports was given to the board of Admiralty. That Board received orders from the Secretary of State and re-issued them to the Navy Board to provide the

[8] Instructions to Peter Crisp, 30 Dec. 1708, TNA, SP.42/120, fos. 126–33, used by Watson. Crisp was appointed agent for transports in the Mediterranean on 29 Nov. 1708 after his predecessor fell ill at Genoa.

transports. The Navy Board, already burdened with the maintenance and manning of the navy, adopted convenient rules of thumb. For transatlantic voyages it hired ships at the spacious rate of 2 tons per soldier. For cross-Channel voyages it packed soldiers into ships at the rate of three men to every ton of shipping. To simplify the chartering of ships, the board hired a considerable number of them through brokers, for example forty-nine in 1742 from John Major. For space and shallow draught it preferred colliers between 200 and 300 tons, which were hired according to the time they were required as well as their tonnage. They were inspected and measured by Deptford dockyard officers and embarked their troops from points along the Thames.[9]

Meanwhile the task of supplying food for the land forces in the Caribbean was transferred from a Treasury contractor to the Victualling Board. That board thus supplied the army as well as the navy in the West Indies, making no distinction in the provisions sent out. It freighted supplies from North America to the West Indies using ships hired in England or hired locally by the Governor of the colony from which the provisions came. The Governor selected the ships and paid for them with bills of exchange drawn on the Victualling Board. By the mid 1740s these arrangements were relatively routine.[10]

During the Seven Years' War, the Navy Board continued to hire ships on behalf of the Treasury to carry troops overseas. The Treasury again also turned to the Victualling Board to supply provisions for the troops serving in North America and the West Indies. However, the Treasury Board also hired ships to transport army provisions in European waters; the Navy Board hired transports to carry naval stores to the dockyards and to depots overseas; the Victualling Board hired vessels to supply the navy with food; and the Ordnance Board hired ships to carry artillery men, ammunition, arms, ordnance and engineering stores and equipment.[11]

It is clear that the colonies in North America supplied resources – both provisions and ships – which supplemented those available in England and Ireland, and that the process of making them available during hostilities was as reliable as that which existed in England.[12] The terms of the charter parties issued in America were similar to those issued in England,

[9] D. Syrett, 'Towards Dettingen: the conveyancing of the British army to Flanders in 1742', *JSAHR* 84(2006), 316–26.

[10] Crewe, *Yellow Jack and the Worm*, 166–80.

[11] D. Syrett, *Shipping and Military Power in the Seven Years War: The Sails of Victory* (Exeter, 2008), 7–68, of which pp. 42, 46 refer to army supply; Syrett, *Shipping and the American War*, 246–7. See also Syrett, 'The Victualling Board charters shipping, 1775–82'.

[12] For colonial, primarily American, ship building and shipping in the eighteenth century, see Goldenburg, 'An analysis of shipbuilding sites in *Lloyd's Register* of 1776', 419; and Parkinson, *The Trade Winds*, 83.

the owners being paid through bills of exchange.[13] It is equally clear that the loss of these colonial resources in 1776 increased the difficulty of hiring all the ships needed during the American War of Independence, and the difficulties of individual boards were exacerbated by the competition between them, the use of different procedures, the inadequate protection of transports and a lack of control of transports once they were distant from Britain.

Competition between boards

Loss of control of the thirteen American colonies resulted in the boards hiring shipping becoming completely dependent on that available for hire principally in the port of London. At the same time, dispersal and detention of transports in different war theatres threatened to exhaust the shipping available. The shortage pushed up prices, which were enhanced by the competition of the four boards that hired ships – the Navy, Victualling, Ordnance and Treasury Boards[14] – for ship owners naturally looked to hire their ships to the highest bidder. In July 1776, for example, the Victualling Board agreed to hire 4,000 tons of shipping, subject to survey: 'but the next day we were informed by the persons who tendered the ships that the agents for procuring shipping for the army had increased their price from eleven shillings to twelve shillings and six pence per ton; and that the owners and masters of the ships . . . absolutely refused to let them go without an advanced price'.[15]

The problem was exacerbated by depletion of the pool of ships available for hire. Before the American War, in 1775, there were about 6,000 British merchant vessels engaged in foreign trade. Then a proportion[16] was owned in the American colonies and was no longer available after the declaration of independence. But an even greater proportion was lost through capture. It has been estimated that 3,386 British vessels were captured, 495 recaptured, 507 ransomed and 2,384 remained in the possession of the enemy. The number available for hire was thus reduced, probably to fewer than 3,500.[17]

---

[13] Syrett, *Shipping and Military Power in the Seven Years War*, 81–6.
[14] M. E. Condon, 'The establishment of the Transport Board – a subdivision of the Admiralty – 4 July 1794', *MM* 58(1972), 69–84.
[15] NMM, ADM. DP/108, 19 July 1776.
[16] One-third of the ships recorded in the 1776 edition of *Lloyds Register* were built in the colonies, principally in Newfoundland, Massachusetts and the Chesapeake region: Goldenburg, 'An analysis of shipbuilding sites in *Lloyd's Register* of 1776'.
[17] C. Wright and C. E. Fayle, *A History of Lloyd's* (London, 1928), 156–7, cited in Syrett, *Shipping and the American War*, 77.

If falling numbers drove up charter prices, so also did the cost of equipping ships. In 1781 the owners of twelve ships still under charter complained of:

the heavy expense of fitting out our ships occasioned by the enormous price of all kind of naval stores and provisions, which within these twelve months have advanced in sundry articles near 20 per cent, and which added to the high discount on Navy bills and the premiums on bills abroad seldom less than 15 per cent, renders it impossible to us to continue our ships in the service at the present freight.

They thus wanted an increase of about 8 per cent per ton.[18]

These owners had to consider their own interest, for most who hired ships to the boards possessed small numbers of vessels. In 1776, 121 vessels were hired by the Navy Board from fifty contractors. Most owned 1 or 2 vessels. But one, Messrs Wilkinson (perhaps acting as a broker), hired out 57 vessels. Between 1779 and 1782 the Navy Board hired 179 vessels under 72 contracts. Again, most owners hired 2 or 3 vessels; two hired out 10; one – Messrs Wilkinson – hired out 69.[19]

For the naval boards, the obvious solution to rising prices was a reduction in the competition by rationalisation of the number of boards competing for the available shipping. For its part, however, the Treasury Board wanted to dispose of its burden of hiring ships for army victualling, in which it had no expertise. In 1778 it considered creating a new board to carry on this business, putting together a commissioner from both the Navy and Victualling Boards to combine their expertise. In 1778 pressure of business prevented a new board being organised.[20] However, in March 1779, the hiring of army victuallers was shifted to the Navy Board which already assisted the Treasury by hiring and fitting troop transports.[21]

The shift in responsibility for the hire of army victuallers from the Treasury to the Navy Board was timely, for the European War made necessary the victualling of troops in the West Indies and the Mediterranean as well as in North America. It was also logical. For, in addition to conveying the stores demanded by the navy, the Navy Board already hired and fitted the transports that carried the soldiers – infantry, cavalry,

---

[18] TNA, ADM. 108/1B, 24 July 1781.

[19] WLC, Shelburne Papers, vol. 151, 24 Oct. 1782. Syrett, 'The Victualling Board charters shipping, 1775–82'.

[20] M. E. Condon, 'The administration of the transport service during the war against Revolutionary France, 1793–1802', unpub. University of London Ph.D. thesis, 1968, 32–4; Syrett, *Shipping and the American War*, 133–4.

[21] See, for example, 'Transport letters from the dockyard commissioners and officers at Plymouth', 1776–81, TNA, ADM. 108/1A.

horses, camp equipment, army clothing, quartermasters' stores – and in the dockyards the board possessed the skills and facilities to undertake the inspection, hiring and modification of ships.[22]

But it was a difficult job, the Navy Board was already burdened, and the clerks were paid very little for their extra duty.[23] After 1780 the Navy Board found it extremely difficult to charter additional tonnage on account of the length of the 'course' of the navy, which meant that it was constantly outbid for ships by the Ordnance and Victualling Boards. Urging rationalisation, in March 1782 Sir Charles Middleton proposed that both the Victualling Board and the Ordnance Board should permit the Navy Board to perform their hiring.[24] He argued that those boards were managed by men unfamiliar with the shipping business, that they had thus been imposed on, and their loss was aggravated by their failure to employ agents 'to control the conduct of the masters'. He claimed that when the whole was 'under the direction of one board, the several branches will be made to co-operate and assist each other. The business will be managed frugally for the public, and from there being no longer any competition the owners must be satisfied with a reasonable profit.'[25]

But neither Middleton's arguments nor temporising expedients had any real effect. In March 1782 the Victualling Board agreed to hire ships on the same terms as the Navy Board. But in June, finding itself unable to hire enough ships, it again raised its freight rate without informing the Navy Board. Middleton at the Navy Board could only despair 'that no price will satisfy the owners of shipping in our employ while other boards give a higher one, and that the consequence of the Navy Board's advancing theirs under the present circumstances can have no other effect than obliging the other offices to do the same'.[26] The war therefore ended with no solution to the problem of the competition.

Differences in procedure and terms of hire

The problem was complicated by the different procedures and terms of hire employed by the boards. Their general practice was the same: after

---

[22] Syrett, *Shipping and the American War*, 138, 181.

[23] *Ibid.*, 78, 132–6. The commissioners of the navy each received £300 a year for undertaking the extra work; the Board Secretary £150; and the clerks together £400. The latter objected that their recompense was not commensurate with the work but, threatened with discharge, the clerks 'sunk under the great hand of power'.

[24] Condon, 'The administration of the transport service', 33–4, 36; Syrett, *Shipping and the American War*, 101.

[25] NMM, MID.7/1, labelled by Middleton 'No. 3 – Re the Transport Service' and headed 'Transports'.

[26] NMM, ADM. BP/3, 26 June 1782.

the offer of tenders, they tentatively selected the ships to be hired, agreed charter parties and had the ships inspected, the reports to the Navy and Victualling Boards determining which of the ships would actually be chartered. But in detail their respective practices varied.

For the Navy Board, ships were inspected at the dockyards by the master attendant, master shipwright and clerk of the survey. They reported on the age, condition of hull and rigging, on the space between decks, the gross tonnage of the vessels, and provided a complete inventory of equipment.[27] The work added considerably to the burden on the yard officers: at Plymouth the master house carpenter helped the master shipwright, managing to survey about one ship a day, which included work on Sundays.[28]

For the Victualling Board, each ship was inspected by its hoytaker in the River Thames. He too inspected the condition of each ship. If, according to the hoytaker, a ship was old and unfit for service, the ship would not be chartered. There is no indication that ships' bottoms were inspected and bored, as required by the Navy Board. But the Victualling Board's hoytaker, carpenter, sailmaker and ropemaker did appraise each ship for tonnage and value, which served as the basis for any payment of compensation should a ship be captured or otherwise lost during government service.

The inspection was important not only to establish the worthiness of ships in view for hire, but because some owners or masters exaggerated the carrying capacity of their vessels, which could have been both inconvenient and uneconomic. A shortfall in the troops or cargo carried could mean another vessel had to be hired and another cargo despatched. For the Navy and Ordnance boards, which both hired ships to carry troops as well as materials, carrying capacity was indicated by gross tonnage. This and the number of months needed were used as the principal components by which payment was made for ships.

But the means by which gross tonnage was calculated was a source of contention. For the ship owners used one method of calculating tonnage – keel length multiplied by breadth of beam, multiplied by half of beam, all divided by 94. Meanwhile the dockyard officers used another method – keel length less three-quarters of breadth, multiplied by breadth, multiplied by half the breadth, all divided by 94. Conflict was inevitable. For the latter method produced a figure that was about 81 per cent that of

---

[27] Syrett, *Shipping and the American War*, 108–16; for copies of surveys, see *British Naval Documents*, 452–3.

[28] TNA, ADM. 108/1A, 22 Feb. 1778; the volume contains reports from Plymouth to the Navy Board, 1775–84.

the ship owners! The yard officers' figure thus reduced by about one-fifth the amount of freight the Navy Board had to pay for the hire of each vessel.[29]

Meanwhile the Victualling Board hired ships not according to their gross tonnage but according to the tonnage of the cargo they could carry and the length of time it would be carried. That is, the Victualling Board paid freight according to the quantities of the provisions that would be loaded on a ship. It was a practice that had long usage and was fully accepted by ship owners. In the mid eighteenth century, 8 barrels of pork, 30 bushels of pease, 20 hundredweight of flour, or 12 bags of bread were each considered to form a single ton of freight. In 1745 an ideal freight totalled 325 tons and comprised 4,510 gallons of oil, 400,000 pieces of beef, and 180,000 pieces of pork.[30]

The Navy and Victualling boards operated the same policy with regard to manning. Both required owners to man the ships at 'the rate of seven men to every hundred tons' and in May 1780, when the shortage of seamen was critical, both reduced that number to six men. These seamen were issued with protections from impressment.

However, the Navy and Victualling boards differed in their respective requirements for ships under hire to be equipped with guns. The Navy Board required ships to be armed with 'at least six carriage guns of six pounders or less bore as the board shall think proper according to the size of the ship'; the owner also had to provide twenty shot for each gun. From January 1776 the Navy Board even offered 'gun money' of £5 per gun to induce owners to provide more than six guns. And in November 1779 carronades, not less than 12-pounders, were permitted instead of 'common guns'. The Victualling Board, on the other hand, made no mention of guns or gun money in its charter parties; and it may be assumed that, even during the American War of Independence, victuallers were not usually required to be armed.[31]

Upon favourable report on ships from the inspecting officers, the boards drew up their hire contracts. The Victualling Board charter parties were similar to contracts of hire used by the Navy Board. The contract set out the obligations of the owner as well as responsibilities of the hiring board. Contracts remained virtually unchanged between 1741 and 1776 and specify that a ship was 'To commence pay on the first day of July next, upon producing a certificate of the ships being completed, fitted,

[29] Syrett, *Shipping and the American War*, 110–12.
[30] Syrett, 'The Victualling Board charters shipping, 1739–48'.
[31] D. Syrett, 'The Navy Board and merchant shipowners during the American War, 1776–1783', paper presented to 'The Shipowner in History' symposium at NMM, 1984, 7; Syrett, 'The Victualling Board charters shipping, 1775–82', 222.

victualled, manned and provided with all proper necessaries and stores for the ship's company and ready to sail'.[32]

The owners had to equip the ships with sails, blocks, cables, anchors, etc., to supply provisions for the crew, and to man them so that they were ready for employment. The date of payment was used as an incentive and deadline for the ship owner to get his ship properly equipped: as the Deptford dockyard officers observed during the American War, 'it is in the owners' interest to get down [river] and in pay as early as possible [and] we presume there is not unnecessary delay'.[33]

To ensure ship owners had fulfilled their obligations with regard to the above points, both boards demanded a final inspection and report on ships coming under hire. Approval gave rise to the issue of charter parties,[34] the beginning of payment for the ships, and the despatch of warrants for the loading and sailing of vessels.[35] Troop transports sometimes had to be fitted with cabins at the dockyards. At Plymouth in February 1776, cabins were fitted by contractors who were instructed by house carpenters who travelled from Deptford for that purpose. Using some contract labour, the master shipwright reported he could prepare fourteen transports in six days.[36] Once fitted, they were ready for the troops, their victuals and despatch.

Protection of transports

By the 1740s charter parties usually specified that victuallers bound overseas to the Mediterranean or the West Indies must sail in convoy. These convoys invariably congregated at Spithead, that anchorage being preferred in winter to the Downes. It was convenient to gather ships together before they sailed, rather than collecting ships in passage. Ships with provisions from Ireland thus often joined at Spithead rather than being collected on the outward voyage. Ships sent to, or hired in, North America to carry victuals like rice to the West Indies were convoyed as a matter of routine, under a general order, by ships proceeding in the same direction. Instructions in 1741 to a captain sent to New York, for example, required him to 'give convoy to all our victuallers from New York on his return'.

---

[32] TNA, ADM. 106/2179, 8 June 1741, quoted in Crewe, *Yellow Jack and the Worm*, 177.

[33] TNA, ADM. 106/3404, fo. 134, quoted in D. Syrett, 'The Navy Board and merchant shipowners', 10; for the copy of a contract for 1776, see *British Naval Documents*, 448–51.

[34] NMM, MID.7/1, fo. 2.

[35] Syrett, 'The Victualling Board charters shipping, 1739–48'; Crewe, *Yellow Jack and the Worm*, 178.

[36] TNA, ADM. 108/1A, 22 Feb. 1776.

There were exceptions. Some owners of large, well-armed ships contracted to sail without convoy. In 1744 an owner chartered two ships, each of 550 tons, manned with sixty men and armed with twenty carriage guns, to be escorted down the Channel into the Atlantic but then to sail alone to Jamaica. The arrangement suited the Victualling Board as the freight rates were not much higher than those for ships tied to convoys and the arrangement continued until 1747.[37]

During the American War of Independence, the large number of army victuallers prompted the Navy Board in 1779 to attempt economies in the number of escorts needed by shipping army provisions in two large consignments each year, otherwise making use of routine trade convoys. To begin with, however, the Admiralty failed to provide escorts in time to prevent delays in the despatch of food to the army in America. The Admiralty was stretched to provide frigates and sloops. Yet it also made the escort of victuallers a low priority and under-estimated the number of escorts required. In 1779 victuallers at Cork, loaded by 4 September, did not sail for America until 24 December owing to delays in the appearance of escorts. In 1780, victuallers that should have sailed in May and July did not arrive in New York until November; those intended for Canada did not get there.[38]

Protests from the army's Commander-in-Chief, its Commissary-General and the Secretary of State for America prompted a review of procedure. In September 1780 the Treasury decided 'to have the supplies sent out in less quantities, and by more frequent convoys', and the Navy Board responded with a proposal for four convoys a year. The Admiralty acknowledged 'the importance of this service in its utmost extent' and agreed 'to supply convoys when applied for tho' there is reason to apprehend it will interfere with other very material services'. Subsequently, between 1781 and 1783, the Admiralty generally did provide escorts in time to permit victuallers to sail on schedule.[39]

Control of transports

Once transports were despatched, there was the problem of getting them back. During the American War of Independence, in 1779–80, the Navy Board had just enough ships under hire to fulfil its engagement to the army to ship its victuals. However, that sufficiency depended on ships

[37] Syrett, 'The Victualling Board charters shipping, 1739–48'; Crewe, *Yellow Jack and the Worm*, 176–9.
[38] NMM, ADM. BP/1, 26 Sept. 1780.
[39] Syrett, *Shipping and the American War*, 154–60.

being recycled. The Navy Board's transport agents were directed not to let any ships remain in American waters. But by the late summer of 1780 the Navy Board was unable to report the return of any of the victuallers despatched in 1779. Threatening 'fatal' consequences for the army and 'enormous' expense to the public, that October the Navy Board forcefully represented the

necessity of their Lordships interposing their authority with the officers commanding abroad, and that it may be recommended to them in the strongest manner not to detain more transports than what are absolutely necessary for the use of the army, nor to divert the storeships or victuallers, as has been the case with the Pacific for upwards of two years past, to any other purpose than those for which they are sent out, and to furnish the agents who conduct them with convoys to Europe as soon as they are cleared or discharged.[40]

While the Treasury directed the army's Commander-in-Chief in America to ensure his officers returned ships promptly, the Admiralty directed naval officers to ensure the vessels were given convoy promptly. Yet these joint efforts failed. By February 1781 the Navy Board estimated that it had under hire one-third more tonnage than it actually needed, had ships been returned promptly to Britain.

The shortage of ships continued to the end of the war. Small numbers of ships were removed from the cycle of replenishment because of detention in North America or the West Indies by customs officials, because they missed their convoy date, or because their crews were impressed and they had difficulty raising another, especially in American waters where seamen were scarce. A more serious cause was due to ships not being unloaded sufficiently quickly. This was because the ports were sometimes too shallow to receive them or because quays or warehouses were damaged or lacking. Ships were often detained to act as floating warehouses, hospitals or prisons. In September 1780 some victuallers had been at New York for eighteen months without even being examined. Their cargoes would inevitably be reduced by vermin. At St Lucia in 1782 there were about fifteen victuallers which remained loaded, and would remain so indefinitely because storage on shore was sufficient to receive the contents of no more than two.[41]

Victuallers were also retained abroad to serve as transports carrying troops north or south from New York and to collect supplies like coal from Cape Breton. But, above all, they remained because they lacked orders. The army Commander-in-Chief blamed the navy for not ordering escorts, while the navy Commander-in-Chief blamed the army for not

---

[40] NMM, ADM. BP/1, 10 Oct. 1780.   [41] Syrett, *Shipping and the American War*, 161–8.

ordering the unloading of vessels quickly enough. Meanwhile, empty vessels remained at New York up to six months, waiting to return to England. The fundamental problem seemed to be a shortage of ships officially designated to act as escorts. Without ships being specifically allotted to this service, local commanders could always find a higher priority for which to use frigates or sloops. Thus, ultimately, it was the Admiralty's reluctance to allot warships to escort duty that lay behind the shortage of victuallers in Britain.[42]

It was a problem that persisted to the very end of the war. By the end of July 1782, there were at least 56 troop transports and victuallers in America, 109 on their way and 81 ready to sail.[43] At the end of 1782, without the ships retained in American waters, the Navy Board could not contemplate shipping all the provisions required by the army overseas and it resorted to hiring neutral ships and freighting provisions in merchant ships. Ultimately, absence of control of the tonnage under hire was at fault, reflecting a lack of comprehension in the British government and a failure to give shipping the priority its management required.[44]

A Transport Board re-instituted

After the American War, Sir Charles Middleton made the recommendation for a separate Transport Board to the Commissioners on fees.[45] Although the commission primarily investigated perquisites, it served as a vehicle for other recommendations. The Commissioners were already relatively well informed about shipping, for they included Francis Baring who had managed the victualling business of the army for the Treasury in the last year of the war. Criticisms of, and recommendations for, the transport service were scattered through both their fifth and eighth reports on the Navy and Victualling Offices.[46]

The fifth report echoed Middleton's criticism of several boards taking up ships for transports. The eighth report noticed abuses that had occurred under the management of the Victualling department's hoy-taker, who hired ships in which he or other members of the department had a financial interest. It also noticed the different systems by which the various boards measured the tonnage of ships and the competition

[42] *Ibid.*, 168–72.
[43] NMM, MID.7/1, fo. 3. Figures given in tonnage, which are converted to number of transports at the rate of 270 tons to a transport.
[44] Syrett, *Shipping and the American War*, 171–80.
[45] For his command of logistics, see his memorandum of March 1782 printed in *Letters of Lord Barham*, II, 47–50.
[46] Condon, 'The administration of the transport service', 35.

for ships, whereby the owners gained at a cost to the public, 'which they could not possibly accomplish if the whole of the transports required for Your Majestys's Service were to be hired exclusively by one board'.[47]

In 1793 Sir Charles Middleton was out of office but still communicating by letter with Pitt, and Henry Dundas of whom he was a distant relation.[48] A note in Dundas's papers states that the inefficiency of not having a Transport Board 'operated so much on the mind [of Middleton] that it induced him to submit to Mr Pitt the absolute necessity there was for its existence'. The note claims that Pitt 'was so convinced of it as immediately to put it into existence'.[49] Hence in July 1794, at the same time as Henry Dundas was made Secretary for War and Colonies, a new Transport Board was formed.[50]

The Order in Council of 4 July maintained that the Navy Board could no longer conduct the transport service for the army 'without great detriment and inconvenience to the more immediate duties of their office', and approved the formation of a specialist board under the direction of the Treasury commissioners. The board comprised two sea captains and a civilian. Both the former[51] became rear-admirals in 1795 and were replaced by other captains, one being Rupert George who remained at the board until the end of the Napoleonic War. The civilian, Ambrose Serle, remained at the Transport Board until about 1809; he was a friend of Middleton's and assisted him as a Commissioner of naval revision between 1805 and 1808 when he wrote the report on the Transport Board.[52]

The function of this board was then defined as the hiring and appropriating of ships and vessels for the conveyance of troops and baggage, victualling, ordnance, barrack, commissariat, naval and military stores of all kinds, convicts and stores to New South Wales, and a variety of miscellaneous services.[53] The duty of hiring and managing ships to convey the army and its food supply remained central to the functions of

---

[47] *Ibid.*, 36, 44; 8th Report of Commissioners on fees, 25.
[48] *Letters of Lord Barham*, II, 386–409.
[49] NLS, Melville Papers, 1044, fo. 107, cited in Condon, 'The administration of the transport service', 48.
[50] Condon, 'The establishment of the Transport Board'.
[51] Sir Hugh Cloberry Christian and Philip Patton.
[52] See NMM, MID.1/168, 10, confidential letters from Serle, 1789–90; Condon, 'The administration of the transport service', 49–50; Morriss, *Royal Dockyards*, 201, 203.
[53] The Transport Office was twice significantly enlarged. In 1795 it received two commissioners and office staff from the Sick and Wounded Office, when the duty of caring for prisoners of war in health was shifted to the Transport Board. In 1806 it received a physician-commissioner and more clerks when the Sick and Wounded Board was dissolved and the care of the sick and wounded seamen was placed under the Transport Board.

the board until 1816 when it was dissolved. The Transport Office then reverted to a sub-section of the Navy Office.[54]

The board in operation

The new board commenced management of the transport service from 1 September 1794. It completely relieved the Treasury, the Navy Board, the Victualling Board and the Ordnance Board of the task of hiring transports. This alone eradicated competition, standardising terms and conditions of hire.[55] To remove the burden of inspecting ships tendered for transports from the Navy Board's dockyard officers, it appointed its own professional staff to survey, measure, value and report on ships offered for hire. The Victualling Board's hoytaker was also relieved of his duty of inspecting and hiring ships to act as victuallers. The only duties remaining to him were the occasional hire of small craft for shipments of provisions to the outports, attendance upon the issue and return of food into store, and account-keeping of provisions on board ships in the Thames.[56]

From the beginning, the Transport Board established a reputation for efficiency. Even before taking full control, it ruled in August 1794 that any employee having any interest, direct or indirect, in any ship hired as a transport would be dismissed forthwith.[57] The ruling came five days after the master of a naval victualler owned by the former hoytaker claimed freight for £1,409 which the Victualling Board would not sanction, having no knowledge of the terms upon which part of the sum was earned.[58]

With this new beginning, the new board assumed an authority that was independent of the existing naval boards. Responsible to the Treasury, the Transport Board received the King's commands directly from the Secretary for War and Colonies rather than from the Admiralty. Indeed, though a formality, it was the Transport Board which reminded the Admiralty of Cabinet decisions at the same time as it requested convoy for the vessels, the Transport Board having already by then directed the masters of transports where to embark troops, ordered its own local

---

[54] Condon, 'The administration of the transport service', 52.
[55] For records of ships hired, 1806–11, see TNA, ADM. 108/150.
[56] 11th Report of the Commissioners of naval revision, 18, cited in Condon, 'The administration of the transport service', 62.
[57] TNA, ADM. 108/31, 26 Aug. 1794, cited in Condon, 'The administration of the transport service', 62.
[58] NMM, ADM. DP/14, 21 Aug. 1794; the ship was the *Crescent*, James Loring master, owned by Messrs St Barbe, Green and Bignall.

agents for transports to manage these embarkations, and instructed its agents to await orders from the Admiralty regarding convoy.[59]

The authority of the new commissioners permitted the Board to transact business with the other boards on equal terms. Such independence was vital to the efficiency of its coordinating role, for, from 1794, the Transport Board hired army and naval victuallers as well as the transports required to carry stores, troops and ordnance. When ships were needed, application was made to the Transport Board for shipping to convey a specified tonnage of provisions or number of troops and the Transport Board responded with ships that were available at the requisite port, making up the necessary tonnage as more were hired or returned from abroad.[60]

Many of the officers appointed to work under the Transport Board were already experienced in the transport business, having worked for the Navy Board or in the Navy Office. Naturally enough, the business as run by the Transport Board much resembled the organisation run previously by the Navy and Victualling boards, and by the previous Transport Board before 1717. To coordinate hired transports, embarkations of troops or provisions and despatch in convoy, the Transport Board employed local agents – to be found at Deptford, Woolwich, Gravesend, Dover, Deal, Portsmouth, Cowes, Southampton, Plymouth, Bristol, Guernsey, Waterford, Cork, Liverpool and Leith.[61] They ensured ships under hire were clean, fitted and stored for the services they were to perform. From 1795 standards of cleanliness for troops were set by Sir Jerome Fitzpatrick, Inspector of Transports, appointed by the Secretary of State for War. In addition there were agents afloat, for example at Spithead, and with large contingents of transports. Because the floating agents were sometimes of higher rank or seniority to those on shore, a standing rule was that the authority of resident agents took precedence over that of the agents afloat. To ensure observance of correct procedures, agent business had to be reported to the Transport Board in regular abstracts.[62]

The transport service was highly centralised. No agent was permitted to draw upon the Transport Board for money to conduct any part of

---

[59] TNA, ADM. 1/3730, 6 Sept. 1794, cited in Condon, 'The administration of the transport service', 60–1.

[60] See, for example, ADM. DP/32B, 30 July 1812, for ships required to convey provisions to the north coast of Spain.

[61] The main locations are given each year in the *Royal Kalendar*. For the work of Captain Daniel Woodriff, transport agent at Portsmouth and Southampton, 1795–7, see his in-letter book, TNA, ADM. 108/28.

[62] TNA, ADM. 1/3741, 19 Aug. 1801, cited in Condon, 'The administration of the transport service', 51, 55.

the service without first submitting the necessity and justification to the board, whose approval and consent had to be received prior to any issue of bills.[63] Of course, especially overseas, agents could be instructed to hire transports by a Commander-in-Chief. But, even then, the agent was required to receive written orders from that Commander-in-Chief, and to arrange hire terms as near as possible to those of charter parties issued by the Transport Board, at rates as cheap and reasonable as circumstances permitted. The documentation of these transactions, with justifications, had to be returned to the Transport Board at the first opportunity.[64] All this required that the agents corresponded regularly with the Transport Board.[65]

To supply provisions to the troop transports and the army victuallers, the Treasury worked as closely with the Victualling Board as it did with the Transport Board. It corresponded with both boards simultaneously to ensure food supplies and transports were coordinated.[66] To save repeated directions to the Victualling Board, in September 1794 the Secretary for War requested the Admiralty to order the Victualling Board to comply automatically with demands for the supply of provisions made directly to it by the Transport Board. This the Admiralty did, to which the Victualling Board had no objection.[67]

However, an important demarcation dispute immediately arose over which of them, the Transport or the Victualling Board, should direct and control the transports once they were laden with provisions. The Victualling Board based its claim on previous directions from the Admiralty; that is, the situation before August 1794. Henceforward, however, the Admiralty permitted the Transport Board to direct transports when and where to proceed at the same time as it requested convoy. The task of the Victualling Board was accordingly reduced to just that of supplying the appropriate provisions to troop transports and victuallers.[68]

---

[63] TNA, ADM. 108/31, 29 Sept. 1794, cited in Condon, 'The administration of the transport service', 56.

[64] Article XII of instructions to agents, TNA, ADM. 1/3737, 14 Jan. 1799.

[65] TNA, ADM. 108/31–70.

[66] For an example of this coordination, see TNA, ADM. 109/102, 19 Sept. 1794, cited in Condon, 'The administration of the transport service', 63.

[67] TNA, ADM. 1/4162, 3 Sept. 1794, cited in Condon, 'The administration of the transport service', 64.

[68] Having laden the transport *Active* and given her master directions to proceed to Spithead, the Victualling Board claimed 'a right to send out provisions, and by that right to give their directions for the movements of the transports, not admitting that the Transport Board had any power to do more than to hire in the first instance'. Meanwhile, the Transport Board claimed 'the victualling service should extend no further than...receiving...the species of provisions they had occasion to embark': TNA, ADM. 1/3730, 17, 20 Sept. 1794; ADM. 108/31, 17 Sept. 1794.

The reasoning applied to the Victualling Board was also applied to the Ordnance and Navy boards. The Transport Board provided shipping according to the ordnance and naval stores that had to be shipped. And it was the Transport Board that informed the Admiralty when they were ready to sail under the auspices of one of their agents and needed convoy.[69]

The Transport commissioners were expected to meet as a board five days a week. Yet from the first they found it necessary to meet on Saturdays, and thereafter it maintained a six-day week throughout the French Revolutionary War. When transports were needed by any of the other boards, application was made to the Transport Board, which invariably provided the necessary tonnage without delay. This became a regular and generally trouble-free routine. Occasionally, exceptional arrangements or expenses had to receive the sanction of the Admiralty or Treasury boards. More usually the Transport Board kept the Admiralty informed of what transports were employed and where. Indeed, from October 1795, it transmitted to the Admiralty weekly a list detailing all the victuallers and storeships in employment, their ports of departure and intended destinations. In this way the transport service came under the view of the Admiralty, which was able to use its ships, as charters permitted, when they were needed.[70]

The scale and speed of the service

The Transport Board's command of all hired shipping worked well. From 1794 the elimination of chartering by a multitude of boards not only gave greater control of ships but reduced competition in hiring them, and reduced expense arising from that competition. The routine of providing the tonnage, once established, reduced inter-departmental communication and the time that took. The Navy, Victualling and Ordnance boards were thus able to concentrate to a greater extent on their own specialist duties. At the same time the specialisation of the Transport Board brought greater efficiency and speed of operation to the hire and despatch of transports.

The Transport Board brought a range of improvements in efficiency. Although the Navy Board had hired transports from as many as 72 contractors, one-third of the vessels from one broker,[71] by 1801 the Transport Board had contracts with over 300 owners. Hence it was more able

---

[69] TNA, ADM. 1/3730, 20 Oct. 1794, cited in Condon, 'The administration of the transport service', 66–7.

[70] Condon, 'The administration of the transport service', 67.

[71] WLC, Shelburne Papers, vol. 151, 24 Oct. 1782. Syrett, 'The Victualling Board charters shipping, 1775–82'.

to pick and choose its ships and contractors. Registration of shipping having begun in 1787, the process of hire was simplified by the decision to take the registered tonnage of ships to indicate their size instead of measuring their tonnage. In consequence, the cost of surveying vessels was much reduced; the cumulated saving was estimated at nearly £1,222,000 by 1815. The Transport Board's agents became more professional in the sense that, although already commissioned sea officers, they had to undergo examination by one of the transport commissioners. Changes and additions were also made to charter parties: the product of experience, including regulations for the guidance of ship masters.[72]

Specialisation and regulation standardised procedure, and helped the Transport Board to hire ships more quickly. Such was the pace of hire of ships for the large-scale expeditions to the West Indies in the French Revolutionary War, they were taken up even before the Transport Board had decided the nature of the freight they should carry. They were allotted to carry troops, victuals or stores as the board's surveyors recommended after hire. Whereas in 1776 ships to convey 27,000 troops, their provisions and ordnance to America had taken nine months to hire and despatch, in 1795 the ships needed to convey a greater number of troops on the first Christian–Abercromby expedition to the West Indies were hired and despatched in only three months; 14,000 went in ships drawn from the East and West India trades.[73]

Four years later, the Transport Board achieved even greater pace. The Helder expedition in 1799 demanded the conveyance of 46,000 men from England and the Baltic to north Holland. Preparations were begun at the end of June 1799. Within a month the Transport Board had 44,000 tons of transport, three-quarters of which had been newly hired into government service. After three months the board had collected, fitted and despatched 90,000 tons of shipping; that was nearly 6 per cent of all the tonnage available, taken up at a rate of 30,000 tons, or 100 ships, a month.[74] At the rate of 2 tons of shipping per soldier, the Transport Board was conveying 15,000 men on average every month. Of course the Helder was closer to England than the West Indies and in 1799 the shorter distance permitted some transports to make three successive trips carrying troops.[75]

Striving for speed of operation, the Transport Board managed to eliminate the interference of customs officers in the West Indies. In England, ships loaded with provisions did not usually clear out through customs

[72] Condon, 'The administration of the transport service', 312–14.
[73] Ibid., 315–16, 319. See also Duffy, *Soldiers, Sugar and Seapower*, 184–96.
[74] Calculated at 270 tons a ship.
[75] Condon, 'The administration of the transport service', 317.

houses. In the West Indies, however, the ships were often stopped and delayed by customs officials. At Martinique in 1798 a victualler, the *Intrepid*, which failed to report to the customs house, was seized and sold! By then, however, under pressure from the Transport Board, the Treasury had ordered customs officers in England to make monthly returns of transports arriving and sailing. At the same time, to enhance the value of this information, the Transport Board instructed the masters of all vessels operating under its auspices to answer fully all the questions put to them by customs officials.[76]

The new board's capacity for speed of operation was demonstrated in 1803 at the declaration which began the Napoleonic War. On 18 May 1803, the new Secretary of State for War and Colonies, Lord Hobart, ordered the evacuation of the Hanoverian army from the River Elbe. While a transport commissioner was sent to the Elbe to prepare craft to ferry the troops down river, the Transport Board was directed to prepare sufficient transports. Seventeen transports, totalling 6,959 tons, were prepared in twenty-four days at freight of £10,840: 'All the ships were hired, fitted, ballasted, victualled, watered, rigged and manned, and had sailed from Deptford by the 11th June.' Only one of the ships was 'rejected, not being completed in time'. The orders for this fleet of transports to sail from the Nore were dated 12 June. Two days later, news was received in London of the capitulation of the Hanoverian army on 3 June. However, as the government was subsequently able to demonstrate, the means of its evacuation could hardly have been prepared faster.[77]

The scale of operations also developed under the Transport Board. The first sailing of the Christian–Abercromby expedition to the West Indies in 1795 involved over 300 transports; the second at the end of the year took 199 transports and merchant vessels.[78] During the Napoleonic War, the confinement of most large-scale amphibious operations to the continent of Europe facilitated short-term hire arrangements and the provision of transports on a large scale. In 1807 the expedition to Copenhagen took 377 transports totalling 78,420 tons.[79] In 1809 the Walcheren expedition took 352 transports.[80] At the same time, the financial challenges of

[76] *Ibid.*, 284.
[77] 'A list of the transports hired by this board for a particular service', Transport Office, 30 Apr. 1804, DRO, 152M/c1804/ON, account number 17.
[78] Duffy, *Soldiers, Sugar and Seapower*, 184–96, 206.
[79] *Naval Papers respecting Copenhagen, Portugal and the Dardanelles presented to Parliament in 1808* (London, 1809), 5.
[80] G. C. Bond, *The Grand Expedition: The British Invasion of Holland in 1809* (Athens, GA, 1979), 14, 17.

the war demanded economy in routine operations. Thus, although the Admiralty retained 17 victuallers to supply the Channel fleet from Plymouth, it was obliged to keep to a minimum the number retained 'for any contingent service'.[81]

The hire of shipping

Economy was achieved through flexibility of hire arrangements. These were adapted to the needs of the transport service but also to the economics of the shipping industry. The hire arrangements made by the Transport Board were thus much like those that were employed by the separate boards before 1794. Ships were hired on terms and conditions of three different types. Cargo space was hired; ships were hired for a single voyage; ships were hired on long-term agreements. For both cargo space and ships that were hired, the Transport Board recorded contract details in ledgers that still survive. Those for the ships include the dates of hire, of entry to payment and of discharge, the rate of payment per ton per month, and the freight earned.[82]

Cargo space for freight

In the eighteenth century some ships sailed to destinations with little outward cargo, and the value of their homeward cargoes covered the cost of their outward voyage. This particularly applied to ships of the East India Company which was anyway partly subject to government, benefited from its garrisons, and tended not to carry much cargo out to India. The East India Company thus normally granted free space to the carriage of troops and stores.[83] Ships also sailed to the West Indies half-loaded, and it made sound sense for their owners to add to the value of their outward voyages. The government thus hired hold space for troops and stores on these outward voyages. Space was also hired in some ships destined for the Mediterranean and Canada.

During the American War of Independence, the use of West India ships to despatch troops to the Caribbean lasted from 1779 until 1782, in which time they carried some 7,500 troops. The arrangement was made after the Secretary of State, Lord George Germain, discussed the idea with the secretary of the West India Committee, whose chairman

---

[81] NMM, ADM. DP/30A, 24 Feb. 1810; ADM. DP/32B, 30 Sept. 1812.
[82] TNA, ADM. 108/148–51, ships' ledgers, 1793–1815; ADM. 108/158–61, freight ledgers, 1795–1818.
[83] Syrett, *Shipping and the American War*, 71–6.

then called a meeting of the merchants and masters trading to the West Indies at the Jamaica Coffee House. There the latter decided upon their terms.

In 1779 the troops were to be paid for at the rate of £6 per head, their beds and bedding being supplied by the Navy Board. Their food, water and eating utensils were supplied 'in the usual manner as on transports' – by the owner. From forty-eight hours after the troops embarked, the owners received demurrage for delays in European ports at the rate of 1 shilling per soldier per day. Demurrage was also paid at 15 shillings per day for ships detained in West India ports. Personal luggage was carried free, but camp equipage was paid for at the rate of 40 shillings a ton.

The use of West India ships to carry troops out to the Caribbean was approved in government because they were large, well manned and generally well armed. It had the advantage for government of releasing merchant tonnage for hire on other terms. But for ship owners it had the disadvantage of ignoring the time a cargo might take to deliver. In August 1775 space in the *Brown Hall* was hired for a consignment of naval stores from Deptford to Halifax, but she was detained by the need to sail in convoy, then diverted to Antigua by adverse winds, and did not reach Halifax until the summer of 1776. In this case, the owners claimed they should have been paid by the month, and the Navy Board actually agreed to pay over half what they claimed.[84] However, the fact that government was able to continue hiring hold space suggests that such unforeseen circumstances were exceptional rather than the rule.

To ensure the masters of hired ships adhered to requirements for freight, payment for a ship's services could be delayed or reduced. Freight money was not usually paid until a voyage was completed and the relevant vouchers, bill and accounts had been cleared in the Navy or Victualling Office. That practice was only broken in 1782 when ships were scarce and owners had to be allowed a small amount of freight money in advance. Nevertheless, masters who failed to take advantage of a convoy could still have a temporary 'stop' put on their payment, while the value of provisions that had gone missing could be deducted from the freight. Indeed, to deter pilfering from cargoes, the owners of victuallers were charged by the Victualling Board double the value of what was not delivered. The Navy Board was equally meticulous – for example, charging the owners of one vessel £7 6s 0d for the loss of eight pairs of stockings from box number 4396.[85]

---

[84] *Ibid.*, 65–9; Syrett, 'The Navy Board and merchant shipowners', 4–7.
[85] Syrett, 'The Victualling Board charters shipping, 1775–82', and 'The Navy Board and merchant shipowners', 15.

After 1794 the Transport Board regularly shipped army provisions as freight or general cargo. Thus, after the initial expedition to the West Indies, victuallers were no longer sent after the transports, West India ships being hired for freight at a rate per ton simply until provisions were delivered, the risk from capture and the sea falling wholly upon the owners. When their cargoes were delivered, they became free of government service and were able to pick up other cargoes for a further or return voyage. East India ships were used to supply the garrison at the Cape of Good Hope on a similar basis. Likewise, provisions were freighted to the continent and the Mediterranean.[86] Ships taken up 'on freight' were used to victual the garrisons at Gibraltar, Minorca and Malta. Though sometimes despatched in small batches of two or three, most went in large convoys. Thirty reached Gibraltar in 1796, all after April owing to the severe winter in England; twenty in 1797; but only twelve in 1798, owing to the British evacuation of the Mediterranean. Following the battle of the Nile, the formation of the second coalition against France and the creation of new garrisons, twenty-five entered the Mediterranean in 1799, and about thirty-eight in 1800.[87]

Ships for a single voyage

There were times when entire ships had to be hired, either for a single voyage or for a period, usually because a destination was not on a normal trade route – for trading ships did not usually enter war zones, especially if no return cargoes offered. During the American War of Independence, private mercantile cargoes did not emanate in any quantity from New York; ships were not attracted there and cargo space to that region was thus not generally available. For deliveries of troops or stores, the Navy Board thus had to hire ships on short-term contracts for the specific voyage, or use ships on long-term charter.

Of these options, the short-term contracts were higher in cost *per diem* and did not have the flexibility of the long-term charters. They were less economical and thus less approved. Nevertheless, on occasions there was no choice, and specific tasks had to be performed on this basis. Hence, during the American War, cargoes of military clothing and oats went out under single-voyage contracts. The freight rate was agreed for, say, a single month. Yet in these brief periods delays could and did occur, and ship owners on these contracts did not hesitate to charge demurrage.[88]

---

[86] Condon, 'The administration of the transport service', 276–8.
[87] TNA, ADM. 1/3730–42, Oct. 1795–Dec. 1801, cited by Condon, 'The administration of the transport service', 293.
[88] Syrett, *Shipping and the American War*, 69; and 'The Navy Board and merchant shipowners', 7.

Ships under long-term contracts

During the American War of Independence, the Navy Board was the board which hired most shipping on a long-term basis. In July 1776, at the beginning of the war, the Navy Board had under long-term charter 416 ships serving as troop transports, and the number of these vessels never fell below about 150 ships until 1782. Between 1780 and 1783, the Navy Board also had under hire at least 97, rising to 212, army victuallers.[89] It should be noted that this excluded the ships hired by the Navy Board to carry naval stores between the dockyards, yards overseas and the fleet; by the Ordnance Board to carry ordnance equipment and stores; or by the Victualling Board to carry food to the fleet. The overall number of ships under hire was thus much larger.[90]

Ships hired under long-term agreements were hired for 'six months certain', but that time-period was usually over-ridden in the contract by a clause specifying the ship would remain in service until 'receiving notice of discharge'.

The hire of shipping on this basis by the Navy Board was triggered by instructions from the Secretary of State to the Admiralty. At the beginning of hostilities these instructions were often secret, for two reasons. Firstly, the equipment of transports to carry troops could serve as warning to the enemy of an impending expedition. Secondly, were they to discover the scale of the government's demand for shipping, ship owners would be likely to raise their charter price. But once hostilities began, the need for shipping was disseminated publicly, either by word of mouth to ship owners, brokers and underwriters in the City of London, by posting notices at such gathering places as the Lloyds and West India Coffee Houses, or by an advertisement in newspapers. In 1778 and 1779 the Navy Board placed the same advert for about two weeks in three daily newspapers: that it was 'ready to treat, on Monday the 4th of next Month, at noon, for the Hire of TRANSPORTS for foreign service'.[91] With some minor differences, the Victualling Board did the same.[92]

At the same time as charter prices were agreed, so too was the voyage or period of service, and the date and place from which the ship

---

[89] Syrett, *Shipping and the American War*, 249–50; and 'The Navy Board and merchant shipowners', 2.

[90] Similarly, during the War of 1812, 991 ships were hired on long-term charter, only 29 for a single voyage: NMM, ADM. BP/46B, 17 May 1826.

[91] The Navy Board advertised in the *Daily Advertiser*, *Morning Chronicle and London Advertiser* and the *Public Advertiser*: Syrett, *Shipping and the American War*, 82.

[92] The Navy Board also placed adverts for ships in the *London Gazette*, but the Victualling Board does not appear to have done so. See Syrett, 'The Victualling Board charters shipping, 1775–82'.

should be ready to sail.[93] Under the terms of a Navy Board contract, the ship could be directed to whatever port the board should direct to load 'soldiers, horses, women, servants, arms, ammunition, provisions and store as shall be ordered to be put on board her'. The region of the world to which the ship would be destined was specified, where the charter party permitted unloading and reloading with like cargo. In this task, the master and seamen were instructed to assist with the aid of their boats.[94] The charter parties kept open a variety of uses for the ships under hire. Hence the Navy Board could use a vessel to carry naval stores, troops as directed by the Treasury or, from 1779, army victuals. The completion of service was also left deliberately open, vague provision being made for the return voyage to England or to another part of the world.[95]

After 1794 the Transport Board adopted the best features of past practice in forming long-term charters. Most ships were hired for six months certain. The greatest number under hire at the end of the Napoleonic War comprised 991 vessels under long-term charter and 29 hired for a single voyage.[96]

Function, destination and size

One great benefit of a large number of transports under hire was the ability to use the ships flexibly according to their suitability for a particular task, in particular according to their cargo and draught. In 1782 thirty-nine ships totalling 14,225 tons were hired by the Navy Board to carry victuals to the army in North America and the West Indies (see table 7.9). They had a mean individual tonnage of 365 tons. However, their size tended to vary according to their destination. (see table 8.1). This was more obvious in the size of 377 transports totalling 78,420 tons employed in the Copenhagen expedition of 1807. Each had an average tonnage of 208 tons, perhaps on account of the shallow waters they had to penetrate.

Care is necessary in generalising from the numerous lists of transports that exist, for the different boards before 1794 calculated tonnages differently: gross for the Navy Board, carrying burden for the Victualling Board. After 1794 the Transport Board maintained the Navy Board's ledgers, from which precise figures for individual tonnages, prices of

---

[93] Syrett, 'The Victualling Board charters shipping, 1739–48'; Crewe, *Yellow Jack and the Worm*, 176–8.
[94] Syrett, 'The Navy Board and merchant shipowners', 12.
[95] TNA, ADM. 106/2179, 8 June 1741, cited in Crewe, *Yellow Jack and the Worm*, 177.
[96] NMM, ADM. BP/46B, 17 May 1826.

Table 8.1 *The tonnage of transports despatched to Charlestown, New York and the West Indies in 1782*[a]

Destination	Total transports	largest	smallest	average
Charlestown	15 vessels	512 tons	168 tons	293 tons
New York	18 vessels	538 tons	107 tons	379 tons
West Indies	6 vessels	634 tons	272 tons	500 tons[a]

[a] NMM, ADM. BP/3, 13 Apr. 1782.

hire and payments may be obtained.[97] There was also continuity in the sizes of ships used by the Navy and Transport boards. Twenty-three horse transports under hire to the Navy Board in 1758 ranged between 200 and 345 tons but had a mean size of 270 tons. Over half a century later, between 1812 and 1815, the average tonnage of ships under hire to the Transport Board for all purposes was 269 tons.[98] During this half-century the size of ocean-going ships had grown. The boards' figures suggest, however, that convenience, flexibility and continuity of cargo sizes may have been more important in determining the hire of ships.

The service at the time of the War of 1812

By the time of the War of 1812, the British transport service had reached a stage of maturity. It had an established capability for meeting the need for fleets of transports. It employed the greatest number of merchant ships ever taken up as transports and was capable of serving military forces on both sides of the Atlantic. It was a measure of the service's power of projection that British strategy against the United States was to attack her coastal cities. It still suffered from problems that existed in the American War of Independence. However, in quality it was a different service to that which existed thirty years earlier. Unified, centrally regulated, it had become a professional arm of the state, with expectations of high commitment from all employees, ship masters as well as agents.

The challenges of scale

By the end of the Napoleonic War the fleet of transports managed by the Transport Board reached 1,020 vessels, with a combined size of 276,554

[97] For the Navy Board and Transport Board, TNA, ADM. 108/148–51, 1793–1815.
[98] NMM, ADM. BP/109 [no day or month given], 1758; ADM. BP/46B, 17 May 1826.

tons.[99] In 1815 the British-registered fleet contained 21,861 sailing ships amounting to 2,477,000 tons. The British government thus employed almost 5 per cent of the available ships and, having hired the larger vessels, 11 per cent of the available tonnage (see table 3.2).[100]

However, for all the shipping at its disposal, the British transport service still suffered from pressure on its resources. Demands on the ships available still exceeded their supply and partly because ships were now adapted to certain purposes, which gave rise to shortages of particular types. These revealed themselves in short-term inconveniences such as that encountered by the Victualling Board in July 1812. That board wanted a transport to carry provisions to Sir Home Popham's squadron on the north coast of Spain; the Transport Board had no vessel at Plymouth that was suitable for that purpose so it gave the Victualling Board a smaller one at Portsmouth. Yet the latter board was assured that 'a large fleet of transports are expected very shortly to arrive from Lisbon, and the first vessel fit for the conveyance of provisions will be appropriated for conveying the residue of the provisions proposed to be sent'.[101]

A year later, the shipping shortage had become more serious. The war with the United States of America, in addition to that in Europe, absorbed more transports and for longer voyages. In June 1813, when the Victualling Board wanted 3,300 tons of transport – twelve vessels – to convey provisions to the Mediterranean, the Transport Board could supply only two-thirds of them. The chairman of the latter board explained that

Notwithstanding every exertion on the part of the Transport Board, the unexampled demands which have been made upon it for army services, the great scarcity of disposable shipping applicable to the Transport Service, and the detention of a considerable proportion of the transports which have been sent out to foreign stations, render it impracticable to procure ships with sufficient expedition to comply with the unbounded demands which have been and continue to be made.

The Victualling Board could only point out the Mediterranean fleet had only a month's supply of provisions in hand and was thus liable to 'serious inconvenience'. The Admiralty Board was obliged to step in and inform the Transport Board that the supply of the Mediterranean fleet was 'so paramount an object that they desire that transports for this purpose be provided in preference to any other'.[102]

[99] NMM, ADM. BP/46B, 17 May 1826.

[100] Ville, *English Shipowning during the Industrial Revolution*, 153, agrees with this percentage.

[101] NMM, ADM. DP/32B, 30 July 1812.   [102] NMM, ADM. DP/35A, 2 June 1813.

The complaint of the Transport Board that transports were being detained on foreign stations echoed that of the Navy Board during the American War of Independence. Although the service was now managed by one board, the Transport Board was still unable to hasten the return of vessels or prevent other branches of military service from retarding them. Of course their detention on one station prevented their use on another. Late in August 1813 the Victualling Board was still trying to get seven vessels despatched urgently to the Mediterranean fleet. But then they wanted convoy, another problem of the war against the American colonies.[103] Indeed in 1814, to economise on escorts, the Admiralty reverted to the practice of the Navy Board from 1779 in attempting to concentrate as many as possible transports destined for the same parts of the world in the same convoys.[104]

Shortages of escorts and the detention of transports overseas were recurring complaints. After 1812 the main cause was the war with the United States which once again required armies to be shipped and relocated 3,000 miles from Britain. The disposition of the British army overseas in 1813 and 1815 (see table 2.3) shows that British troops in Europe, the Mediterranean and the West Indies were reduced to provide soldiers for Canada and the mobile force which attacked America's coastal cities in 1814. How did the transport service meet the challenge of shifting large bodies of troops around the world?

Its principal tool was the corps of professional agents who worked under a body of regulations that gave the single board control of the detail of transport management. A product of the Commission of naval revision and long experience at the Transport Board, the regulations were comprehensive both for ship masters and board agents. They were drafted by Ambrose Serle, a transport commissioner from 1794 until 1809 who, having been with the British army in America in 1776–8, knew at first hand the importance of an effective transport system. Under the instructions he drafted, ship masters as well as the agents were expected to perform duty to the state, a requirement backed by sanction as well as exhortation.

Although civilians hired for only 'six months certain', the masters were left in no doubt they would be 'severely punished' for any 'improper conduct'. They were 'strictly required' to conform to regulations from which no deviations were allowed 'under any pretence whatever' without authority. 'No excuse' would be admitted; 'no indemnification' was permitted. On any infringement of duty the board would 'not fail to deduct

---

[103] NMM, ADM. DP/33B, 23 Aug. 1813; ADM. DP/34A, 12 Mar. 1814.
[104] NMM, ADM. DP/34B, 6 Dec. 1814.

from the pay of their vessels to the full extent allowed'. Deductions would 'infallibly take place'![105]

Less threatening, but more exacting, were the instructions for agents of transports. Serving afloat as well as on shore, they accompanied transports whenever sufficient sailed together to warrant supervision, just as they had over 100 years earlier. The naval officer commanding the convoy, known as the convoy captain, had charge of all matters regarding the sailing and protection of the convoy. The transport agent had charge of the transports in so far as their economy of operation, utilisation, safety and cleanliness were concerned. In addition, he acted as the representative of the Transport Board, for which he served as its eyes and ears.[106] In some respects their instructions echoed those of the early eighteenth century. But equally evident were recent concerns, dating from the American War of Independence, now priorities for a state that had to conserve its resources and had a duty to those in its service.

Fleet management afloat

The first duty of the transport agent was to ensure the owners and masters of ships under hire complied in full with the terms of their contracts. He was expected to examine the charter party of each ship to check that each carried the full complement of men and proper proportions of stores. Shortages were to be made up by the masters at the expense of the owners. He was also to investigate defects and damage to ships. In the case of defects for which the Transport Board was not liable, the faults were to be surveyed by three naval carpenters and, if incapable of proceeding, the agent was 'to protest against her' so that the Transport Board might terminate payment for the ship. In the case of damage for which the Transport Board might be liable, the agent and master were to hasten repair. But as time under repair was lost to freight, any neglect on the part of the master that delayed repair was deemed sufficient for the agent 'to protest against her'. Moreover, if sent home by an agent, any unwarranted delay on the part of a master was thought sufficient cause for the board to discharge the vessel. On the other hand, the masters of those ships that performed satisfactorily, were 'from time to time' to receive certificates which,

---

[105] From eighteen clauses in *Regulations to be observed by all Masters of ships and vessels employed in His Majesty's Transport Service*, TNA, ADM. 108/182.

[106] *Regulations to be observed by Agents employed in His Majesty's Transport Service* (1814), TNA, ADM. 108/182.

when presented by the ships' owners, would permit them to receive payment.

To make full use of the transports, the agent was instructed to ensure ships turned around quickly at ports in and at the end of each voyage. When the transports reached a port, all seamen and stores had to be mustered and checked once a week. From a British port, the agent had to despatch a return to the Transport Board once a week to reach the Transport Office 'regularly every Monday'; from a foreign port, the return was expected about once a fortnight. The return had to be accompanied by a report of the ships ready for further service, and of those not, with the reasons for their state, and the time when expected to be ready. The agent was exhorted to be particular about every defect and deficiency in case the Transport Board was obliged to make a reduction in the freight paid on this account or because of the inactivity of their masters.

To ensure the safety of the ships in his charge, before they sailed the agent was expected 'to obtain satisfactory information respecting the competency of the masters'. This related not only to their qualities as seamen, but to their knowledge of the waters and coasts to which the convoy would be heading. Should, in the agent's view, pilots be necessary, the masters were to obtain them, charging the cost to the ships' owners. If masters failed to get them, the agent had to do so, transmitting the cost to the Transport Board for deduction from the earnings of the owners. If a change of master was deemed necessary, the agent had to set out his reasons so that they could be 'clearly understood by all parties'. Safety from capture by the enemy was equally important. The agent was to note 'neglects or breaches of orders of any kind' that entailed risk: failing to set sail or to keep station, for example, or to anchor within reach of escorts should they need protection.

Before, during and after each voyage, the agent was to attend to matters affecting the health of troops that were carried. He was to attend their embarkation, see each ship's master was issued with sufficient food for the length of the voyage, see that it was issued to soldiers in the correct amounts and investigate any complaints. Deaths, desertions and absences of soldiers were to be reported to the Transport Board. While troops were on board, each morning hammocks were to be lashed up and brought on deck, weather permitting. At disembarkation, the master (who was accountable for beds and stores) had to claim back the hammocks from the soldiers, get them scrubbed and dried, and have the beds and blankets shaken, well aired, and stored in a dry room ready for re-issue. Meanwhile carpenters were to lift and wash the floor boards of the cabins and clean out the spaces beneath. While the transports were unoccupied, these 'bottom boards' were shifted in turn to permit a 'free circulation of air' into deck and floor spaces.[107]

The conduct and discipline of transport crews were considered important too. None were to be permitted to engage in private trade or freight: only on the Commander-in-Chief's direction was the agent to permit privately owned goods on board transports. Privately owned, runaway slaves were not to be accommodated either. Seamen from the transports were to be discouraged from 'improper straggling and misconduct' on shore, especially 'at unseasonable hours'. In this respect, those who disobeyed orders and were subsequently impressed were to be given no protection from naval service.

Of all these matters, the agent afloat was required to keep a log book containing every 'particular'. Accounts of stores lost or expended were to be included. A copy of the log was to be transmitted to the Transport Board every six months and at the termination of every voyage. More immediately, the agent had to report arrivals and sailings of the transports in his charge, the numbers of troops embarked or landed, the types and quantities of stores shipped or unloaded and the terms of the freight. Private 'intelligence' was to be marked 'private' to distinguish it from public correspondence, but both forms of correspondence were to be sent to the board through the local resident or senior transport agent on shore.

In this way, the Transport Board gathered information valuable to the economic management of the transport fleet. By 1815, the Board and its agents were integrated in a highly regulated, centralised branch of government. Before 1794 there had been Navy Board agents like George Teer who served the public faithfully. But there had also been others, like the Victualling Board's hoytaker at Deptford, who had a vested interest in ships that were hired. After 1794, loyalties, ethics, regulations and even the language[108] of the transport service changed. In this respect, the transport department was representative of a state that was itself changing in response to war and new standards of efficiency.

Because the Transport Board made use of shipping, the primary vehicle of seaborne trade, the board directly interacted with the maritime economy. No other department of government, among the number examined here, so directly bridged the gap between the state and the sea. Its ability to draw upon the maritime economy partly accounted for

---

[107] The agent for transports at Portsmouth reported in 1796 that one ship had to 'be whitewashed and fumigated after the nasty fellows that have been on board of her': TNA, ADM. 108/28, 1 Apr. 1796.

[108] For example, by December 1815, the transport service included the term 'troopships': NMM, ADM. DP/35B, 28 Dec. 1815.

the scale and speed of its operations, which contributed much to the success of British amphibious operations after 1794. But its knowledge and practice derived from the past and from the mercantile world. In terms of hire it complied with the shipping industry. In organisation, it drew upon antecedent experience dating back to the first Transport Board. At the same time, the great use made of agents and the employment of an Inspector of Transports derived from contemporary bureaucratic practice. The product was a service that was capable of projecting the state wherever in the world it required to go and an organ of government central to British maritime expansion.

Britain's military operations overseas demanded an organisation to meet the local needs of an army on campaign or serving as a garrison. This gave rise to the Treasury's Commissariat. At the beginning of the Seven Years' War the Commissariat did not exist; by the beginning of the Napoleonic War it was virtually an established department of government. Although the Commissariat was wound up at the end of each war, methods of management developed in the Seven Years' War were revived in each succeeding conflict. The American War of Independence taught many lessons in the provision of transport and the shipment of food which were put into practice in the French Revolutionary War. Individual campaigns nevertheless posed their different challenges, and supply to the soldier in the field was the product of attention to detail on the part of innumerable clerks, commissaries, contractors, and army and navy officers. The successful supply of the British army and its allies in the Egyptian campaign and in the Peninsular War resulted from their concerted efforts. It was helped too by the development of a global economy, a factor that affected all forms of supply, especially those of food and finance. By the second half of the Napoleonic War, the work of the Commissariat in the Peninsula resembled that of central government on a smaller scale.

The mid-century infrastructure

Until the late eighteenth century the supply of money to the army was based on two presumptions. The first was that a standing army was a threat to civil liberties. This dictated that government should so stint the funding of its army at home that it could not become a threat. The second was that troops on campaign should live off the land with little or no assistance from civil authorities. This ensured that, even during wartime, the funding and organisation needed to supply the army were both kept to a minimum, and during peace they shrank to what was politically acceptable: 'Political prejudices were thus accommodated at the expense of both military and administrative efficiency.'[1] The effect was to deprive the British army of employment, money and basic facilities.

When armed force was needed, there was preference for reliance, if possible, on local militia and on the employment of the forces of Hanover, Hesse-Cassel and Brunswick. Those regular forces the British government did maintain were allowed enough money by the Treasury for colonels of regiments to supply their soldiers' basic uniforms and clothing but little else, forcing the soldiers to purchase anything more at their own expense.[2] Deductions from the pay of soldiers for extra articles preserved the fiction that the army was self-supporting. But the pay of soldiers could not support all the deductions. Hence, in addition to its regular funding, the army in fact needed supplementary allowances – contingencies or 'extraordinaries'.[3]

The supply of basic facilities was no more straightforward. Because a standing army was regarded with opprobrium, the Treasury did not itself provide barracks. Nor, as a matter of course, did the Treasury provide anything so dangerous as the means of moving the army's baggage and food supplies. The Crown's right of purveying transport was abolished by statute at the Restoration. In Britain, army commanders were thus required to apply to the justices of the peace of counties through which they were to pass to impress wagons and horses in the parishes on the route. From 1692 recompense beyond a statutory scale was made from the county rate. Though this safeguarded the constitution, the army's supply of transport was inadequate, even at the time of the 1745 Jacobite risings when England was invaded.[4] As a result, for the supply of necessaries, billets and transport, the army relied on a variety of unofficial and semi-official sources, including shop keepers, inn-keepers, householders and merchants, to provide its accommodation and transport.[5]

When the army did go on campaign, facilities were more forthcoming. But under-funding affected organisation because, instead of being deliberately created, it developed piecemeal across a number of departments as expedience demanded. Until the Seven Years' War, this organisation was complex and confused, growing and shrinking with the onset and cessation of hostilities. This prevented the concentration of power in too

---

[1] M. Van Creveld, *Supplying War: Logistics from Wallenstein to Patton* (Cambridge, 1977), 27–9, 38.

[2] Barnett, *Britain and her Army*, 143–4.

[3] H. M. Little, 'The Treasury, the Commissariat and the supply of the Combined Army in Germany during the Seven Years' War (1756–1763)', unpub. University of London Ph.D. thesis, 1981, 25.

[4] R. C. Jarvis, 'Army transport and the English constitution with special reference to the Jacobite risings', *Journal of Transport History* 2(1955–6), 101–4.

[5] Little, 'The Treasury, the Commissariat', 8–9.

few hands but demanded a great deal of communication between departments and handicapped coordination and concentration of supplies at the geographical point where the army most needed them.

As campaigns demanded, the board of Ordnance looked after barrack accommodation, its furniture and fuel, as well as the supply of munitions. The board appointed its own commissaries for overseas campaigns, who received and issued ordnance stores and handled money needed locally to purchase materials and pay employees.[6] The Secretary at War provided some personal camp equipment, uniform and medical supplies. The Treasury then provided what might be missing, under the head of 'contingencies' or 'extraordinaries'. For example, the Treasury provided bread wagons as they were required, and contracted for the supply of bread and biscuit when the army was in camp or on campaign. Straw, forage and firewood might also be supplied within Britain. Foreign troops taken into British pay were similarly supplied by the Treasury, but not by the Secretary at War, so that they had to supply their own medical supplies, uniform and camp equipment. To ship these troops overseas, the Navy Board supplied the transports, the Victualling Board fed them while at sea, and the Sick and Wounded Board cared for those who fell ill on the voyage. Once they were overseas and on campaign, the Commander-in-Chief employed the quartermaster-general as a chief of staff to manage marches, camps, equipment and maintenance as far as sutlers and foraging parties could meet their needs.[7]

For supply purposes, this patchwork of departmental providers needed coordination, and that task fell to the officials acting for the Treasury. This was natural enough, for when army expeditions ventured into places contractors were unwilling to supply, the Treasury was obliged to enlarge its 'extraordinaries' to include complete rations of food and drink.[8] For this, the Treasury relied on a small number of commissaries, who were responsible for supplying, distributing and accounting for the goods and services consumed by the army. Under two Comptrollers of Army Accounts, the commissaries made contracts, ensured supplies reached the army, and examined accounts, signing certificates for the sums due to the army's creditors. Their contract arrangements, however, were subject to the approval of the Treasury Board which (with the Commander-in-Chief of the army) authorised warrants for payment by the Paymaster General or his deputy. Where remittances of specie were insufficient to

---

[6] The papers survive of Ordnance commissary Richard Veale, 1759–63: HRO, 109M91/CG19, 82. For Veale's career, see HRO, 109M91/MIS2, fo. 8.

[7] Little, 'The Treasury, the Commissariat', 12–17; J. A. Huston, *The Sinews of War: Army Logistics 1775–1953* (Washington, 1966), 3–4.

[8] Binney, *British Public Finance*, 177; Baker, *Government and Contractors*, 22.

make payments, bills of exchange could be drawn on the Treasury or its bankers.[9]

With tolerance of inefficiency, these arrangements sufficed for the scale of the land forces employed during the first half of the eighteenth century. However, with the growth of forces employed outside Britain during the Seven Years' War, the Treasury was obliged to develop a substantial bureaucracy to arrange the necessary supplies. In April 1758 the Treasury became responsible for maintaining an enlarged continental army. Whereas previously it had been limited to Hessians and Prussians, it was enlarged by troops belonging to Hanover, its allies and a British contingent. By the end of 1758, the Treasury had become responsible for the maintenance of over 60,000 soldiers on the continent; by the end of 1759, the number had risen to 70,000; between 1760 and 1762, there were over 95,000, which, with supporting workers, amounted to over 100,000 men. Specifically, the Treasury undertook to provide forage, grain and flour, the bakery and hospital trains for the British troops and the artillery train for the German troops. The Treasury thus undertook supply and transport on a large scale. It met this responsibility by placing the business under the Superintendent of Extraordinaries who was permitted to control payment and develop specialised functions with an enlarged number of commissaries.[10]

The Seven Years' War and the Treasury Commissariat

This new organisation was set up under Thomas Orby Hunter, a respected Member of Parliament and junior Lord of the Admiralty, who was appointed Superintendent of Extraordinaries.[11] In 1759 the Superintendent of Extraordinaries was given authority to employ as many subordinates as commissaries as he judged necessary for efficiency, to issue warrants for payments by the Deputy Paymaster, and to draw bills of exchange on government bankers, without referral to the Treasury Board. Despite a brief hiatus in the delegation of this authority in 1760–1, this new power established a single officer who controlled both supply in the field and payment for the provisions and services required.[12]

Reflecting an understanding that decisions were better made in the field, from 1759 the Treasury was even prepared to delegate to the

[9] Little, 'The Treasury, the Commissariat', 15–16; H. M. Little, 'The emergence of a Commissariat during the Seven Years' War in Germany', *JSAHR* 61(1983), 201–14.
[10] Little, 'The Treasury, the Commissariat', 265–6.
[11] L. Namier and J. Brooke, eds., *The History of Parliament: The House of Commons, 1754–1790* (3 vols., London, 1964), II, 656.
[12] Little, 'The Treasury, the Commissariat', 51, 77.

Superintendent's commissaries full authority to form and conclude contracts for what was needed. The Treasury Board was simply to be kept informed of steps taken to maintain supply. For eighteen months the Treasury limited its interventions to comments on what it considered minor blemishes in the terms of contracts.[13] The commissaries were able to issue warrants for immediate payment by Deputy Paymasters who were appointed to attend detachments of troops in the field. By 1761 the commissaries were accompanied by cashiers with small reserves of specie and were thus able to pay cash for goods and services purchased in small quantities.

The growth of specialisation and scale

Early in the Seven Years' War, the commissaries had been jacks of all trades. From 1760, they were appointed to perform specialist functions within one of three main areas of supply, control and account. Great emphasis was placed on professional performance, and commissaries specialised in specific areas of this work, for example the supply of forage, transportation, bakeries, and so on.

At the same time, the number of commissaries grew to match the growing size of the army for which they had to provide. In 1757–8, the Hessians and Prussians had been supplied by one commissary. In July 1758, the British contingent brought another. By the end of 1759, there were thirteen; by 1760, eighteen; by 1761, twenty-seven; and by 1762, forty-one. The growth was partly the result of complaints of inefficiency from Prince Ferdinand. Staffing ratios were thus improved. The ratio of commissaries to troops declined from 1 to 5,416 soldiers in 1759 to one to 2,362 in 1762.[14] They were assisted by deputies, assistants, inspectors, storekeepers, clerks, craftsmen and labourers, all comprising a considerable bureaucracy and labour force which grew in proportion to the army. For example, 60 magazine keepers and subordinate officials in 1759 had grown to more than 400 in 1761. By 1763 this bureaucracy was regarded as equal to the Excise in terms of the variety and extent of its business and the number it employed.[15]

By the end of the Seven Years' War, the Treasury had thus developed an organisation better able to deal with the growing scale and greater complexity of supplies demanded by the army. Before that war, the word 'commissariat' was hardly used. By the end, the Treasury had replaced a small and haphazard collection of commissaries by a 'recognisable institutional structure'. This employed a greatly increased staff, of whom

[13] *Ibid.*, 73.   [14] *Ibid.*, 55–6.
[15] *Ibid.*, 56; Little, 'The emergence of a Commissariat', 205–6.

the leading officers were well rewarded, and undertook increasingly specialised duties, for the execution of some of which authority was delegated
from board level to those operating in the field. In effect, the Treasury
had created a new department to manage the supply of Britain's army
overseas. Its only disadvantage was that it was temporary. As soon as
peace was concluded, the new department was virtually dissolved.[16]

Nevertheless, while war lasted, the Treasury's Commissariat supervised the supply of troops with their foodstuffs and necessities as required
by their military commanders. The foodstuffs were obtained by a variety of means: from contractors and agents, from local residents and
from government officials. There were two staples: bread and forage,
the fuels of soldier and horse respectively. Just these two commodities
were enough to entail the maintenance of numerous teams of horses and
wagons, attached to depots and magazines, which ensured that bakeries
were supplied with flour and firewood, and the troops and horses were
supplied with foodstuffs.

The scale of these supplies matched those of the navy. In Germany
in 1761 the army maintained at British expense contained over 100,000
men and 60,000 horses. Between them, they consumed more than 66
tons of oats, hay and straw and nearly 90 tons of bread each day. Their
supply was from 60 to 70 magazines, employing between 400 and 500
staff. As the field of operations shifted, so the commissaries had to find
barns and granaries to serve as storehouses. Their receipts had to be
checked and certified, as did issues *pro rata* to those who were entitled to
receive them. The grinding of grain had to be organised in local mills, and
the flour supplied to the bakeries. The latter employed several hundred
staff, while some 2,500 wagons and carts employed about 17,000 horses
and 5,600 wagon-masters, drivers, farriers and labourers.

Despite such quantities of food being dispersed among so many magazines, using a staff that was largely German and paid on a parsimonious
scale, effective supervision and accountant control were both generally
wanting. In 1761 irregularity, confusion, malpractice, theft and fraud
accounted for waste on a wide scale. Thomas Pownall, responsible from
1760 for ensuring supplies reached the army in Germany, catalogued a
series of irregularities. Receipts at magazines were accepted though neither quantities nor qualities complied with the terms of contracts. Issues
were pilfered, grain and flour disappearing between magazines, mills and
bakeries. Wagons were not maintained, horses not properly cared for.
Rations were issued to anyone who maintained a pretence of belonging
to the army, including camp-followers: Pownall claimed that 'the whole

---

[16] Little, 'The emergence of a Commissariat', 207–14.

of the issue is a kind of irregular scramble for what every one can get and such a one as cannot be controlled'.

The confusion was amplified by the tendency of contractors and local authorities to assume titles, such as 'commissary', 'bailiff', 'inspector', and to give orders regarding deliveries at magazines which suited themselves rather than the army. This was possible because documentation had not been standardised; full and exact specifications were not prescribed; certificates of receipt were not checked against issues; accounts were not examined; and responsibilities were not subject to audit. In the absence of these procedures, pilfering flourished, frauds took root, and army officers were induced to connive at, indeed participate in, corrupt practices.

Control and accountancy

The Treasury was convinced that, to cure this state of affairs, more commissaries were needed, and appointed a special group specifically to improve control and accounting. Thomas Pownall was one of this group. He had been a clerk for eleven years at the board of Trade before going to America to fill a variety of administrative posts that, in eight years, culminated in the governorship of Massachusetts Bay between 1757 and 1759. During this last appointment he made arrangements for the supply of the colony's troops during their participation in the conquest of Quebec. He thus had first-hand experience of provisioning at long distance. In May 1760 Pownall succeeded Hunter as Superintendent of Extraordinaries. Initially Pownall despaired of reforming practice in Germany. But this gave way to a determined and logical programme of improvements which established new standards.

First of all, despite a backlash, Pownall laid down that only those officials with authority from the British Commissariat could interfere in supply arrangements to troops for which Britain had responsibility. The employees of the Commissariat then received detailed instructions which laid down their personal responsibilities and mode of procedure. Those to magazine keepers, for example, ran to twenty-six paragraphs and ensured that they checked receipts for both quantity and quality against contract specifications. Issues of rations were to be made according to monthly lists of effective numbers of troops. For receipt and issue, official weights and measures were established and their patterns circulated to all magazines. Mills and wagon trains were similarly treated to new regulations.[17]

---

[17] TNA, T.1/413, fos. 364–9.

With their duties clarified, staff under the control of the Commissariat became subject to closer supervision. Large magazines were divided into five geographical areas, to each of which a deputy commissary and inspector were appointed. Small magazines were subject to sub-inspectors and their assistants. Again, bakeries and trains were similarly treated.[18]

Local control was maintained through the requirement for magazine, mill and train managers to keep detailed and systematic records. Magazine keepers now had to keep seven books of accounts detailing receipts and issues, origins and destinations, and all cash transactions.[19] So too the managers of mills, who were required to complete printed forms stating quantities of bread baked and the amounts drawn by individual regiments. Printed forms were also used to maintain a close record of the horses, wagons and drivers available for trains. All these accounts had to be returned to the regional headquarters of the Commissariat where they were examined, checked and collated under the eyes of the commissary exercising the office of check and the commissary general of accounts. They kept two master records which stated the supply situation for the whole region. The Grand Ledger stated monthly the quantities of all supplies in each location; and the Journal recorded all movements of provisions and transports.[20]

According to H. M. Little, the historian of these developments, they made an immediate impact: 'As it now became possible for the first time to check the accounts of magazines, bakeries and trains, to establish the legitimacy of suppliers' demands for payment and accurately to charge those who had consumed provisions, the opportunities for undetected inefficiency, negligence and fraud were significantly diminished.'[21]

Overall control in London was maintained by the establishment in 1762 of a central department of financial control and account. This demanded thorough accounts, which were subject to exacting examination. Four commissaries managed the accounts: two examined vouchers for payments, one liquidated accounts that were passed, and the fourth recorded those settled.[22]

The Treasury's army commissaries have, in the past, been accused of being 'rascally and ignorant' and 'a shady crew'. But H. M. Little observes that among the fifty-nine superior Commissariat officers who served during the Seven Years' War, 'there is more evidence of experience, ability and integrity than of the amateurish incompetence and

[18] TNA, T.1/420, fos. 118–19.   [19] TNA, T.1/413, fos. 365.
[20] See TNA, T.38/806–11.
[21] H. M. Little, 'Thomas Pownall and army supply, 1761–1766', *JSAHR* 65(1987), 92–104.
[22] Little, 'The Treasury, the Commissariat', 69–73.

dishonesty which have frequently been uncritically accepted as the hall-
mark of commissaries in this period . . . a significant number had enjoyed
the best possible experience of service in other wars or in other fields'.[23]
As the commissaries were not permitted perquisites, they were necessarily paid well by the Treasury, given accommodation and other allowances,
and came to enjoy a status and prestige unknown to their predecessors.

The American War of Independence

The work of the Commissariat in America during the War of Independence is well documented by Treasury records which are naturally concerned with the financial matters.[24] But it was the logistical demands
which taught important lessons in London. For the war in America highlighted the two principle challenges of meeting the needs of the army at
long distance: the arrangement of contracts for the supply of food and
the hire of shipping to carry those supplies.

Treasury contracts for food

At the outset of the American War of Independence, the government
assumed that their armed forces would be able to obtain food and forage
supplies from within North America. However, in 1775 the Americans lay
siege to the British garrison at Boston and cut off all supplies of provisions
from the local region. In mid May 1775 General Gage had to request
that provisions be sent immediately from Europe because 'all ports from
whence our provisions usually come, have refused suffering any provisions or necessaries whatever to be shipped for the King's use . . . and all
avenues for procuring provisions in this country [are] shut up'.[25] It was
a continuing experience. Around Boston, the King's troops were rarely
able to control more territory than the ground upon which they stood.
The Treasury tried to arrange contracts for supply but Americans would
not undertake them and, with a decline in regular trade, fewer British
merchants were ready to perform them.

When the Treasury found that insufficient provisions would be delivered by the contractors into America, the shortfall had to be shipped
from Britain by the Treasury itself. It was obliged to arrange contracts
for supply to commissaries within the British Isles and to hire ships to

---

[23] *Ibid.*, 98.
[24] TNA, T.64/101–19. For a summary tabular account of the financial responsibilities
for the Commissary-General of the army in America, see TNA, WO.60/19, account
May–December 1782 for Brook Watson, signed 31 Dec. 1782.
[25] Syrett, *Shipping and the American War*, 121.

convey the victuals to the army. In the autumn of 1775 it despatched more than thirty victuallers to Boston. However, principally owing to captures and adverse weather, only eight of these reached their intended destination. In March 1776, with the Americans bombarding its positions, the British garrison of 10,000 men at Boston was obliged to evacuate the town by ship to Halifax. Reinforced and guided by a strategy of dividing the American colonies, the main army in North America shifted in August to Long Island where it enlarged its zone of occupation to the whole of eastern New Jersey. This territory seemed to solve the provision problem, but in January 1777 American positions forced the British to withdraw from that territory.

The inability of the British army to hold more than a narrow bridgehead around New York forced the Treasury to supply the great bulk of its provisions, forage and fuel from the British Isles. By 1777, the British army was 30,000 men; by 1779, it comprised over 43,000 men; and in 1781–2, about 92,000. On top of that number of soldiers, in 1782 the Treasury had to fund rations in America for civilians and Indians equal to 28,000 men more than the army's strength, in other words for 120,000 men.[26]

Even at the height of the American war, the Treasury Board met only about 100 times a year. The business of army supply was thus overseen by the Secretary to the Treasury, John Robinson. He soon developed a list of contractors who made deliveries to shipment points in England and Ireland.[27] In January 1781 twenty individual merchants or partnerships were prepared to supply a portion of the overall demand (see table 9.1). Each contractor received a standard letter from Robinson, a copy of which went to those responsible for shipping the provisions. That to Messrs Stephenson and Blackburn on 9 January 1781 required them

to deliver with all possible dispatch and by the 10th of March next 12 month's supply of provisions of the dry articles to be furnished in complete rations for 5,000 men to his Majesty's Agent Victualler at Cowes, and also to deliver the like supply of the wet articles of provisions in complete rations for the same number of men to his Majesty's Agent Victualler at Cork by the 25th of March next, that is to say, for each person to be victualled for 7 days successively 7 pounds of flour of the first quality made from wholly kiln dried wheat, 7 pounds of beef or in lieu thereof 4 pounds of pork (one fifth part of the meat to be in beef and the remainder in pork both of the first quality), 6 ounces of butter or in lieu thereof 8 ounces of cheese, three pints of pease (which has been kiln dried) and half a pound of oatmeal all of the first quality,[28] and my Lords are pleased to direct that the beef and pork shall be put in barrels & half barrels and in the proportion of

---

[26] Baker, *Government and Contractors*, 4.   [27] *Ibid.*, 6–34.

[28] For discussion of these quanitities, see Baker, *Government and Contractors*, 22.

Table 9.1 *Contractors providing food supplies to the army in North America and the West Indies in January 1781*[a]

Contractors' names	No.	Service
Adam Drummond		
Moses Frank		
Jno. Nesbitt	12,000	
Jno. Henecker,		
Wm. Deveynes		
Wm. Smith	12,000	NY and dependency 37,000
Benj. Smith		
W. Fitzhugh		
and Peacock	12,000	
Paul Cox	1,000	
Messrs James, Robert Smith		
Atkinson	15,000	Canada
Antho. Richardson	6,000	
John Durand	4,000	11,000 Leeward Islands
Chas Potts	1,000	
Messrs Stephenson & Blackburn	3,000	Pensacola
Do	5,000	Halifax for Nova Scotia, St Johns and Penobscot
Jno Whitelock	1,500	
Hender Mason	1,000	
J. Boyle French	1,500	
Robert Mayne	4,000	15,000 Charles Town for the Carolinas
R. V. Sadler	2,400	
Ed. Lewis	2,300	
Geo. Brown	2,300	
		6,000 Georgia, E. Florida Bermuda and Bahamas

[a] TNA, ADM. 108/4A, 8 Jan. 1781.

200 half barrels to every 300 tons of provisions, and that four good iron hoops be put on both the whole barrels and on the half barrels.[29]

Contractors had agents who provided the provisions they had undertaken to supply. Rarely did the agents supply exact quantities. The Treasury kept a precise record of excesses and deficiencies. Not one of the agents in 1779 submitted precise quantities; nor did any of them err consistently in excesses or deficiencies. Overall, throughout that year, taking all deliveries into account, only pease and butter were deficient in

[29] TNA, ADM. 108/4A, Treasury to Navy Board, 17 Jan. 1781.

quantity – by 9 tons and 31 tons, respectively. All the other commodities were over-delivered – beef by 112 tons, flour by 154 tons[30] – which was probably encouraged by payment for excesses as much as by desire to avoid penalties for under-delivery.[31]

The Navy Board and shipments

During the American War of Independence, the supply of food to shipment points in Britain proved adequate to the growing number of troops overseas. However, the food then had to be shipped to them, and in this lay the principal challenge for the Treasury, for it lacked knowledge of the shipping industry, technical staff and facilities for fitting vessels that had to be hired. In 1776 the Treasury thus approached the Admiralty to hire, lade and despatch army victuallers, but the Admiralty declined the invitation. In February 1776, the Treasury thus contracted with the company Mure, Son and Atkinson to act as its shipper, and the company performed its task well. By early 1778 the Treasury had a fleet, chartered by Mure, Son and Atkinson, of 115 ships totalling 30,052 tons.[32]

However, in mid 1777, the Treasury discovered that, while the Navy Board was hiring ships at 11 shillings a ton, the Treasury was paying 12s 6d. In November 1777, the Treasury thus attempted to get the Navy Board to hire ships for the Treasury, and even had Mure, Son and Atkinson discharge the ships under hire through them which cost more than 11 shillings. However, the Navy Board already had 78,000 tons of shipping under hire and it refused to undertake the Treasury's task. The Treasury consequently reversed its directive to Mure, Son and Atkinson and continued as before until early 1779, when it again requested that the Navy Board undertake the hire of ships to carry army provisions. This time the Navy Board responded positively, partly because it had undergone a change of personnel. In August 1778, Charles Middleton had become Comptroller of the Board and he was, in the words of David Syrett, 'nakedly ambitious . . . a bureaucratic imperialist'.[33]

The Navy Board decided that the army victuallers would sail from Cork under convoy. To minimise the number of warships needed to escort the vessels, the Navy Board intended to maximise the provisions conveyed in each convoy, and to despatch the victuallers with regular trade convoys. Thus food for the soldiers in Florida, Georgia, New York and Rhode Island would be despatched in one or two large convoys each year. New York would act as the redistribution point; from there, ships for the West

---

[30] TNA, T.64/200, 2 Dec. 1779.
[31] For a reflection on the contractors' probity, see Baker, *Government and Contractors*, 142.
[32] Syrett, *Shipping and the American War*, 121–31.   [33] *Ibid.*, 23.

Indies could sail with the warships going south, those for Quebec going with ships destined for Newfoundland and the St Lawrence. The plan demanded coordination between the government departments: the Treasury to have the different provisions stockpiled ready for shipment, the Navy Board to supply the victuallers at the appropriate place and time, and the Admiralty to appoint warships to provide convoy when the victuallers were loaded. However, to achieve efficiency, from the beginning the Navy Board had to move into several other areas of Treasury jurisdiction to ensure that supplies were received and loaded into victuallers speedily and in correct proportions.

The Treasury's main department for army provisions was already at Cork. The Treasury's contractors delivered the provisions here ready for loading. To Cork, in April 1779, the Navy Board appointed an agent for transports, Stephen Harris.[34] Harris very soon discovered the provisions were not being conveyed to the victuallers as he needed them. In June 1779 he complained to the Navy Board that, though there were fifty-two lighters in the port, the Treasury's commissary, Robert Gordon, was employing only about six a day, at most eleven, and that delays in loading would result in shortages in Canada. Also, instead of sending off half-as well as whole barrels so that ladings could be a mixture, the contractors were delivering either all whole or all half-barrels, frustrating attempts to mix and maximise the cargo carried by the victuallers. Then Harris and Gordon argued about the best place to load the provisions from the lighters into the victuallers.[35] The conflict had to be resolved. At the end of July 1779 the Treasury relieved their commissary of his part in loading the victuallers and gave the Navy Board's agent full control.[36]

Henceforward, shipments of pork and beef would be despatched in mixed ladings of whole and half-barrels, the bulk of the vessels in convoy for New York. The first 'division' of army victuallers supervised by Harris began loading on 1 June 1779 and completed its cargo on 16 July. It consisted of twenty-eight vessels, totalling 8,045 net tons, carrying 8,157 tons of provisions. Nineteen of the vessels were destined for New York; the other nine for Georgia.[37]

Meanwhile, the Navy Board was reviewing the whole organisation of supply. Since May, Harris had complained that the casks containing dry provisions – flour and peas – arriving at Cork from England were so badly made that they had to be repaired; yet the coopers undertaking the repairs

---

[34] TNA, T.64/200, fos. 13–14, 3 Apr. 1779.
[35] TNA, T.64/201, fo. 24, extract of letter from Commissary Gordon, 16 July 1779.
[36] Syrett, *Shipping and the American War*, 139–42.
[37] Together they carried 1,231 barrels and 1,674 half-barrels of beef; 10,222 barrels and 9,042 half-barrels of pork; 2,860 firkins of butter; 24,150 barrels of flour; 9,279 barrels of pease; and 991 barrels of oatmeal: TNA, T.64/200, 30 July 1779.

had not employed extra staff but kept ten or twelve ships waiting for their lading up to eight weeks.[38] The Navy Board responded by deciding to ensure contractors supplied provisions in sound casks, to time limits, at more lading points. In mid July 1779 it required that provisions for the east coast of New York and east Florida be despatched from Cork; those for Canada, the West Indies and west Florida, from the River Thames. To provide balanced cargoes, enough English dry provisions would have to be shipped to Cork for loading; and enough wet Irish provisions would have to be shipped to Rotherhithe for loading. In every 300 tons of provisions there had to be 200 half-barrels. The Treasury was perfectly amenable; indeed, it required contractors to comply with any directions the Navy Board might make to them.

To manage this new arrangement, in August 1779 the Navy Board appointed an agent victualler, George Cherry, in the River Thames. Cherry had been agent victualler to the fleet under Lord Howe in North America and knew the requirements for supply that were necessary to the preservation of provisions. He was to receive all provisions from contractors and to ensure they complied with the terms of their contracts. At Cork in October, complying with the Navy Board, the Treasury dismissed Robert Gordon, whose duties were all transferred to Stephen Harris. As an agent for transports, he was not fully suited to the task of overseeing the supply of victuals and in December 1779 Harris was replaced by John Marsh as agent victualler, with the same duties, powers and recompense as George Cherry in the Thames.[39]

Meanwhile, the Navy Board had come to realise that the River Thames was not the best place to load army victuallers. This was a laborious process for, without sufficient docks and wharfs, victuallers had to be loaded by lighter while at anchor in the river, which was already busy. The premises at Rotherhithe were anyway too small to cope with the quantities of provisions being delivered there. The ships had then to force their way downstream, for convoy from the Nore, usually no farther than Spithead, before joining a major convoy. With adverse winds and tides, there were many delays and progress was difficult to monitor (see table 9.2). On the suggestion of George Cherry, the Navy Board decided that army victuallers would be more efficiently loaded from premises at Cowes. Here, in March 1780, large premises, capable of holding 40,000 barrels, were offered to the Navy Board by Mr MacKenzie. Cowes was adjacent to the anchorage at Spithead, from where most convoys proceeded to North America. It was also closer to Cork whence the wet

---

[38] TNA, T.64/200, 27 Dec. 1779.
[39] Syrett, *Shipping and the American War*, 142–4.

Table 9.2 *State and situation of army victuallers on 6 May 1780*[a]

For What Contract Employed	Ships' Names	Tonnage	Remarks
West Florida	Baltic Merchant Polly	608	Dispatched in October 1779
	Anne & Elizabeth	310	Sailed with Commissary
	Love & Unity's	364	Walsingham
	Total	1,282	
West Indies	Benjamin & Ann Grace	731	Dispatched in October 1779
	Tryal	538 ⎫	Sailed with Commissary
	Britannica	317 ⎬	Walsingham
	Perseverance	286 ⎭	
	Eliza	360 ⎫	at Spithead laden
	Lord Sandwich	317 ⎭	
	Friendship	220	Nearly laden in the River
	Brilliant	229	To load
	Total	2,998	
Nova Scotia & Newfoundland	Camel	293 ⎫	
	Friends	216 ⎪	At Spithead and intended
	Two Brothers	203 ⎬	to sail with the
	Jane	327 ⎪	Newfoundland Fleet
	Four Sisters	170 ⎭	
	Duke William	423	Nearly laden in the River
	Juno	226 ⎫	Cannot complete their
	Enterprize	226 ⎭	lading for want of Pork
	Total	2,084	
Canada	Ocean	300 ⎫	Sailed with the Dana
	Hercules	250 ⎭	
	Mary 1st	322 ⎫	
	Howden	205 ⎪	
	Argo	332 ⎬	Laden, and at Spithead
	Prosperous Armilla	382 ⎪	
	Nancy/Thomson	245 ⎭	

(*cont.*)

Table 9.2 *(cont.)*

For What Contract Employed	Ships' Names	Tonnage	Remarks
	Amphitrite	441	
	Nancy (younge)	283	
	Isabella (2d)	362	
	Spring	378	Under convoy of the
	Sophia	237	Atlanta
	Wier	281	
	Holmton	676	
	Bridgwater (has on bd)	125	At Gravesend to take Troops
	Peerith	301	Will complete their
	Valiant	341	ladings by the 13th of
	Providence Increase	234	May
	Total	5,695	

G. Cherry, Agent for the Army Victualling on the River Thames

[a] 'Account of the shipping employed for the purpose of transporting the provisions supplied on the several contracts mentioned ... with their state and situation as nearly as can be ascertained this 6th of May 1780', TNA, T.64/200, 8 May 1780.

provisions came. The Navy Board thus adopted the recommendation of George Cherry and moved its Rotherhithe depot to Cowes in summer of 1780.[40]

At both Cork and Cowes, the Navy Board proceeded to raise standards of delivery. Taking the Victualling Board's contract for the supply of ships abroad as its model, in August 1780 the Navy Board had the Treasury insert penalty clauses in its contracts. Its agents then refused to accept late deliveries and provisions unfit for consumption or packed in sub-standard casks. Contractors or their agents who failed to accept these measures were subject to further sanctions. Contractors were limited to one agent at each port; specifications were established for the preparation of provisions supplied; inspections were introduced and warranties of six months running from date of delivery. The Treasury was never to be entirely free of complaints from the army. Nevertheless, in December 1780 the Commissary General at New York reported provisions 'remarkably good and

[40] TNA, T. 64/200, fos. 80–3, proposal of MacKenzie 13 Mar. 1780 and advice of Cherry 23 Mar. 1780; Syrett, *Shipping and the American War*, 144–5.

the packages much better than sent out before'. When deficiencies and shrinkage persisted, in February 1781, again at the instigation of the Navy Board, the Commissary Generals overseas were required to open and sample one barrel in ten and to report to the Navy Office any short-ages of casks on board a victualler.[41] The new measures had effect. In August 1783 the Commissary General in Canada reported provisions that appeared 'sweet, sound and wholesome [and in] every way fit for issuing to His Majesty's troops, the packages in good condition being the strongest and best that have yet come to this province'.[42]

The French Revolutionary War

The experience of the American War of Independence was not lost. The Commissary General in America, Brook Watson, was re-employed in 1793. The junior Secretary at the Treasury in 1782–3, George Rose, remained as senior Secretary from 1783 until 1801. George Cherry was appointed to the Victualling Board. Frustrated by the slow pace of reform, Sir Charles Middleton resigned from the Navy Board in 1790 but provided advice to Pitt and Dundas in 1793–4 and was appointed to the board of Admiralty in 1794. These men became key players in the reorganisation of army supply in the French Revolutionary War.

The Victualling Board and army supply

Although before 1793 the Treasury was responsible for feeding the army overseas, the Victualling Board was accustomed to supplying some pro-visions to soldiers. As the navy's agent victualler at New York in 1775–6, Evan Nepean – Secretary of the Admiralty twenty years later – had main-tained a mutually beneficial exchange of food with the army's commis-sary. The same happened elsewhere during the war. A whole sequence of transactions was brought to summary account in 1785.[43]

Soldiers shipped on board warships and transports were also victualled by the navy. Accounts of soldiers victualled were kept by naval pursers, and demands made on the War Office at the rate of 3 pence per man a day. The Paymaster of the Army stopped the money from the payment

[41] Syrett, *Shipping and the American War*, 148–53.
[42] Nathaniel Day to Treasury, TNA, T. 1/580, 4 Aug. 1783, quoted in Syrett, *Shipping and the American War*, 154.
[43] NMM, ADM. DP/5, 6 May 1785.

of regiments and redirected it to the Treasurer of the Navy. On account of the need to provide detailed accounts, the debt incurred during the American War was still being recovered in 1792.[44] Troops were generally victualled at two-thirds the quantities allowed seamen.[45] Women attached to the soldiers were victualled at the same rate, but reduced to half in 1799 when children were settled at one quarter the whole allowance.[46]

The Treasury continued to be charged for victualling troops under transportation.[47] Meanwhile the Admiralty remained conscious of its authority over the provision of supplies through the Victualling Board.[48] The system of orders to the subordinate board for issues to troops on board transports eventually became the same as that to pursers for issues to seamen on board warships. Regulations governing these procedures were rationalised by the Commissioners for revising and digesting the civil affairs of the navy in 1808.[49]

The provisioning of all troops overseas was a far greater matter. In 1781 Lord North was informed the Treasury probably paid a great deal more for pork and beef than the naval Victualling Office. North subsequently wrote to John Robinson, Secretary to the Treasury, 'to consider whether we shall not save considerably to the public by putting the victualling of the army under the [victualling] commissioners'.[50] At that time nothing was done. But in October 1793 the Treasury formally delegated responsibility for provisioning troops overseas to the naval Victualling Board.[51] Initially this entailed the purchase of provisions and the hire of army victuallers.[52] On 26 October 1793 George Rose, with ten years'

---

[44] NMM, ADM. DP/12, 14 Feb. 1792.

[45] Syrett, *Shipping and the American War*, 189.

[46] M. E. Condon, 'Living conditions on board troopships during the war against Revolutionary France, 1793–1802', *JSAHR* 49(1971), 14–19. There were 135,000 soldiers fed on these allowances in transports going overseas between 1793 and 1802. See also NMM, ADM. DP/32A, 24 Feb. 1812.

[47] See for example, the Secretary of State's directive to the Admiralty, 21 Feb. 1793, for the Victualling Board to supply provisions for three battalions of Guards being sent to Holland for the defence of that Republic. The Admiralty observed 'as that is not an naval service you are to keep a particular and distinct account of the expense thereof and to solicit the Lords Commissioners of the Treasury to defray the same': NMM, ADM. C/679, 22 Feb. 1793.

[48] Watson, 'The Commission for victualling', 173–8.

[49] TNA, ADM. 114/71, 14 Sept. 1808.

[50] Condon, 'The administration of the transport service', 34.

[51] In view of the extra duty, the chairman of the Victualling Board was given an extra £500 a year and each of the other commissioners an extra £250. Officers and the clerks under them were given allowances that dated back to 24 June 1793.

[52] 32nd Report of the Select Committee on Finance, 1798, 3; TNA, ADM. 109/102, 24 Oct. 1793; ADM. 109/110, 24 Oct. 1793.

experience as Secretary to the Treasury, was also appointed Storekeeper of Army Provisions at the Victualling Office.[53]

The administrative re-arrangement was logical. The rations of the British soldier overseas in 1793 were much the same as those consumed by the seaman, comprising bread or flour, pork or beef, oatmeal, pease and rice. Troops on the continent of Europe received an allowance of butter, which was altered for sugar and cocoa oil in the West Indies. They also received a quantity of spirits or wine; in 1794, for the West Indies, wine was thought better for health. One pint of wine was considered equal to half a pint of spirits. By 1801 wine was issued to soldiers three days a week instead of rum.[54]

With rations of a similar nature, the new arrangement promised economies. For the Victualling commissioners could, if need be, engage for army supplies with contractors who already dealt with the Victualling Board and feed the army in much the same way as it fed the navy.[55] From the start, the Victualling Board aimed at maintaining stocks of provisions overseas that would last land forces three months. In 1793, before war was declared, land forces were maintained in New South Wales, Canada, Nova Scotia (including New Brunswick and Newfoundland), the West Indies and Gibraltar. In 1794, as garrisons were reinforced, supplies were altered in quantity under the directions of the Treasury as advised by the Comptroller of Army Accounts.[56]

Regularity of supply to troops overseas was aided by the establishment of facilities dedicated to army supply. The dedication of the yard at St Catherine's and the Hartshorne brewery near the Tower of London to army supply separated it from naval victualling.[57] During the 1790s the shipment of provisions from the Pool of London became difficult as the Thames became more congested. However, Cowes on the Isle of Wight continued to serve as trans-shipment depot, where cargoes were combined with supplies from Ireland.[58] Only in 1816 was the army victualling establishment at St Catherine's run down, but it was still maintained until at least 1821.[59]

[53] TNA, ADM. 109/102, 24, 26 Oct. 1793; Condon, 'The administration of the transport service', 19–20.
[54] TNA, T/27/44, 17 Dec. 1793, cited in Condon, 'The administration of the transport service', 282–3.
[55] For letters and orders to the Victualling Board from the Admiralty and Navy Board, 1694–1819, see NMM, ADM. G/773–98; for the same from the Treasury, 1793–1816, see TNA, ADM. 109/102–10.
[56] TNA, ADM. 109/102, 16 Mar. 1793, 8 Oct. 1794, 22 Apr. 1795.
[57] NMM, ADM. DP/35B, 5 Dec. 1815.   [58] TNA, ADM. 109/110, 25 Jan. 1794.
[59] Coad, *The Royal Dockyards 1690–1850*, 283; NMM, ADM. DP/35B, 5 Dec. 1815; ADM. DP/38A, 11 Feb. 1818; ADM. DP/41A, 20 Mar. 1821.

The Transport Board and equipment supply

For nearly a year, the Navy Board assisted the Victualling Board in the hiring and fitting of ships to serve as army victuallers. The Navy Board also supplied the troop transports needed by the army. However, this was rendered unnecessary in July 1794 when the hire of victuallers and transports was shifted to the newly formed Transport Board. As Norman Baker observed, these two new Board responsibilities may 'be viewed as a progression towards the application of specialist knowledge to the problems of army provisioning'.[60]

The responsibility of the Transport Board for the supply of shipping to all government departments has been examined in chapter 8. In addition to the supply of army transports and victuallers, however, the board was soon being used by the Treasury to supply clothing, equipment and forage to forces overseas. By the end of 1795, the Home Secretary, Secretary for War and Colonies, and Secretary at War were each giving lists of articles for supply to the army directly to the Transport Board. It thus came to coordinate supply of the army's equipment just as the Victualling Board did the food to the state's land forces overseas.[61]

The transport service acted as a natural intermediary between the suppliers of food, equipment and forage and the army Commissariat. Receipts, bills of lading and charter parties accompanied the cargoes to their destinations where army commissaries or their deputies gave receipts for deliveries, which were returned to the Victualling and Transport Offices where the boards sanctioned payment for their freight. The whole charge was then recouped from the Treasury.[62]

The Commissariat and local supply

In 1793 the Commissary-General, Brook Watson, accompanied the army on campaign. He went with the Duke of York's army to Flanders where he oversaw the supply of provisions and forage to the army.

In 1796 a recent deputy commissary, Havilland LeMesurier, published *A System for the British Commissariat*, espousing the arrangements for the Commissariat established by Brook Watson. LeMesurier maintained that neither the Commander-in-Chief of an army nor the military departments at home could be expected to focus on the supply of every article demanded by an army, and that this 'civil administration' was

---

[60] Baker, *Government and Contractors*, 252–3.
[61] TNA, ADM. 108/4B, Treasury letters to the Transport Board, 5 Aug. 1794–1 Dec. 1795.
[62] Condon, 'The administration of the transport service', 281–2.

the function of the Commissariat. He argued that, though appointed only for the duration of hostilities, the commissaries performed an indispensable service, to which there was no alternative, and that it could be extended wherever the army demanded local supply. LeMesurier went on to describe the commissaries' methods of supply, finance and account, appending instructions and specimen returns.[63]

The treatise appears to have made an impact. It was followed by an extension of the service. In 1797, under the threat of invasion, the country was divided into districts, and commissaries were appointed to supply the troops in each district. In 1798 the authority of the Commissary-General was extended over all these districts. The Commissariat consequently came to serve not only the army overseas but the forces at home as well.

Garrisons, expeditions and campaigns

Despite the restructuring of central government to support the operations of the army overseas, until the very end of the French Revolutionary War garrisons and expeditions suffered from difficulties of supply. From the outset, the Treasury demonstrated its readiness to act pragmatically in doing all it could to ensure supplies reached garrisons, for example in the West Indies. The expeditions to that region in 1795–6 demanded supplies from Britain in great quantities. Yet, as the campaign in Egypt in 1800–1 demonstrated, it was not only the central boards and their agents who were important for the sustenance of an army overseas: it was local commanders who had to improvise local supplies.

Overseas garrisons

From the onset of hostilities in 1793, the Treasury recognised that some overseas garrisons might be deprived of supplies from Britain and that it needed to act pragmatically to ensure troops overseas were supplied. This was evident in attempts to prevent the army garrison at Grenada going short in the autumn of 1793 when the Governor of the island was directed to purchase provisions locally. To facilitate the supply of these provisions, in October 1793 the Treasury decided to relax the Navigation Laws, permitting American ships of one deck to carry cattle and grain from the United States into Grenada.

Initially this was only for so long as it was 'necessary for the re-establishment and subsistence of the said island' or until other measures

[63] Printed as Appendix B in Glover, *Peninsular Preparation*, 267–305.

were taken to achieve the same level of commercial activity as in other British islands. However, the policy of relaxing the Navigation Laws was extended by degrees to the whole of the West Indies. In December 1793 the Treasury also sanctioned vessels from Spanish America to import cattle, salt beef and pork, flour, staves, hoops and lumber. These vessels were permitted to depart with sugar, molasses, rum and coffee to the net value of the imports. Then, in March 1794, a British contractor was permitted to send British ships to America for livestock, beef and flour, these being delivered to troops in Canada as well as the West Indies.

The Navigation Laws were waived as other foreign West Indian islands were occupied: St Domingue in July 1794, and Martinique in January 1795. The same applied at Jamaica. Because trade there languished with the onset of hostilities, army commissaries on that island were permitted to land provisions shipped from America. However, as trade revived, and the army was assured of supplies from Britain, in October 1796 the import of provisions into Jamaica from the United States was prohibited once more.[64] This policy of expedience was effective in preventing deprivation. In October 1796 the British army Commander-in-Chief in the West Indies observed: 'It is fair to say that the supplies have been regular and there has scarcely been a complaint.'[65]

A similar combination of supplies freighted from home with back-up from elsewhere applied at the Cape of Good Hope. The East India Company was glad to freight provisions out in their ships. However, after the poor harvests in Britain in 1795, which subsequently reduced the wheat and pease available for sowing, in 1797 the Victualling Board arranged for a supply of flour to be shipped from India.[66] This usage of home and local supplies as convenience dictated owed much to experience in victualling the navy, which was fed by a similar policy.

Expeditions overseas

The supply of large-scale expeditions posed a different challenge from that of garrisons. They could not rely on local supplies. Expeditions demanded that great stocks of food be ordered at least four or five months in advance, that large quantities accompany the soldiers, and that these be followed by further consignments for as long as the expedition lasted.

[64] Condon, 'The administration of the transport service', 285–7.
[65] TNA, T.1/733, 19 Oct. 1796, quoted in *ibid.*, 288.
[66] Condon, 'The administration of the transport service', 289.

In the case of the expedition to the West Indies in 1795, the Victualling Board ordered over 2½ million pounds of flour, over a million pounds of pork, with beef, pease and pork in proportion, at the beginning of 1795. The quantities for despatch were advised by the Comptroller of Army Accounts on 16 March and the amounts were authorised by the Treasury on 20 April. Despatch was then hastened to precede, rather than follow, the hurricane season, with further consignments to follow that season. The supplies were to be shipped by transport and consigned to the Commissary General at Martinique, John Jaffray.

In the event, in July scarcities of wheat and pease delayed the despatch of the full order. In August, the Treasury ordered the beef, pork and butter be despatched with part quantities of flour and pease, but the part quantities were only despatched in October.[67] In September, thirteen army victuallers were ready to sail with Admiral Christian. He sailed on 16 November 1795 but met a succession of storms and did not reach Barbados until 21 April 1796.[68] The storms, adverse winds and captures took their toll on the transports. Between October 1795 and April 1796 only nine of eighteen victuallers scheduled for the West Indies actually reached their destination.[69]

Nevertheless, the Victualling Board subsequently reported that, between 7 July and 31 December 1795, the quantities of food shipped to Martinique, in terms of the days they would last 25,000 men, were as follows:

Beef	761,252 pounds or 243 days with 1,877 pounds over
Pork	2,947,208 pounds or 235 days with 9,708 pounds over
Flour	3,682,560 pounds or 147 days with 7,560 pounds over
Pease	18,160 bushels or 108 days with 70 bushels over
Rice	239,947 pounds or 134 days with 623 pounds over
Butter	274,165 pounds or 409 days with 339½ pounds over.[70]

However, the Victualling Board was not just catering for 25,000 men. In August 1795, it was expected to provide for 40,000 men in the West Indies; in 1796, for 50,000.[71] To meet these demands, the Victualling Board established a routine of quarterly supply. It also sanctioned

[67] TNA, ADM. 109/102, 12, 13 June, 29 July, 8 Aug., 14 Oct. 1795.

[68] Duffy, *Soldiers, Sugar and Seapower*, 203–16.

[69] Condon, 'The administration of the transport service', 285. From a 200-ship convoy which sailed on 15 November 1795, the storm on 17 November drove 6 ships on to Chesil beach, Dorset, and destroyed 10 or 11 others; see E. Booth, *Christian's Fleet: a Dorset Shipping Tragedy* (Stroud, 2003).

[70] TNA, ADM. 109/102, 15 Sept. 1795.

[71] TNA, ADM. 109/102, 24 Aug., 15 Sept., 2 Nov. 1795.

substitutions, for example rice for the pease which could not be obtained in Britain in 1795–6.[72]

As expeditionary forces of 1795–6 declined into garrisons, local and contract supplies replaced those shipped by the Victualling and Transport Boards. Hence, soon after 1795 the supply of troops in the West Indies again became a patchwork of practices.

The Egyptian campaign

The expedition to Egypt in 1800–1 fell short of the 15,000 men originally allotted to the task of destroying the French army of occupation.[73] But by 1800 Henry Dundas, Secretary for War and Colonies, seems to have become complaisant about what could be achieved by expeditionary forces supported by the Victualling and Transport Boards. He rested his support for the expedition to Egypt 'with just confidence on the continuance of the same exertions which had so successfully contributed to the supply of His Majesty's troops in all the various and complicated services of the present war'.[74] George III was more perceptive of the risks. He informed Dundas in October 1800:

It is with reluctance I consent to the proposal of sending 15,000 troops under the command of Abercromby to Egypt; as that service must probably prove a burial ground for them to as great an extent as St Domingo; for unless the army be supplied from home as amply and as regularly as that in America was by the Treasury in the time of Lord North nothing but famine can attack it.[75]

George III was right to be concerned. The Treasury and Admiralty's scheme for supply was vague, the lines of communication to Egypt were long, and the possibility of their disruption was high. The War Office, Treasury and Commissariat with Abercromby remained in regular communication.[76] But the expedition took far longer than expected and deficiencies had to be met within the Mediterranean.

The plan was to stockpile provisions at Gibraltar, Minorca and Malta; to re-supply those stocks from England by victuallers every two months;

[72] TNA, ADM. 109/102, 21 Oct., 2 Nov. 1795.
[73] P. Mackesy, *British Victory in Egypt, 1801: The End of Napoleon's Conquest* (London, 1995), 15.
[74] Add. MS 40,100, 6 Oct. 1800, fo. 295, quoted by Condon, 'The administration of the transport service', 301.
[75] Add. MS 40,100, 5 Oct. 1800, fo. 293, printed (number 2256) in A. Aspinall, ed., *The Later Correspondance George III* (5 vols., Cambridge, 1962–70), III, 424.
[76] TNA, WO.58/168, War Office and Treasury letters to the deputy Commissary-General with Sir R. Abercromby's troops, 1799–1806.

and to secure fresh foods and grain from friendly ports in the Mediterranean. This had foundation in what was already happening. The British island garrisons and the navy were supplied from North Africa, Sicily and Italy with wheat and fresh foods, live bullocks being purchased at Leghorn. There were mills at Minorca and Malta so that grain might be milled. Wine was despatched from Lisbon by the Victualling Board's agent who supplied Portuguese wines to victuallers destined for the West Indies and the Mediterranean. Yet an army of 10,000 men on the move demanded constant re-supply from its stockpiles and it needed contingency supplies in case of emergency.

The experience of General Abercromby ensured the plan was strengthened. Abercromby, aged sixty-six, had experience dating back to the Seven Years' War; he had served in Flanders in 1793–4; he had commanded the expedition to the West Indies in 1796 and the Helder expedition of 1799. He was accustomed to having his opinion heard.[77]

In May 1800 Abercromby recommended the stockpile at Minorca be set and maintained at a level sufficient for six months' supply, half of which was to be reserved for an emergency. In addition the victuallers were to be hired for six months certain so they could be sent wherever they were needed and, if necessary, serve as warehouses afloat. Abercromby's recommendations were adopted. By the end of September 1800, six months' supply for 6,000 men was established at Minorca, and three months' supply for 6,000 men at Malta. The oldest stocks of perishable foods – flour, biscuit and salt meat – were issued first to ensure progressive turnover. In November 1800, six weeks after the expedition was launched, Abercromby demanded the stock at Malta be enlarged also to six months.

Meanwhile, Lord Keith, Commander-in-Chief in the Mediterranean, permitted the supply of the army to take priority over that of the navy, which had also to rely on the army's magazines of salt meat, bread and flour. However, he supplemented these supplies from England with local purchases of biscuit, flour, fresh meat, wine and vegetables. Buyers were sent on shore in Morocco to buy vegetables, which were dipped in salt water before being loaded; cattle too were purchased there and cut up by butchers carried on warships for that purpose.[78] He purchased shoes and cordage in Sicily. He ordered rope, canvas, spars, fuel, candles and medicines from Lisbon and Naples. And where naval craft or transports

---

[77] Mackesy, *British Victory in Egypt*, 6–12.
[78] A private on board the *Dictator, A Faithful Journal of the Late Expedition to Egypt* (London, 1802), 22; cited in Condon, 'The administration of the transport service', 304–5.

were lacking, he hired small craft and paid for pilots to ship the supplies to the army.[79]

The order launching the expedition reached Gibraltar on 24 October and it sailed three days later. It reached Malta on 19 November and departed for the eastern Mediterranean on 12 December. It put into Marmaras Bay on the coast of Asia Minor at the end of December and did not depart again for Egypt until 22 February. When it landed on the coast of Egypt on 8 March, the expedition had only five days' provisions available, which had also to meet any demands from the navy. This was increased to seventeen days after supplies were landed from warships and transports. Nevertheless, for an army facing stern resistance and immediately fighting for its survival, the margin between starvation and supply was slim.[80]

After all the planning, what had happened? Quite simply, while consuming the rations with which it travelled, the army had left behind the victuallers that carried new supplies. The victuallers carrying supplies to Malta and Minorca had been held at Gibraltar when the expedition left there. By the first week in October 1800, there were seventeen awaiting orders at Gibraltar. Fresh provisions, fruit and vegetables had been purchased and consumed by the army at Malta and Marmaras, but the rate of consumption had been far greater than the rate of replenishment. When the army reached Egypt, the former British consul there, George Baldwin, obtained sheep, poultry and other fresh foods from the Arabs. At first this source of supply seemed abundant but it soon diminished.

Nevertheless, with the army ashore, its location could at last be sent to the victuallers, which eventually began to arrive: those from Gibraltar first, then those direct from England in April. Extra supplies were drawn from the stocks at Malta and Minorca. Even so, the rate of consumption still exceeded the rate of supply. The truth was that, even together, the supplies brought by the victuallers and those obtained locally were insufficient, especially as the navy as well as the army was dependent upon them. Moreover, during May and June 1801, the purchase of provisions and their supply to the army were hampered by difficulties of conveyance within Egypt. Lord Hobart was informed early in June that the army's

chief problem is provisions of all kinds which, tho' they abound in the country, we do not find it easy to get at; indeed the present state of the bar at Rosetta (which is almost impassable) and the low state of the river render the carriage of

[79] NMM, KEI. L/22, 19 Sept. 1801, cited in Condon, 'The administration of the transport service', 305.
[80] Mackesy, *British Victory in Egypt*, 13–19, 68–72; Condon, 'The administration of the transport service', 303–6.

bread and spirits, which we are obliged to get from the fleet, tedious to a degree, and renders it impossible to advance with that degree of rapidity

as would have been desirable.[81]

Inadequacies in the money available to the army also handicapped the purchase of local supplies, especially wheat, of which there was plenty in Egypt. That which was purchased was sent to Malta and Minorca for grinding and manufacture into biscuit. This was organised by Lord Keith, who also despatched warships to Tunis, Algiers and Sicily for food supplies. At the same time, he shipped out rice in large quantities, available cheaply in Egypt, for the supply of the whole station. The role of Keith in arranging extra supplies has been considered critical to the success of the expedition. M. E. Condon observed that 'it was Keith's exertions . . . during the spring and early summer of 1801, in creating a supply system from the Mediterranean to Egypt to supplement the infrequent arrival of victuallers from England, that kept the British expeditionary force alive in Egypt in 1801'.[82]

In the matter of food supply, the expeditions of the French Revolutionary War demonstrated that there were lessons to learn for the Victualling Board. In campaigns in Flanders and the Helder in 1794 and 1799, respectively, provisions sent from England also proved deficient.[83] Local supplies were necessary, often arranged through diplomacy, and always through the energy and enterprise of commissaries and officers on the spot. The Victualling Board did its best. As demonstrated by the Egyptian campaign, it established stocks and tried to ensure the army had all it needed. But it erred on the side of economy. Thus the speed with which the army consumed its provisions outpaced communication, transportation and replenishment. The board also failed to foresee contingencies and the necessity for extra supplies. In consequence, at times the army was obliged to depend on local arrangements that were, of necessity, on occasions emergency measures.

The Napoleonic War and the Commissariat

Throughout the Napoleonic War, the Commissariat operated under the Treasury much like the Transport or Victualling Offices under the Admiralty. Sir Brook Watson, who had served as a Commissary-General as far back as the American War of Independence, was replaced by Thomas

---

[81] TNA, WO. 1/345, Hutchinson to Lord Hobart, 2 June 1801, quoted by Condon, 'The administration of the transport service', 306–7.
[82] Condon, 'The administration of the transport service', 307–8.
[83] *Ibid.*, 290–2, 295–30.

Aston Coffin in 1806, when the office was extending its formal responsibility for supplying troops throughout the world, for example in Curaçao and the Leeward Islands.[84] In 1809 a Commissary in Chief based in London was appointed (James Willougby Gordon) with authority over all stations overseas except Ireland and the East Indies. From 1809 the Commissary in Chief was made responsible to the Treasury Board for making all contracts for the supply of land forces – both at home[85] and overseas – with bread, oats, forage, straw, fuel, candles and barrack equipment.[86]

While the bureaucracy for the supply of the army was enlarging, the overseas expeditions of the Napoleonic War, at least the large-scale ones, benefited from their concentration within Europe. Short supply routes improved communication and eased transportation, as ships could shuttle back and forth to the ports of unloading. The great advance in the management of supply of land forces during this war was the degree of coordination achieved in operations of a large scale.

The Walcheren expedition

The Walcheren expedition in 1809 involved over 39,000 troops, including 3,000 cavalry, 245 warships (120 cutters and gunboats) and 400 transports each averaging 250 tons. Not all the troops were transported at once: 25,000 went over initially, embarked from the three main eastern anchorages under the commands of the port Admirals at Spithead, the Downes and the Nore. Admiral Campbell in the Downes was ordered to assemble 'every ship and vessel under his command' on 25 June. They included 2 former East Indiamen which could each carry a regiment of 1,000 men after their bulkheads and stores were removed. Ships of the line had their lower-deck guns removed and their main holds prepared for troops, their lower holds for the reception of horses. The Transport Board's agent at Portsmouth provided the forage for the horses.

The Admiralty ordered the Navy, Victualling, Ordnance and Transport Boards 'to give every assistance to Admiral Otway [at Spithead] in the embarkation' on 27 June 1809. Four days later, Otway reported 'that with the number of ships promised him' he could embark '14,660 infantry,

---

[84] TNA, WO.57/1, Commissary-General's in-letters, 1806–9.
[85] Accounts of contracts, with prices and quantities for all commodities, exist for each of the counties and for Ireland. See TNA, WO.58/44, quarterly returns of contracts made by the Commissary in Chief, 1809–12; also WO.57/51, proposals and agreements for contracts in Ireland 1809–19.
[86] T. M. O. Redgrave, 'Wellington's logistical arrangements in the Peninsular War 1809–1814', unpub. University of London Ph.D. thesis, 1979, 13–14.

and 460 horses with their riders or drivers and [a] brigade of guns'. At Spithead, regiments were embarked from 3 July; in the Downes and at Ramsgate on 16 July, where Deal boatmen embarked the soldiers. The Nore 'armament' joined the ships in the Downes on 22 July. The Portsmouth division sailed on the 25th, and the whole force departed on the 28th to reach the island of Walcheren the next day.[87]

Commissariat officers accompanied the Walcheren expedition but hardly had time to get organised. Letters to the officers in 1809 relate to the repair of barracks for the troops, and the despatch of bricklayers, other craftsmen and 1 million bricks.[88] Troops were fed and watered directly from the government yards in England by victualler. On average, each of the victuallers carried enough bread to serve the army about $2\frac{1}{2}$ days; enough meat to last it $4\frac{1}{2}$ days.[89] Allowing time for unloading and distribution overland, at least one victualler had to reach the Scheldt every day to prevent the army going short. In addition, water – 300 tons at a time – was delivered by the Victualling Board.[90]

The Peninsular War

Regular deliveries were also necessary to the supply of land forces serving in the Iberian Peninsula in the second half of the Napoleonic War. By March 1811 that campaign had the support of 256 transports of one type or another. To ensure the security of these vessels, convoys were organised: ninety-eight in 1811 alone, which was nearly one every three days. During that year, on six occasions the number of vessels under convoy amounted to more than 100. Others were smaller, one convoy consisting of just 2 vessels. But so lavish was the provision of supplies to the Lisbon depot in 1811 that, by the summer, owing to lack of storage space, Wellington himself was obliged to request that the flow, for the time being, be quenched. As the war went on, the distance by sea from Britain to these depots became shorter, though the coastline became more dangerous, as depots were formed in the ports of the northern coast.[91]

Yet the Peninsular War lasted five years, and some campaigns were conducted distant from the coast where supply rested to a great extent on the organisation of the Commissariat in the field. As soldiers carried

[87] TNA, ADM.1/6040, 'Precis of reports on Scheldt Expedition', part 2, 'Means of Transport and accommodation for Troops'.
[88] TNA, WO.58/85, letters to commissary officers, 1809–10.
[89] NMM, ADM. DP/29, 6 Nov. 1809.   [90] NMM, ADM. DP/29, 11 Dec. 1809.
[91] C. D. Hall, 'The Royal Navy and the Peninsular War', MM 79(1993), 403–18; NMM, ADM. DP/34A, 3 May 1814.

rations for only three days, supplements had to come from store by mules. Carola Oman, author of the seven-volume history of the war, observed that the 'future of the army in 1809 depended on whether the commissariat department would be able to rise to the height of its duties'. Contemporary opinion suggests that it did.[92]

R. Glover noted that, in the Peninsula, the Commissariat revealed a new ruthlessness in its requisition of supplies in this period. Even so, Wellington refused to forage as vigorously as the French – that is, plunder the produce of local peasantry – and believed in payment for requisitions.[93] Local resources were used to a great extent by the British army, but money and transport ('the field commissary's right and left arm') were key to their provision. Indeed the dependence of the British army on the supplies provided by the Commissariat tied Wellington to his supply routes and affected the strategy he was able to pursue. It also added to the costs of maintaining the British army in the Peninsula, which, at least in 1813 and 1814, placed restrictions on the scale of his campaigns.[94]

Early in his career, Wellington had been praised for his 'judicious and masterly arrangements in respect to supplies', and his awareness of their primary importance ensured the business and finances of the Commissariat received his full consideration.[95] 'It is very necessary', he observed, 'to attend to detail, and to trace a biscuit from Lisbon into a man's mouth at the frontier, and to provide for its removal from place to place, by land or by water, or no military operation can be carried on'. An army without supplies and starving, he wrote, is actually worse than none: 'The soldiers lose their disciplined spirit.'[96] In the Peninsula, Wellington's requirements shaped the bureaucracy which supplied his army, in conjunction with those of the Treasury.

In April 1809 a Commissary-General, John Murray, was appointed for both Spain and Portugal, and he superseded the senior commissary for Portugal, Philip Rawlings. Murray was relieved in June 1810 by Hugh Kennedy, who remained until December 1811 and returned ten months later to remain until the end of 1813. Wellington had a good working relationship with Kennedy, based, it has been suggested, on their 'common love of system'.[97] However, Wellington was also aware of the benefits

[92] Oman, *Wellington's Army 1809–1814*, 319.   [93] Glover, *Peninsular Preparation*, 255–66.

[94] Redgrave, 'Wellington's logistical arrangements', 2, 4–6, 191–2.

[95] J. Weller, 'Wellington's Peninsular War logistics', *JSAHR* 42(1964), 197–202.

[96] G. E. Rothenberg, *The Art of Warfare in the Age of Napoleon* (London, 1977), 185, derived from E. B. Hamley, *The Operations of War Explained and Illustrated* (London, 1907), 18–19.

[97] Redgrave, 'Wellington's logistical arrangements', 38–9.

of continuity in staffing. In January 1810, when Murray complained of instances of inefficiency among his staff in the Peninsula, Wellington denied want of zeal but admitted signs of inexperience.

I must say [he wrote to Gordon the Commissary in Chief in London] that there are instances of want of ability or of activity, or of activity misapplied, owing to want of knowledge of the duties of the employment which these gentlemen fill. In fact the duty of a commissary to an army in the field on any great scale of operations can be but little understood by the officers of the English Commissariat; and it must be difficult to perform them adequately in Spain and Portugal. In these countries the population is excessively small in proportion to their extent; the produce consequently in any particular district but little adequate to support the drain of subsistence of a body of troops. It must be difficult to procure what is required at the same time that the deficiency cannot be made up excepting by drawing supplies from a great distance; and the communications are bad and difficult, and the means of transport in the country are few, not easily procured, of a bad description, and slow in their movements. Your experience will point to you how difficult it must be for the commissary to supply the consumption of the troops in a country of this description; particularly for commissaries whose duties till they joined the army had been confined to the receipt from, or the superintendence of, the delivery by a contractor of a certain specified quantity of supplies; or even upon a larger scale of service, to the delivery to a small number of troops engaged in operations near the sea, of provisions received regularly and daily from the ships. To these circumstances must be added the complicated nature and the extent of the accounts which the commissaries are required to keep at the same time that they have these extensive, laborious, and to them novel duties to perform.

In January 1810, Wellington concluded that the commissaries had 'done their duty well; and they are certainly improved and are improving daily'.[98] There were exceptions, for which Wellington was confident Gordon would apply a remedy. Gordon did, with the appointment of seven new acting assistants.[99] In 1810 the Commissariat in the Peninsula continued to grow. By the summer of 1812 it had 719 employees, excluding local labour. In September 1811 the local employees (Portuguese) numbered 1,254. Of these, 973 were muleteers, drivers, carmen, labourers, servants and herdsmen; but some 41 were clerks, 124 were conductors, and 26 were storekeepers – 'roguish Portuguese under-commissaries' as they were called by one British officer. They were dispersed among the eight infantry divisions, nine cavalry regiments, headquarters and principal depots. There were also innumerable borrowed soldiers and casual locals.[100]

[98] TNA, WO. 57/38 part I, Wellington to Gordon, 30 Jan. 1810.
[99] Ibid., part II, Murray to Gordon, 13 Mar. 1810.
[100] Redgrave, 'Wellington's logistical arrangements', 32.

The challenges of the Peninsula

The importance of the Commissariat in supplying the British army and its allies in the Peninsula is unmistakable. However, its achievement rested on a number of foundations. It had to arrange transportation of supplies to the troops, cater for deficiencies in the commodities required, keep accounts of all movements and issues, and fund local purchases. Each of these operations had their difficulties. The way in which they were overcome reflects both on attitudes within the Commissariat and on the situation in Europe, which was very different to that half a century earlier.

Transportation and local supply

The role of the navy in providing escort to the flow of transports to the Iberian Peninsula has been discussed elsewhere.[101] The role of the navy ceased at the coast where supplies for the army were transferred from ship to shore or lighter. From port and riverside depots, the Commissariat took control of transportation. A large proportion of the local people employed by the Commissariat were used to drive mule trains, for mule transport complemented carts along the supply routes from the coast. In a constant flow, supplies were transported to inland depots from which the army could draw its provisions. For this purpose, from 1810 mule brigades were attached to units of the army: 200 mules to an infantry brigade, 250 to a cavalry regiment. In 1812 Wellington ensured his engineers cleared both the Douro and Tagus for navigation, ordered carts be constructed where they could not be obtained, and persuaded the Portuguese Regency to issue a regulation that made it easier for the British army to obtain carts for transport.[102]

In Portugal at the beginning of 1810, depots for supplies were established at Lisbon, Abrantes, Aldeia Gallega, Elvas and Figueira. In addition storeships were moored in the rivers Tagus and Douro. There were depots at these places both for Quarter-Master General stores – equipment for camps and hospitals – and for provisions including biscuit, flour, salt meat, wine, spirits and forage. In January 1810 provisions were available at six other depots: Santarem, Vitta Nova, Oporto, Almeida, Portalgre and Thomar. Some of these depots for provisions were temporary or transitory, depending on the movements of the army and on the convenience of communications and transport. Hence, for example, though

---

[101] C. D. Hall, *Wellington's Navy. Sea Power and the Peninsular War 1807–1814* (London, 2004), 111–29.

[102] Redgrave, 'Wellington's logistical arrangements', 148–9, 188; Weller, 'Wellington's Peninsular War logistics', 200.

not listed in January 1810, the following month seven new depots for provisions were established at Villa Franca, Abrantes, Rio Mayor, Leiria, Coimbra, Fordao and Vizeu.[103]

The depots for the supply of food may be divided into permanent depots and temporary or forward depots. The former were at the main ports or up-river unloading ports; the forward depots were a few days' march – perhaps 50 miles or more – beyond the permanent depots. According to the location, food was provided by a mixture of large-scale purchases and shipments from the rear, smaller local contracts, and local one-off purchases. The permanent depots, having to provide a predictable number of rations, took advantage of contracts to complement shipments from Britain. Thus troops around Lisbon were supplied by contract established in 1808 and still in force in 1813. Forward depots made local contracts for supplies, especially meat, long forage, alcohol – and soft bread where military units were stationary for a while. Troops advanced beyond these forward depots bought up local produce where and when they could, purchasing many of their provisions in this way.[104]

Local supplies, especially of forage and transport, were not always made available voluntarily. A degree of force was often needed to get what the army needed. The British Commissariat never went as far as the French. Wellington observed that the bayonet was used in every part of their system of supply and that, as the *de facto* government of every province they occupied, the French took what they wanted as a kind of taxation. The British could not act the same way without depriving their allies of resources. They thus entered into financial transactions for all they needed. Nevertheless, the degree of compromise on the British part depended on the power they exercised. In France in 1814, payment was generous, partly to conciliate the French. In Spain, forcible requisition was rarely used. 'I cannot over the heads of the Spanish authorities knock loudly at the doors of the several magistrates as we do in Portugal', Kennedy was informed by one of his officers in November 1812.[105] Yet, as this officer tacitly acknowledged, in Portugal provincial governors and local magistrates were bullied, sometimes ignored, and inhabitants forced to make supplies available, even at times by a ruthless clerk holding a pistol.[106]

---

[103] TNA, WO.57/38, parts I and II, return of Quarter Master General's stores in Portugal, 1 Jan. 1810; returns of provisions in Portugal, 1 Jan., 1 Feb. 1810.

[104] Redgrave, 'Wellington's logistical arrangements', 185–6.

[105] Ogilvie to Kennedy, 1 Nov. 1812, NLS Acc 4370, no. 17879, fo. 104, quoted in *ibid.*, 192.

[106] Redgrave, 'Wellington's logistical arrangements', 191–3.

The provision of main commodities

The supply of every commodity had its particular problems. In the Iberian Peninsula, when water transport was not available, imported commodities had to be transported long distances over difficult terrain. As a result local purchases became more important than shipment from Britain, and they had to be paid for by means that were locally accepted. In consequence the theatre of operations developed its own economy. This is evident from an examination of the supply of the four most important commodities consumed by the British army – meat, forage, alcohol and bread. Without them, morale and discipline would have suffered. Other subsidiary commodities were supplied but were trivial in relative importance.

Fresh meat was the most difficult commodity to supply from Britain. In consequence, both permanent and forward depots purchased live cattle. These must have been driven in from a wide catchment area, for consumption in the closing stages of the war demanded at least 1,000 head a week, and sometimes above 1,500. From late in 1813 a small proportion of this quantity – 100 a week – was shipped from Ireland to Passages and St Sebastian. A much greater supply of salt meat was shipped from Britain: consumption amounting to more than $1\frac{1}{2}$ million pounds a year. However, it had a limited use owing to the difficulties of transporting it inland. In 1810 nearly three-quarters was kept at Lisbon, Oporto and Coimbra, and the other quarter on navigable stretches of the River Tagus at Santarem and Abrantes.[107]

Forage to feed horses was of two types. One was a form of grain: barley, oats and Indian corn (maize) were the most common but some local variety of bean or pulse was used occasionally. The other form was long forage: straw or hay, but shortages compelled reliance on other vegetation. There was a basic ration: 10 pounds of barley or Indian corn or 12 pounds of oats and 14 pounds of straw or hay, but a shortage of one commodity could be made up with more of the other. Generally, 1 pound of corn was regarded as equivalent to 2 pounds of forage, which was equivalent to 4 pounds of green vegetation. The rules of thumb and the provisions made by commissaries varied. There was common agreement, however, that most, if not all, should be obtained locally by foraging parties.

Very little forage was shipped from England. The only shipments that appear in the records of the Commissariat were 3 million pounds of hay in the summer of 1809; 2 million pounds in February 1810; the same

[107] *Ibid.,* 71–3.

a year later; and $4\frac{1}{2}$ million pounds in the winter of 1813–14. In 1810, import was imperative owing to the exhausted state of Portugal.[108] But, much like the salt meat, land transport difficulties confined consumption of these imports close to the coast and navigable rivers. Once the army was accessible from the ports of northern Spain, at a time when numbers of horses were increasing, imports increased. From early 1812, oats, barley and Indian corn were shipped from Ireland; and the following year a contract was made for the import into northern Spain of 12–15,000 quarters of oats or barley each month. Other imports came from North America and even Brazil. In 1812–13, the British ambassador to Brazil arranged '31 cargoes of wheat, flour, rice, Indian corn and beans'.[109]

Alcohol was heavy to transport – 1 pint of wine weighed nearly $1\frac{1}{4}$ pounds – and was available almost everywhere in the Peninsula, so was bought as close to the army as possible: 85,000 pints was bought by one commissary in one month alone in the summer of 1810. However, distilled alcohol – spirits – supplied three rations from every pint and was generally cheaper than wine. A small stock was thus normally kept as a reserve in case the local alcohol supply ran out, or scarcity raised the prices prohibitively high, which was the more likely the longer the army remained in one place. Only in 1811 was a regular quantity of rum shipped from England: 284,000 gallons a quarter, sufficient for 75,000 men for three months. This saved money in the Peninsula where cash was always short. Yet the shortage of transport was even more pressing, and when the forces advanced into Spain the rum was left behind.[110]

Bread was perishable and best eaten close to the place where it was baked. Supplies were ordered from local contractors or obtained through the mediation of local magistrates. Troops on the move were issued with biscuit, of which stocks were held in the forward depots. The biscuit was hard-baked to keep for one to two months and was lighter than bread, the normal ration being only 1 pound, as opposed to $1\frac{1}{2}$ pounds of bread. However, when troops halted for more than one night, bread was wanted for its superior gastronomic qualities and, for this, flour was supplied by the commissaries. Flour was generally ground from wheat, and small quantities of both were imported into the Peninsula from Britain in 1810. Yet, thereafter, bad harvests resolved the Treasury to

---

[108] TNA, WO.57/38, part II, Murray to Gordon, 13 Mar. 1810. In 1810 forage corn was also sought from 'the western Islands', Morocco, Gibraltar and Malta.

[109] Redgrave, 'Wellington's logistical arrangements', 91–4, 101, 104.

[110] *Ibid.*, 85–6.

refuse supply from Britain, as late as October 1812, 'even for the army except under circumstances of absolute necessity'.[111] Consequently, the Commissariat and its contractors looked to the Azores, the Black Sea, Sicily, Morocco, Algiers, Malta, Egypt, Canada and America.[112] In April 1812 a consignment in seventy-one ships was held up by the American trade embargo resulting from the outbreak of Anglo-American hostilities, but all arrived safely later that year. Indeed, the supply from America continued under British licence until the summer of 1813. Only then was that trade terminated, when, consequent upon Napoleon's defeat in Russia, large supplies of wheat became available from the Baltic and were also made available from England.[113]

Numbers and accountancy

The scale of supply expanded as the size of the army managed by Wellington grew. Before the Talavera campaign in 1809, Wellington's British army totalled 26,000 men and 4,000 horses. By the battle of Bussaco in September 1810, these numbers had increased to 40,000 and 6,000. While occupying the lines of Torres Vedras north of Lisbon, with the addition of a Spanish army and certain Portuguese militia units, 70,000 men and 12,000 horses had to be fed. By August 1811, with 'His Majesty's allies in Portugal', the army reached 80,000 men.[114] Wellington maintained the principle that Portuguese units should draw their supplies from British stores but, the army being dispersed, in 1811 the Commissariat catered for only 60,000 men. However, retiring into Portugal in late 1812, numbers grew once more to 70,000 men and 12,000 horses and, by the time of the advance through the Pyrenees, reached 100,000 and 20,000 respectively.[115]

All these men and horses had to be fed each day. With such large numbers, failures of supply did occur, especially when troops were on the move. This is evident in the period 25 December 1811 to 16 January

[111] TNA, WO.58/134, Herries to Kennedy, 15 Oct. 1812, quoted by Redgrave, 'Wellington's logistical arrangements', 58.

[112] Some of the Mediterranean supply was carried in shipping chartered by the Transport Board, but for 'enormous' freight charges unforeseen by one contractor: F. W. C. Philips, to Colonel Gordon, Commissary in Chief, 10, 24 Feb. 1810, TNA, WO.57/38, part 1.

[113] Redgrave, 'Wellington's logistical arrangements', 52–3, 58–61, 66. He repeats figures from W. F. Galpin, 'The American grain trade to the Spanish peninsula', *American Historical Review* (1922): 1803 – 122,410 barrels flour; 1805 – 22,633; 1808 – 41,761; 1809 – 65,149; 1810 – 88,696; 1811 – 529,105; 1812 – 557,218; 1813 – 542,399; 1814 – 4,141.

[114] Christie, *Wars and Revolutions*, 317.

[115] Redgrave, 'Wellington's logistical arrangements', 51.

1812 from stores passing through Albuquerque for a force of 9,500 men and 2,000 horses under Hill: 9,000 bread rations were available each day, but only 4,000 of meat, 7,000 of alcohol and 1,500 or 1,200 of forage corn, depending on whether the allowance was 10 or 12 pounds per horse a day.[116]

The whole task of supply was monitored from behind the lines by the Commissary-General, and in London by the Commissary in Chief. The latter received monthly accounts of the provisions and forage remaining in store at each of the main depots or magazines. These encompassed the quantities of biscuit (Portuguese and English), flour, wheat and rye; the amount of salt meat and live bullocks; the pipes of wine and gallons of rum; and the forage in store – oats, barley, wheat, Indian corn, beans, bran, hay or straw.[117] These were compiled from statements required by the Commissary-General, in the Peninsula, who based them on statements of stocks returned by field commissaries daily and weekly, and on accounts of receipts and issues returned weekly and monthly by depot commissaries. The work of commissaries indeed was heavily dominated by returns and accounts. Altogether, by 1813, there were two daily, six weekly, eleven monthly, and eight occasional returns and accounts demanded from accountant officers.[118]

Some of these returns acquitted officers of public stores or money for which they were responsible, for commissaries were personally accountable for the public property entrusted to them unless they revealed good reason for their exoneration. Accounts of purchases had to be accompanied by receipts from payments and vouchers of fair price certified by two local signatories, expected to be 'magistrates' but often respectable inhabitants. These accounts had to be delivered at Lisbon or headquarters by the hand of a commissariat employee, not entrusted to the post. At headquarters, accounts contributed to the grand account that was submitted to the Treasury and was then subject to minute examination – every error, omission or question being referred back to the commissary on active service. Shortages of clerks, however, meant that in 1813 accounts under examination were dated 1810. The whole process was prolonged because methods of account keeping were not standardised. Double-entry book-keeping was not used, though in use elsewhere in government. In 1813 Kennedy, the Commissary-General in the Peninsula, noticed his Egyptian accounts from 1802 were still not passed and proposed improvements in accounting methods but, to his frustration, in vain.[119]

---

[116] *Ibid.*, 165.   [117] TNA, WO.57/38, parts I and II, 1 Jan., 1 Feb. 1810.
[118] Redgrave, 'Wellington's logistical arrangements', 22–3.   [119] *Ibid.*, 22–4.

Finance and dispositions

Neglect of accounting methods did not denote inefficiency. Rather, it indicated the preoccupation of the senior Commissariat with, in their view, higher priorities. Of these, the principal was supply of money. Without it, the campaign would have collapsed. The Commissary-General was responsible for paying from the military chest all the costs of the campaign, not just those arising from the supply of provisions, although they certainly were the greatest expense. The other costs, in diminishing order of amount, were the Ordinary expenses of the deputy Paymaster, charges for transport by land and water, subsidy payments to the Portuguese government, Ordnance and medical expenses, wages and contingencies, and purchases of clothing and equipment. The first four of these expenses accounted for 90 per cent of disbursements in the Peninsula in 1811–13.

Where was the money obtained to meet these costs? There were two methods of supply. The first and most obvious was to draw cash and specie from the Treasury's accounts at the Bank of England and at the Royal Mint.[120] However, the Bank of England did not always have sufficient cash or specie available for export. The period of the Peninsular War was one of bullion shortage, caused by the revolts in South America which interfered with mining operations and the export of specie. The capture of Spain's quicksilver mines in southern Spain by the French in 1810 exacerbated the situation, mercury being necessary to the extraction of silver from ore. To this shortage was added a disruption of British trading patterns as a result of Napoleon's Berlin and Milan decrees, vigorously enforced by the emperor after the parliamentary Select Committee on the price and circulation of gold bullion revealed the damage his measures had done to the British economy. In spite of these difficulties, in 1810 the Treasury sent £650,000 to the Peninsula, compared to £450,000 in 1809. But by then expenses had grown so much that the larger sum was a small proportion of the amount required.[121]

The second method of supply was to raise money from that which passed naturally through the trading centres of the Peninsula. The cash was paid for with bills of exchange drawn upon the Treasury. Over £92,500 was raised in this way between 10 January and 12 February

---

[120] TNA, WO.60/44, tabular accounts of John Charles Herries, Commissary in Chief, with the Treasury and with HM Mint for bullion, 20 Mar. 1816. The accounts indicate the variety of types of coin and forms of specie, with weight or value, supplied to Herries in 1814.

[121] Redgrave, 'Wellington's logistical arrangements', 107–9, 113.

1810.[122] Even so, at the end of January 1810, Thomas Dunmore, deputy Commissary-General at Lisbon, was so anxious at the 'rapid and alarming decrease of our funds' – the military chest being reduced to about £2,000 in specie but pledged to pay no less than £60,000 over the coming week – that he felt obliged to raise a further £80,000 by offering 'an advantage beyond the course of exchange'. The 'advantage' was kept secret even within his own office and he left it to John Murray, Commissary-General, to explain the higher rate to the Treasury.[123] In June 1810, Commissary-General Drummond was ordered to Lisbon to collect information on the money available and how it might be diverted into British hands. Drummond devised a scheme for selling certificates of entitlement to British Exchequer bills. However, having been impressed by the sale of its own bills, the Treasury shelved Drummond's scheme and relied on its own bills to purchase coin for the military chest.

Yet neither of these two methods of raising cash or specie was adequate. In 1811–12 the Commissariat slipped into increasing debt. In 1810 Wellington ordered that suppliers who could not be paid in ready money were to be given a bill, in effect an IOU, on the Commissary-General. The Commissary bills bore the amount to be paid and, from the shortage of money, came into circulation at a discount (2–3 per cent in 1810 but as much as 25 per cent in 1812) proportionate to the delay before the Commissariat could be anticipated to cash them. Much like Navy or Victualling bills, the Commissary bills bridged the gap between the Commissariat's income and its expenditure. In effect they formed a loan to the British government, the absence of interest compensated for by the ability of suppliers to charge higher prices that took into account the time before the bills would be cashed. Hence, by this means, the Commissariat purchased requirements even in rural locations, and the country people were able to sell the bills to those who could travel to Lisbon or army headquarters to exchange them for cash.

Commissary bills filled a critical gap in army funding between 1810 and 1812, when the shortage of ready money reached crisis point. At the end of that year several consignments of gold were received from South America. Moreover, with Napoleon's disastrous invasion of Russia, his grip on Germany and upon European markets slackened and money became more available from the continent. There, the agency of Nathan Rothschild raised funds sufficient to meet both the British subsidy obligations to allied powers and the expenses of the army in the Peninsula to the amount of £100,000 a month. Rothschild delivered gold to the Mint

---

[122] TNA, WO.57/38, part 1, Thomas Dunmore to Commissary in Chief, 17 Feb. 1810.
[123] TNA, WO.57/38, part 1, Dunmore to Murray, 25 Jan. 1810.

in London and coin to the Commissariat in the Peninsula.[124] For the advance from Spain into France, this extra funding was still not enough, and by 1814 another financial crisis loomed. To avoid a complete failure of credit, Wellington reduced the supplies available to his Spanish auxiliary forces, let his transport system deteriorate, and dispersed his army between Bordeaux, Bayonne and Toulouse.

By 1814 the supply of money thus had a direct effect on the disposition of the British army. Otherwise, managing a mixture of expedients, the Commissariat performed a difficult task of supply that was little appreciated. Indeed, the dual responsibility of the Commissariat to the army's Commander-in-Chief in the Peninsula and to the Treasury in London often placed it in a difficult position. Both Wellington and the Treasury had to be satisfied. The officers of the Commissariat were caught in the middle, with only 'the established custom of the department' to guide them.[125] Meanwhile, just as Wellington and the Commissary-General in Spain could suspect the Treasury was not doing all it could to supply money, the Treasury and the Commissary in Chief could assume Peninsula staff were too dependent on London. In studies of the army in the Peninsula, Wellington's view has received most attention and the government in London has had a bad press. It is accepted now that both the Treasury and the Commissariat did their best in difficult circumstances to meet the army's needs.[126]

Those circumstances demanded the Commissariat in the Peninsula create a local economy which, in its financial, account-based and centralised nature, much resembled the British state on a small scale. The work of the Commissariat in the Peninsula benefited from Wellington's appreciation of the logistical challenges imposed on the Commissariat, and his understanding of the relationship between the success of the Commissariat and the capabilities of his army. But clearly the Commissariat did not enter the Peninsula unprepared. Its operations in previous expeditions had given it, and the Treasury, a valuable depth of experience in planning and scheduling supplies to reach armies on campaign. Before 1793 it benefited from the experience of the Navy Board; after 1794 it benefited from the support of the Transport and Victualling Boards. By that time it had become a state bureaucracy that was regularly projected overseas and systematically supplied by the home departments. The coordinated operation of these government departments made the

---

[124] TNA, WO.60/109, transactions with Messrs Rothschild in April and May 1815.
[125] TNA, WO.57/38, part 1, Philip Rawlings, deputy Commissary-General at Lisbon, to the Commissary in Chief, 9 Mar. 1810.
[126] Redgrave, 'Wellington's logistical arrangements', 114, 117–19, 125, 130, 187.

projection of the British army relatively routine. It permitted the army to take or hold territories in pursuit of victory in Europe. It facilitated the organisation of amphibious operations and the garrisoning of territories in which Britain was interested throughout the world. Without it, however successful the navy was at sea, the British state could have had no lasting command on shore. In this sense, the Commissariat was the branch of the state bureaucracy that ensured Britain remained an imperial as well as a maritime power.

Conclusion

This book has shown how the British state during the second half of the eighteenth century developed the logistical capability to project its military forces throughout the world. It has focussed on the branches of the state that met particular organisational and supply needs. Each of these branches of supply was important in its own right. But the development of each was shaped by three factors: by the insular nature of Britain which gave the sea an important influence on branches of supply; by the partnership of the state and the private sector; and by the ideas and ethics that both united the state and gave its bureaucracy a special administrative culture.

It is the contention of this book that the bureaucracy that served the armed forces developed its competence between 1755 and 1815, to reach its highest level of capability at the time Britain achieved dominance at sea during the first decade of the nineteenth century.[1] It was a capability that served not only the navy but also the army, and brought with it the ability to command territory overseas, and mount campaigns distant from Britain. It was a capability that was enhanced by the growth of a maritime economy in which state spending strengthened domestic demand during wartime and stimulated those very resources – shipping, trade revenues, imports of food and naval stores – upon which the state depended to maintain its armed forces.

This book has included among the state's servants the contractors, the ship builders and ship masters who were employed by the state, but at the heart of the state were the commissioners, secretaries and clerks who took decisions, gave directions, and performed the accounts and calculations that supplied the soldiers and seamen stationed distant from Britain. They are the unsung heroes of British maritime ascendancy in the period 1755–1815. This is not to detract from the credit

---

[1] C. D. Pringle and M. J. Kroll, 'Why Trafalgar was won before it was fought: lessons from resource-based theory', *Academy of Management Executive* 11(1997), 73–89.

earned by military officers, colonial officials and statesmen. But it was the competence of this bureaucracy, at the head of a cascade of agents, officials and merchant contractors who worked to their figures, that was central to the military and naval achievements of Britain. Without the organisation and resources supplied by this bureaucracy, the fighting services of the British state were impotent.

Britain's bureaucracy met the needs of an insular state dependent on the sea. This gave Britain's logistics a special maritime basis. They employed the resources of a maritime economy. Ships were central to any operation beyond Britain's shores. Procurement, distribution, trade and finance were all dependent on the maintenance and growth of the maritime economy. Soldiers were shipped by sea; their transports were merchant ships hired into state service. Half the value of all Britain's naval stores came from abroad. The dockyards built some warships, but during the second half of the eighteenth century an increasing proportion was built in merchant shipyards which, first and foremost, served the mercantile world. The sea served to concentrate resources: guns cast in Scotland, food produced in Ireland, saltpetre imported from India. But then these resources had to be organised and deployed to the benefit of the state, and it was to this end that the competence of Britain's state bureaucracy was turned.

The need to send Britain's armed forces to sea developed in Britain's bureaucracy a special ability suited to that requirement. Infrastructure and processes were developed in each branch of supply. For the navy there were dockyards, workforces and materials to be managed to achieve the greatest output commensurate with contemporary standards. The navy as well as the army needed armaments; they needed men; food; and the transport to maintain supplies to wherever these armed forces were sent. For each of these commodities, a dedicated branch of bureaucracy developed expertise in the hire, purchase, measurement, distribution, management and control of resources.

Administrative ability went hand in glove with methods of production. Each infrastructure was open to developments in contemporary technology and methods of organisation. This contributed to the ability of ships to remain in operation for longer and eventually to command the sea. Examples in the fields of ship construction and fitting, ordnance supply and victualling are the introduction of piecework and of copper sheathing, the demand for the lathe boring of cannon, and use of iron water tanks. Only in the field of recruitment to the navy was there little innovation and little advance in competence, although the Impress Service spread its net wider, and, with an increase in the discharge of invalids, the turnover of men was increased.

For an insular nation, the provision of a navy was fundamental to its defence, and by 1755 Britain had the most powerful navy in Europe with a civil bureaucracy – managed by the Admiralty, Navy and Victualling boards – that was capable of keeping the greater part of the British fleet at sea during wartime. With this fleet, the French navy could be blockaded and, when it escaped, defeated in important encounters. Such battles were, however, few and far between and the main achievement of the British wartime state was to mobilise a large fleet and keep it at sea throughout long war periods. To defend Hanover and support allies, an army too was necessary for campaigns on the continent of Europe. For that purpose, during the Seven Years' War, the civilian Commissariat developed to support the army in the field. As this book has shown, that Commissariat eventually developed into a bureaucracy capable of managing an economy dedicated to the support of the army distant from Britain.

With the ability to maintain an army overseas as well as a navy, offence by sea, as well as defence, was possible, and Britain's military bureaucracy during the mid eighteenth century wars began to develop the expertise needed to project and maintain armed forces a long distance from Britain. The American War of Independence demonstrated that make-shift arrangements by the Treasury for the transport of troops, their food and equipment would not do. Hence in 1793 the naval Victualling Board became responsible for the food shipped to the army overseas as well as the navy; and in 1794 a new Transport Board became responsible for supplying hired ships to all government departments and conveying all supplies from Britain to units of the army and navy serving outside the British Isles.

These bureaucratic developments gave the British state the capability of projecting and maintaining a military presence in almost any part of the world. The structure of this bureaucracy preserved and focussed expertise that was firmly based in experience. In the development of this experience, great seamen such as Lord George Anson and James Cook are well known, but within the state's bureaucracy were numerous unknown clerks and agents such as George Teer, John Payne and George Smith, who contributed expertise. Organising figures such as Sir Charles Middleton, Thomas Blomefield and William Congreve listened to them. As in the case of Middleton, they drew knowledge from merchant seamen engaged in privateering, merchants connected with industry, and City ship owners and financiers. The experience was cumulative and had direct influence on the capability of the state in maritime operations.

The British state alone did not have the power to project armed forces. It needed the assistance, indeed partnership, of the private sector. It needed money, supplies, manpower. The partnership of the Treasury with the Bank of England was crucial to the raising of loans capable of sustaining military operations and subsidies to allies. Crucial too was the raising of revenues by the Customs and Excise services and assessors of property and income taxes. The cooperation of the British people was partly due to subtle management but also due to respect in the population for the demands of the state. Taxpayers understood that revenues paid for defence and were partly expended on materials, stores, provisions and equipment that British industry, trade and shipping supplied.

Contractors were indispensable to the supply of these commodities. Each branch of supply had its large and small merchants. Before and during the American War of Independence, when the management of overseas supply was still in its infancy, merchants could take advantage of the state, as the Brymer–Champlin correspondence in 1773 reveals. However, just as the Victualling Board learned to manage its contractors – weeding out those who were unreliable, fostering those who provided steady service – so too the Treasury, Navy, Ordnance and Transport boards did the same. Payment was, of course, important, and the introduction of ninety day bills in 1796, by limiting delays before payment of bills to three months, can only have helped relationships with merchant suppliers. Indeed then, given that credit of up to nine months was common in private trade, business with the state was probably even more profitable than private transactions.

Developments in the commercial environment helped to promote relations between the state and its contractors. Central was the growth of equity law. As is evident in the financial management of food contractors, the growing precision of contract specifications did much to eliminate inefficiency and abuse. But disputes did arise, especially in the 1780s when loopholes in contracts were being discovered. Traditionally, dealers resorted to arbitration by specialists like themselves in their field. The latter usually favoured the views of contractors rather than those of the state. The ability of the latter to appeal to the law courts for rulings on fair dealing achieved judgements of greater independence and objectivity. The development of equity law thus altered the context in which the contracts were drawn up and observed.

Equity in principle was equally important for the working artificer and seaman, both free agents in the private sector, for the principle determined the fairness with which they were treated by the state. Perquisites persisted until the very end of the eighteenth century, sanctioned on account of deficiencies in payment, and only banned on improvement in levels of pay. The principle was similarly evident in the provision of short-allowance money to seamen as recompense for deprivation, or inability to consume, full rations. Even more was it evident in the instruction to pursers of naval vessels to ensure that officers also went on short allowance when it became necessary for the seamen. Agreement to the demands of seamen in 1797 for abolition of the purser's eighth, by which they thought themselves deprived, equally demonstrated the readiness of the state not just to conciliate, but to respect an appeal for justice. The provision of food was a vital medium in the unspoken relations between the state and its seamen.

On a far greater scale, the desire for equity underlay economic reform in attempting to ensure the public received value for the money it provided for state services. Economic reform in turn promoted the application of utilitarian values to state service. The effect was an administrative culture which was unique to British bureaucracy and was calculated to promote efficiency in the performance of duties. Economic reform had an impact on three fundamental aspects of state service: on the recompense of office holders, on their sense of responsibility and on their knowledge of their duties.

Before about 1796–1801, the greater part of state bureaucracy was dominated by a culture in which perquisites were normal, and in which ambitions were focussed on posts that provided the highest rates of private income. The removal of perquisites and the establishment of adequate salaries, graded according to the importance of the work to the public, focussed ambitions upon the posts that carried the greatest responsibilities. In effect, public payment and public duties completely replaced private reward for personal services which may not have conformed to public priorities. The eradication of any potential private interest created the civil service, which survived in recognisable forms into the late twentieth century.

The uncompromised focus on public duty was accompanied by a new emphasis on industry, integrity and trust in the office holder. The outspoken championship of these qualities lay at the heart of the political struggle between Sir Samuel Bentham, the Inspector General of Naval Works, and the Navy Board, between 1796 and 1807. While the former advocated individual responsibility, the latter stood for collective responsibility which had been the dominant organising principle in public administration since the seventeenth century. Although the Navy Board appears to have won that contest, the principle of individual responsibility became rooted in the naval departments, to flower in the 1820s. The period around 1800 was thus an important turning point in the culture of British bureaucracy. With less political fuss, the principle had already been established in the Ordnance departments, where Inspectors were created during the 1780s, to be responsible for the quality of cannon, for manufacturing gunpowder and for the construction of gun carriages. After 1793 the Victualling and Transport departments also expanded their range of agents working at a distance from London, often abroad with fleets. The principle was thus advocated at an appropriate time.

Accompanying this proliferation of agents and inspectors was the necessary up-dating and revision of orders relating to duties. The Admiralty's regulations and instructions to the navy's Sea Service had been revised almost every decade from the middle of the eighteenth century. But those to the civil departments had been neglected, despite an awareness of the need for the revision of standing orders since the 1760s. After 1780, the state bureaucracy managing the logistics of the armed forces was subject to a series of investigative commissions and committees which urged revision. The new salary scales of bureaucrats from about 1796 gave rise to some new instructions, but the full revision eventually occurred between 1805 and 1809 by the Commission for revising and digesting the civil affairs of the navy, which had its counterpart in the Commission of army revision.

The state's military bureaucracy was thus shaped after 1796 by new values, new thinking and new instructions. It gave state service a new ethos. The capability that had existed during the eighteenth century was enhanced, and necessarily so – for the demands placed on the military bureaucracy after 1800 exceeded anything required of it before that. The achievement of British maritime ascendancy during the Napoleonic War corresponded with the establishment of this new ethos. Confidence at sea was complemented by the new capability on shore. Moreover, the latter was concerned not only with the maintenance of a navy of unprecedented size, but with the projection of armed forces throughout the world. Standards of organisation, management and control of resources had to be of the highest level, and they were. The logistics that determined British maritime power stemmed from the bureaucracy at the very heart of the state, and the credit for their effectiveness must be placed where it was earned.

The state's bureaucracy had a military and imperial impact. It also had an economic and social one too. It was, after all, the military bureaucracy which, through spending, added to domestic demand and to the money supply during wartime.[2] Before, and even after, 1796, when ninety day bills were introduced, the short-term debt of the military departments in bills grew at times to several million pounds before being partly cleared. The Treasury was responsible for engineering the reduction of the short-term as well as the long-term debt. But, during the eighteenth century,

---

[2] O'Brien, 'Central government and the economy, 1688–1815', 234–40.

the Treasury was tentative about interfering in the internal procedures of government departments, which were responsible for their own estimates, accounting and audit of expenditure. This gave them relative independence to increase financial requirements and expenditure as war demanded.

The stimulus of the military departments to the economy was naturally directed through the medium of their contractors. During wartime, about three-fifths of state expenditure went on the armed forces, and nearly half of naval expenditure on the supply of materials, especially stores, food and shipping. This not only enhanced the capacity of merchants and financiers to undertake further contracts, but stimulated the industries upon which they drew. The role of the state in sparking innovation has been acknowledged for other periods, and was the same in the eighteenth century.[3] Peaks in patents for inventions coincided with the final years of each major war between 1755 and 1815 – in 1763, 1783, 1801–2 and 1813.[4] War brought about gradual economic, and thus social, change. In contributing to this process of change in society, the influence of the bureaucracy complies with the form of 'military revolution' looked for by Jeremy Black, one which transforms the society from which the military forces emanate.[5]

The process of state and economic transformation accelerated after 1780. There was a relationship, in that state spending on the armed forces grew even greater, and sums put into the hands of contractors, ship owners, importers and industrialists also grew. Attitudes to the benefit derived from state debt changed. Fear of the interest charges turned to pleasure at the investment they provided. As Patrick Colquhoun observed in 1815, the interest of the domestic public debt was 'the seed sown to produce a bountiful harvest of newly created property every year'.[6] It was war, and the spending of the military bureaucracy, which was largely responsible for this bounty. Diminished fear of debt enhanced spending. By the end of the Napoleonic War, the British state had become a very different state from that at the end of the American War of Independence or Seven Years' War. In this change, the departments concerned with procurement and supply played a major part, helping to create what may best be called a military-bureaucratic state.

---

[3] Kennedy, *Rise and Fall of the Great Powers*, 105.

[4] T. S. Ashton, 'Some statistics of the Industrial Revolution in Britain', originally printed in *The Manchester School* 16(1948), 214–34, extracts repr. in *Science, Technology and Economic Growth in the Eighteenth Century*, ed. A. E. Musson (Bungay, Suffolk, 1972), 115–20.

[5] See chapter 1.

[6] Colquhoun, *Treatise on the Wealth, Power and Resources of the British Empire*, 284–5.

The logistics of maritime ascendancy thus helped to transform British society as well as to defeat Napoleon on the continent of Europe and to extend the territory Britain could control throughout the world. The bureaucracy of the state lay at the heart of this change. It was the capability of the military bureaucracy which assured Britain of victory at sea and of an effective army in its land campaigns. The success of this state bureaucracy in turn depended upon reform in government, partnership with the private sector, and experience in managing support for the armed forces by sea. These circumstances gave British state bureaucracy a distinct culture which permitted Britain to achieve and maintain maritime ascendancy between 1755 and 1815.

Bibliography

MANUSCRIPT SOURCES

BRITISH LIBRARY

Report of John Rennie	Add. MSS 27,884
Papers of John Jervis, Earl of St Vincent	Add. MSS 31,163, 31,167
Papers of Horatio, Viscount Nelson	Add. MSS 34,906
Papers of Henry Bathurst, Earl Bathurst	Loan 57/107

DEVON RECORD OFFICE

Papers of Henry Addington, Viscount Sidmouth   152M/c1803–1804/ON

HAMPSHIRE RECORD OFFICE

Papers of Portsmouth Ordnance yard   109M91/CG, CO, MIS

LIBRARY OF CONGRESS, WASHINGTON DC

Papers of Sir George Cockburn	Container 13
Papers of Henry Dundas, first Viscount Melville	

NATIONAL ARCHIVES OF SCOTLAND

Papers of Henry Dundas, first Viscount Melville   GD51/2

NATIONAL MARITIME MUSEUM

Admiralty in-letters

From the Navy Board	ADM. B/161, 214, 228
	ADM. BP/1–50, 105–119
	ADM. Y/1, 2, 4, 5
From the Victualling Board	ADM. D/39–43
	ADM. DP/1–42, 102–112, 201

Navy Board in-letters

From the Admiralty ADM. A/2596, 2953, 2990
From the Inspector General of ADM. Q/3320
Naval Works

Victualling Board in-letters

From the Admiralty ADM. C/679; G/773–98

Private papers

Papers of Cox and Sons, navy MS81/165
agents
Papers of Sir William Cornwallis COR/42
Papers of Sir John Duckworth MS88/090
Papers of Sir George Grey GRE/8
Accounts of Andrew Lindgren MS87/079
Papers of Sir Charles Middleton, MID. 1–10, 14
first Baron Barham
Papers of Sir Alexander Milne MLN 153/3/41
Papers of John Montagu, Earl of F5/2
Sandwich
Papers of Sir George Murray MS84/057
Papers of Sir Evan Nepean NEP/7
Papers of Charles Sergison SER/126

Single

volumes CAD. A/10, B/8, B/10, D/14;
LOG. W/3

RHODE ISLAND HISTORICAL SOCIETY LIBRARY

Papers of Christopher Champlin

ROYAL NAVAL MUSEUM, PORTSMOUTH

Admiralty Library MSS. 1/3, 1B/12, 35, 51, 82, 305
Miscellaneous volumes 321/73, 322/73, 1973/320, 1984/373, 2003/77
Papers of Thomas Peckston

SHEFFIELD CITY LIBRARY

Wentworth Woodhouse Muniments R108–137-1.

THE NATIONAL ARCHIVES OF THE UNITED KINGDOM

Admiralty Office records
In-letters
From Channel Fleet ADM. 1/117
From Lieutenants ADM. 1/2934

From Impress Service	ADM. 1/3663–4, 5117–19
From Transport Board	ADM. 1/3730, 3737
From Secretary of State	ADM. 1/4172
From Ordnance Board	ADM. 1/4015
From Regulating Captains	ADM. 1/5118/10
Petitions	ADM. 1/5126
Courts martial	ADM. 1/5402, 5425
Scheldt expedition	ADM. 1/6040

Registers, lists and indexes

Transactions at Cape of Good Hope	ADM. 7/3/2
Weekly accounts of ships fitting	ADM. 7/60
Convoy lists sent to Lloyds	ADM. 7/64
Lists of ships and vessels under convoy	ADM. 7/65
Admiralty Board Room journal	ADM. 7/257
Memorials and reports	ADM. 7/344
Daily returns, ships' stations	ADM. 7/502
View of progress of navy	ADM. 7/567
Weekly returns, ships on home service	ADM. 7/652
Ships' ordnance	ADM. 7/677
Convoys	ADM. 7/782, 796
List Books of ships' stations	ADM. 8/30, 35, 53, 56, 58, 79, 89, 99, 100.
Digests and Indexes	ADM. 12/1–174

Accounts and accounting records

Bill Books	ADM. 18/104–26
Succession book, impress service	ADM. 30/34
Abstracts of contracts	ADM. 49/32
Estimates of debt	ADM. 49/38
Comparisons of debt	ADM. 49/39, 40, 49, 50
Instructions	ADM. 49/59
State and condition of ships	ADM. 49/100, 102
Statements of debt	ADM. 49/173
Estimates for building warships	ADM. 95/8

Navy Office records

Sheerness dockyard letters	ADM. 106/1844
Portsmouth dockyard letters	ADM. 106/1883
Plymouth dockyard letters	ADM. 106/1937
Navy Board to Admiralty Board letters	ADM. 106/2227–37
Standing orders to yards	ADM. 106/2507, 2513–15, 2534
Ships' armaments	ADM. 106/3063–7

Committee on stores   ADM. 106/3574
Secret papers   ADM. 106/3575
Progress and dimensions books   ADM. 180/6, 10, 9–11

Transport Office records
Plymouth officers' letters   ADM. 108/1A, 1B
Treasury Board letters   ADM. 108/4A, 4B
Transport Board out-letters   ADM. 108/28
Transport Board minutes   ADM. 108/31
Ships' and freight ledgers   ADM. 108/148–51
Freight ledgers   ADM. 108/158–61
Rules for agents and masters   ADM. 108/182

Victualling Office records
In-letters   ADM. 109/102–3,110
Out-letters   ADM. 110/1, 5, 64
Board minutes   ADM. 111/1, 126, 191
Abstracts of contracts   ADM. 112/140–3
Contracts ledgers   ADM. 112/162–212
Information of the Attorney General   ADM. 114/3
Plymouth, Rotherhithe, Woolwich   ADM. 114/41
Average prices of corn   ADM. 114/70
Previsioning for transports   ADM. 114/71

Works Office records
Maps and plans   ADM. 140/555, parts 14, 18

Ordnance Office records
Upnor depot letter-books   ADM. 160/2–14, 56
Proportions of stores   ADM. 160/150

Treasury Office records
Treasury Board papers   T. 1/413, 420, 580
Commissariat finances   T. 38/806–11
Navy Board correspondence   T. 48/89, 64/101–19, 200, 201,
                                       207

War Office Records
Troops at home and abroad   WO. 17/2814
Ordnance stores, navy   WO. 55/1745
Commissary-General's in-letters   WO. 57/1
Commissary in Chief's in-letters   WO. 57/38, 51
Quarterly returns of contracts   WO. 58/44
Letters to commissary officers   WO. 58/85
Letters to Egypt expedition   WO. 58/168
Commissariat accounts   WO. 60/19, 44

Subsidies to foreign states   WO. 60/45
Transactions, Messrs Rothschild   WO. 60/109

WELLCOME INSTITUTE, LONDON

Papers of Horatio, Viscount Nelson   MS 3678

WILLIAM L. CLEMENTS LIBRARY, UNIVERSITY OF MICHIGAN

Papers of Henry Dundas, first Viscount Melville
Papers of Robert Dundas, second Viscount Melville
Papers of William, second Earl of Shelburne

WILLIAM R. PERKINS LIBRARY, DUKE UNIVERSITY, NORTH CAROLINA

Papers of Captain Sir Andrew Hamond
Papers of Lloyd Kenyon, first Baron Kenyon, 1782

PRINTED SOURCES

PARLIAMENTARY REPORTS, PAPERS AND PROCEEDINGS

5th, 6th, 8th and 9th Report of the Commissioners appointed to inquire into fees, gratuities, perquisites and emoluments which are, or have been lately, received in the several public offices, 1786–8, *Commons Reports* 1806, VII, also repr. *House of Commons Sessional Papers of the Eighteenth Century*, ed. S. Lambert (repr. Wilmington, DE, 1975).

11th Report of the Commissioners appointed to inquire into the state and condition of the woods, forests and land revenues of the Crown, 1792, *Commons Journals*, 47.

17th, 24th, 31st and 32nd Reports of the Select Committee on finance, 1797–8, *Reports of Committees of the House of Commons*, 1797–1803, 1st series, XII–XIII.

3rd, 6th and 9th Report of the Commissioners appointed to inquire into irregularities, frauds and abuses practiced in the naval departments, *PP* 1803–4, III; 1805, II.

3rd Report of the Commissioners for revising and digesting the civil affairs of His Majesty's navy, *Commons Reports* 1806, V.

6th Report of the Select Committee on finance, *PP* 1817, IV.

8th Report of the Select Committee on finance, 1818, *Commons Reports* 1818, III.

PP 1805 (152), VIII, 217.

PP 1805 (192), VIII, 277.

PP 1805 (193), VIII, 485.

PP 1806 (2), XI, 665.

PP 1826 (164), XX, 505–11.

The Parliamentary Debates, ed. T. C. Hansard, 2nd series, X; 3rd series, II.

PRINTED PRIMARY SOURCES

An Eighteenth Century Secretary at War: The Papers of William, Viscount Barrington, ed. T. Hayter (Army Records Society, London, 1988).

British Naval Documents 1204–1960, ed. J. B. Hattendorf, R. J. B. Knight, A. W. H. Pearsall, N. A. M. Rodger and G. Till (NRS, 1993).

Calendar of Treasury Books and Papers, 1742–45, ed. W. A. Shaw (London, 1908).

Historical Manuscripts Commission 14th Report. Appendix Part V: The Manuscripts of J. B. Fortescue, Esq., Preserved at Dropmore (London, 1894), II.

Landsman Hay: The Memoirs of Robert Hay 1789–1847, ed. M. D. Hay (London, 1958).

Letters and Papers of Charles, Lord Barham 1758–1813, ed. J. Knox Laughton (3 vols., NRS, 1907–11).

Naval Administration, 1715–1750, ed. D. A. Baugh (NRS, 1977).

Naval Courts Martial, 1793–1815, ed. J. Byrn (NRS, 2009).

Naval Papers respecting Copenhagen, Portugal and the Dardanelles presented to Parliament in 1808 (London, 1809).

Papers relating to the Blockade of Brest, 1803–1805, ed. J. Leyland (2 vols., NRS, 1899, 1902).

Pering, R., *A brief inquiry into the causes of the premature decay in our wooden bulwarks* (1812).

Portsmouth Dockyard Papers 1774–1783: The American War, ed. R. J. B. Knight (Portsmouth, 1987).

Regulations and Instructions relating to His Majesty's Service at Sea (London, 1787).

Select Documents in Australian History 1788–1880, ed. G. M. H. Clark (Sydney, 1950).

Selections from the Correspondence of Admiral J. Markham during 1801–4 and 1806–7, ed. Sir C. Markham (NRS, 1904).

Shipboard Life and Organisation, 1731–1815, ed. B. Lavery (NRS, 1998).

Surgeon's Mate. The diary of John Knyveton, surgeon in the British fleet during the Seven Years War 1756–1762, ed. E. Gray (London, 1942).

The Channel Fleet and the Blockade of Brest, 1793–1801, ed. R. Morriss and R. C. Saxby (NRS, 2001).

The Dispatches and Letters of Vice Admiral Lord Viscount Nelson, ed. Sir N. H. Nicholas (7 vols., 1844–6).

The Economy of His Majesty's Navy Office (London, 1717).

The Eighteenth Century Constitution 1688–1815, ed. E. N. Williams (Cambridge, 1970).

The Health of Seamen: Selections from the Works of Dr James Lind, Sir Gilbert Blane and Dr Thomas Trotter, ed. C. Lloyd (NRS, 1965).

The Letters of Lord St. Vincent 1801–4, ed. D. Bonner Smith (2 vols., NRS, 1921, 1926).

The Life and Adventures of John Nichol, Mariner, ed. Tim Flannery (1822, repub. Edinburgh, 2000).

The Manning of the Royal Navy: Selected Public Pamphlets 1693–1873, ed. J. S. Bromley (NRS, 1974).

The Nagle Journal: A Diary of the Life of Jacob Nagle, Sailor, from the Year 1775 to 1841, ed. J. Dann (New York, 1988).

The Private Papers of George, Second Earl Spencer, 1794–1801, ed. J. S. Corbett and H. W. Richmond (4 vols., NRS, 1913–24).

The Private Papers of John, Fourth Earl of Sandwich, 1771–1782, ed. G. R. Barnes and J. H. Owen (4 vols., NRS, 1932–8).

The Royal Navy in the River Plate, 1806–1807, ed. J. D. Grainger (NRS, 1996).

The Saumarez Papers: Selections from the Baltic Correspondence of Vice-Admiral Sir James Saumarez 1808–1812, ed. A. N. Ryan (NRS, 1968).

The Siege and Capture of Havana 1762, ed. D. Syrett (NRS, 1970).

The Tomlinson Papers: Selected from the Correspondence and Pamphlets of Captain Robert Tomlinson, R.N., and Vice Admiral Nicholas Tomlinson, ed. J. G. Bullocke (NRS, 1935).

The Vernon Papers, ed. B. McL. Ranft (NRS, 1958).

BOOKS AND ARTICLES

Acerra, M. and Meyer, J., *Marines et revolution* (Rennes, 1988).

Albion, R. G., *Forests and Seapower: The Timber Problem of the Royal Navy, 1652–1862* (Cambridge, MA, 1926).

   'The timber problem of the Royal Navy, 1652–1852', *MM* 38(1952), 4–22.

Aldridge, D. D., 'The victualling of the British Naval expeditions to the Baltic Sea between 1715 and 1727', *Scandinavian Economic History Review* 12(1964), 1–25.

Alsop, J. D., 'Royal Navy victualling for the eighteenth century Lisbon station', *MM* 81(1995), 468–9.

Anderson, J. L., 'Aspects of the effect on the British economy of the wars against France, 1793–1815', *Australian Economic History Review* 12(1972), 1–20.

Ashton, T. S., 'Some statistics of the Industrial Revolution in Britain', *The Manchester School* 16(1948), 214–34; excerpts repr. in *Science, Technology and Economic Growth in the Eighteenth Century*, ed. A. E. Musson (Bungay, Suffolk, 1972), 115–20.

Aspinall, A., ed. *The Later Correspondence of George III* (5 vols., Cambridge, 1962–70).

Astrom, S.-E., 'North European timber exports to Great Britain, 1760–1810' in *Shipping, Trade and Commerce: Essays in Memory of Ralph Davies*, ed. P. H. Cottrell and D. H. Aldcroft (Leicester, 1981), 81–97.

Aylmer, G. E., *The State's Servants: The Civil Service of the English Republic 1649–1660* (London, 1973).

Bagby, P., *Culture and History: Prolegomena to the Contemporary Study of Civilisation* (Berkeley and Los Angeles, 1958).

Baker, H. A., *The Crisis in Naval Ordnance* (NMM Monograph 56, Greenwich, 1983).

Baker, N., *Government and Contractors: The British Treasury and War Supplies 1775–1783* (London, 1971).

Bamford, P. W., *Forests and French Sea Power 1660–1789* (Toronto, 1956).

Banbury, P., *Shipbuilders of the Thames and Medway* (Newton Abbot, 1971).

Barnett, C., *Britain and her Army 1509–1970: A Military, Political and Social Survey* (London, 1970).

Barrow, T., 'The Noble Ann affair' in *Pressgangs and Privateers*, ed. T. Barrow (Whitley Bay, 1993), 13–22.

Bateson, C., *The Convict Ships 1787–1868* (Sydney, 1983, repr. 1985).

Baugh, D.A., *British Naval Administration in the Age of Walpole* (Princeton, NJ, 1965).

'Great Britain's "Blue Water" policy, 1689–1815', *International History Review* 10(1988), 33–58.

'Why did Britain lose command of the sea during the war for America?' in *The British Navy and the use of Naval Power in the Eighteenth Century*, ed. Black and Woodfine, 49–69.

'Maritime strength and Atlantic commerce: the uses of "a grand marine empire"' in *An Imperial State at War*, ed. Stone, 185–223.

Bayly, C.A., *Imperial Meridian: The British Empire and the World 1780–1830* (Harlow, 1989).

'The second British empire' in *OHBE* V, 44–72.

Beaglehole, J. C., *The Life of Captain James Cook* (Hakluyt Society, London, 1974).

Beveridge, W., *Prices and Wages in England from the 12th to the 19th century* (London, 1939).

Binney, J. E. D., *British Public Finance and Administration, 1774–92* (Oxford, 1958).

Black, J., 'Naval power and British foreign policy in the age of Pitt the Elder' in *The British Navy and the Use of Naval Power in the Eighteenth Century*, ed. Black and Woodfine, 91–107.

A Military Revolution? Military Change and European Society 1550–1800 (Basingstoke, 1991).

'British naval power and international commitments: political and strategic problems, 1688–1770' in *Parameters of British Naval Power 1650–1850*, ed. Duffy, 39–59.

'Military organisations and military change in historical perspective' *JMH* 62(1998), 871–92.

A System of Ambition? British Foreign Policy 1660–1793 (Stroud, Glos., 2000).

Rethinking Military History (London, 2004).

'Was there a military revolution in early modern Europe?' *History Today* 58, 7(2008), 34–41.

Black, J. and Woodfine, P., *The British Navy and the Use of Naval Power in the Eighteenth Century* (Leicester, 1988).

Blackman, H., 'Gunfounding at Heathfield in the eighteenth century', *Sussex Archaeological Collections* 67(1926), 39.

Bond, G. C., *The Grand Expedition: The British Invasion of Holland in 1809* (Athens, GA, 1979).

Booth, E., *Christian's Fleet: A Dorset Shipping Tragedy* (Stroud, 2003).

Bowen, H. V., *War and British Society, 1688–1815* (Cambridge, 1998).

The Business of Empire: The East India Company and Imperial Britain, 1756–1833 (Cambridge, 2006).

Breen, K., 'Gibraltar: pivot of naval strategy in 1781', *Transactions of the Naval Dockyards Society* 2(2006), 47–54.

Breihan, J., 'William Pitt and the Commission on Fees, 1785–1801', *HJ* 24(1984), 59–81.

Brewer, J., *The Sinews of Power: War, Money and the English State, 1688–1783* (London, 1989).

'The eighteenth-century British state: contexts and issues' in *An Imperial State at War*, ed. Stone, 52–71.

Bromley, J. S., 'Prize Office and Prize Agency at Portsmouth 1689–1748' in *Hampshire Studies*, ed. J. Webb, N. Yates and S. Peacock (Portsmouth, 1981), 169–99.

Brook, J., *The House of Commons, 1754–1790: Introductory Survey* (Oxford, 1964).

Browne, D. K., 'The structural improvements to wooden ships instigated by Robert Seppings', *Naval Architect* 3(1979), 103–4.

Bruce, C. M., 'The Department of the Accountant-General of the Navy', *MM* 10(1924), 252–66.

Buchet, C., 'The Royal Navy and the Caribbean, 1689–1763', *MM* 80(1994), 30–44.

Burroughs, P., 'The Ordnance Department and colonial defence, 1821–1855', *Journal of Imperial and Commonwealth History* 10(1981), 125–49.

Byrn, J., *Crime and Punishment in the Royal Navy: Discipline on the Leeward Islands Station 1784–1812* (Aldershot, 1989).

Cain, P. J. and Hopkins, A. G., *British Imperialism: Innovation and Expansion 1688–1914* (Harlow, 1993).

Canny, N., Marshall, P. J. and Winks, R. W., eds., *Oxford History of the British Empire*, I, II, V (5 vols., Oxford, 1998–9).

Cardwell, M. J., 'The Royal Navy and malaria, 1756–1815', *Trafalgar Chronicle* 17(2007), 84–97.

Caruana, A. B., *The History of English Sea Ordnance 1523–1875* (2 vols., Rotherfield, East Sussex, 1994–7).

Christie, I. R., *Crisis of Empire: Great Britain and the American Colonies, 1754–1783* (London, 1966).

'The Cabinet in the reign of George III, to 1790' in *Myth and Reality in Late Eighteenth Century British Politics*, ed. Christie (London, 1970), 55–108.

'Economical reform and "The influence of the Crown", 1780' in *Myth and Reality in Late Eighteenth Century British Politics*, ed. Christie (London, 1970), 296–310.

Wars and Revolutions: Britain 1760–1815 (London, 1982).

Stress and Stability in Late Eighteenth Century Britain: Reflections on the British Avoidance of Revolution (Oxford, 1984).

Clammer, D., 'The impress service in Dorset, 1793–1805', *Maritime South-West* 21(2008), 101–13.

Clark, D. M., *The Rise of the British Treasury: Colonial Administration of the Eighteenth Century* (Newton Abbot, 1960).

Clowes, W. L., *The Royal Navy: A History from the Earliest Times to 1900* (7 vols., London, 1897–1903, repr. 1997).

Coad, J., 'Historic architecture of H.M. Naval Base Portsmouth, 1700–1850', *MM* 67(1981), 3–59.

'Historic architecture of H.M. Naval Base Devonport 1689–1850', *MM* 69(1983), 341–92.

Historic Architecture of the Royal Navy: An Introduction (London, 1983).

The Royal Dockyards, 1690–1850: Architecture and Engineering Works of the Sailing Navy (Aldershot, 1989).

'The development and organisation of Plymouth dockyard, 1689–1815' in *The New Maritime History of Devon*, ed. Duffy *et al.*, I, 192–200.

Cole, G., 'The Office of Ordnance and the arming of the fleet in the French Revolutionary and Napoleonic Wars', unpub. University of Exeter Ph.D. thesis, 2008.

Colledge, J. J., *Ships of the Royal Navy* (London, 1987).

Colley, L., *Britons: Forging the Nation 1707–1837* (New Haven and London, 1992).

'The reach of the state, the appeal of the nation: mass arming and political culture in the Napoleonic Wars' in *An Imperial State at War*, ed. Stone, 165–84.

Collinge, J. M., *Navy Board Officials 1660–1832* (London, 1978).

Colquhoun, P., *A Treatise on the Commerce and Police of the Metropolis* (London, 1800).

Treatise on the Wealth, Power and Resources of the British Empire (London, 1815).

Condon, M. E., 'The administration of the transport service during the war against Revolutionary France, 1793–1802', unpub. University of London Ph.D. thesis, 1968.

'Living conditions on board troopships during the war against Revolutionary France, 1793–1802', *JSAHR* 49(1971), 14–19.

'The establishment of the Transport Board – a subdivision of the Admiralty – 4 July 1794', *MM* 58(1972), 69–84.

Congreve, W., *A statement of facts relative to the savings which have arisen from manufacturing gunpowder at the royal powder mills; and of the improvements which have been made in its strength and durability since the year 1783* (London, 1811).

Conway, S., *War, State and Society in Mid-Eighteenth Century Britain and Ireland* (Oxford, 2006).

Cookson, J. E., *The British Armed Nation, 1793–1815* (Oxford, 1997).

Corbett, J. S., *England in the Mediterranean, 1603–1714* (2 vols., London, 1904).

England in the Seven Years' War (2 vols., London, 1907, repr. 1992).

Cordingly, D., *Nicholas Pocock 1740–1821* (London, 1986).

Crewe, D., *Yellow Jack and the Worm: British Naval Administration in the West Indies, 1739–1748* (Liverpool, 1993).

Crimmin, P. K., 'Admiralty relations with the Treasury, 1783–1806: the preparation of naval estimates and the beginnings of Treasury control', *MM* 53(1967), 63–72.

"'A great object with us to procure this timber... '": the Royal Navy's search for ship timber in the Eastern Mediterranean and Southern Russia, 1803–1815', *IJMH* 4(1992), 83–115.

'Letters and documents relating to the service of Nelson's ships, 1780–1805: a critical report', *HR* 70(1997), 52–69.

'John Jervis, Earl of St Vincent 1735–1823' in *Precursors of Nelson*, ed. Le Fevre and Harding, 325–50.

Crocker, A. G., Crocker, G. M., Fairclough, K. R. and Wilks, M. J., *Gunpowder Mills: Documents of the Seventeenth and Eighteenth Centuries* (Guildford, 2000).

Crosby, A. W., *America, Russia, Hemp and Napoleon: American Trade with Russia and the Baltic, 1783–1812* (Ohio, 1965).

Crouzet, F., 'Wars, blockade and economic change in Europe 1792–1815', *Journal of Economic History*, 24(1964), 567–88.

Crowhurst, P., *The Defence of British Trade 1689–1815* (Folkestone, 1977).

The French War on Trade: Privateering 1793–1815 (Aldershot, 1989).

Crozier, R. D., *Guns, Gunpowder and Saltpetre: A Short History* (Faversham, 1998).

Cullen, L. M., *Anglo-Irish Trade 1660–1800* (Manchester, 1968).

Davies, D., 'The birth of the imperial navy? Aspects of English naval strategy c. 1650–90' in *Parameters of British Naval Power 1650–1850*, ed. Duffy, 14–38.

'Gibraltar in naval strategy c.1600–1783', *Transactions of the Naval Dockyards Society*, 2(2006), 10–18.

Davis, R., *The Rise of the English Shipping Industry in the Seventeenth and Eighteenth Centuries* (Newton Abbot, 1962, repr. 1972).

The Industrial Revolution and British Overseas Trade (Leicester, 1979).

Dawson, K., *The Industrial Revolution* (London, 1972).

Deane, P., 'War and industrialisation' in *War and Economic Development*, ed. J. M. Winter (Cambridge, 1975), 91–102.

Derrick, C., *Memoirs of the Rise and Progress of the Royal Navy* (London, 1806).

Derry, J. W., *Castlereagh* (London, 1976).

Dickinson, H. W., *Educating the Royal Navy: Eighteenth- and Nineteenth-Century Education for Officers* (Abingdon, 2007).

Dickson, D., *Old World Colony: Cork and South Munster 1630–1830* (Cork, 2005).

Dickson, P. G. M., *The Financial Revolution in England: A Study in the Development of Public Credit, 1688–1756* (London, 1967).

'War finance, 1689–1714' in *New Cambridge Modern History*, vol. VI: *1688–1725*, ed. J. S. Bromley (Cambridge, 1970), 284–93.

Dinwiddy, J., *Bentham* (Oxford, 1989).

Doe, H., 'Enterprising women: maritime business women, 1780–1880', unpub. University of Exeter Ph.D. thesis, 2007.

Donaldson, D. W., 'Port Mahon, Minorca: the preferred naval base for the English fleet in the Mediterranean in the seventeenth century', *MM* 88(2002), 423–36.

Doorne, C. J., 'Mutiny and sedition in the Home commands of the Royal Navy 1793–1803', unpub. University of London Ph.D. thesis, 1997.

Drayton, R., 'Knowledge and empire' in *OHBE*, II, 231–52.

Duffy, M., 'The foundations of British naval power' in *The Military Revolution and the State 1500–1800*, ed. Duffy (Exeter, 1980), 49–90.

Soldiers, Sugar and Seapower: The British Expeditions to the West Indies and the War against Revolutionary France (Oxford, 1987).

'Devon and the naval strategy of the French wars 1689–1815' in *The New Maritime History of Devon*, ed. Duffy *et al.*, I, 182–91.

'The establishment of the Western Squadron as the linchpin of British naval strategy' in *Parameters of British Naval Power 1650–1850*, ed. Duffy, 60–81.

'The creation of Plymouth dockyard and its impact on naval strategy' in *Guerres Maritime 1688–1713* (Vincennes, 1996), 245–74.

ed., *Parameters of British Naval Power 1650–1850* (Exeter, 1992).

Duffy, M., Fisher, S., Greenhill, B., Starkey, D. and Youings, J., eds., *The New Maritime History of Devon* (2 vols., London, 1992, 1994).

Duggan, J., *The Great Mutiny* (London, 1966).

Eames, A., *Ships and Seamen of Anglesey 1558–1918* (NMM, Greenwich, 1981).

Eastwood, D., '"Amplifying the province of the legislature": the flow of information and the English State in the early nineteenth century', *BIHR* 62(1989), 276–94.

Ehrman, J., 'William III and the emergence of a Mediterranean naval policy, 1692–4', *HJ* 9(1949), 269–92.

The Younger Pitt: The Years of Acclaim (London, 1969).

The Navy in the War of William III, 1689–1697 (Cambridge, 1953).

The Younger Pitt: The Reluctant Transition (London, 1983).

Elvin, J. G. D., *British Gunfounders 1700–1855* (privately printed 1983, NMM library aquisition N9869).

Emsley, C., *British Society and the French Wars 1793–1815* (London, 1979).

Engerman, S. L., 'Mercantilism and overseas trade, 1700–1800' in *International Trade and British Economic Growth from the Eighteenth Century to the Present Day*, ed. P. Mathias and J. A. Davis (Oxford, 1996), 182–204.

Ertman, T., 'The *sinews of power* and European state-building theory' in *An Imperial State at War*, ed. Stone, 33–51.

Evans, C., Eklund, A. and Ryden, G., 'Baltic iron and the organization of the British iron market in the eighteenth century' in *Britain and the Baltic: Studies in Commercial, Political and Cultural Relations 500–2000*, ed. P. Salmon and T. Barrow (Sunderland, 2003), 131–56.

Evans, D., *Arming the Fleet: The Development of the Royal Ordnance Yards 1770–1945* (Gosport, 2006).

Evans, E. J., *The Forging of the Modern State: Early Industrial Britain* (Harlow, 1983, 3rd edn, 2001).

Farnell, J. E., 'The Navigation Act of 1651: The First Dutch War and the London merchant community', *EconHR* 2nd ser. 16(1963–4), 439–54.

Fayle, C. E., *A Short History of the World's Shipping Industry* (London, 1933).

'The employment of British shipping' in *The Trade Winds*, ed. Parkinson, 72–86.

Fedorak, C. J., 'The Royal Navy and British amphibious operations during the Revolutionary and Napoleonic Wars', *Military Affairs* 52(1988), 141–6.

Ferns, J. L., 'Missing cannons: the Walker Company of Rotherham', *The Local Historian* 17(1986), 236–41.

Fincham, J., *A History of Naval Architeture* (London, 1851, repr. 1979).

Fisher, F. J., 'The development of the London food market, 1540–1640', *EconHR* 5(1935), 46–64, repr. in *Essays in Economic History*, ed. E. M. Carus-Wilson (London, 1954), 135–51.

Flinn, M. W., 'Sir Ambrose Crowley and the South Sea Scheme of 1711', *Journal of Economic History* 20(1960), 51–66.

Foord, A. S., 'The waning of "the influence of the Crown"', *EHR* 62(1947), 484–507.

Fry, M., *The Dundas Despotism* (Edinburgh, 1992).

Furber, H., *Henry Dundas, 1st Viscount Melville, 1742–1811* (London, 1931).

Gale, W. K. V., *Ironworking* (Aylesbury, 1981).

Gardiner, R., ed., *The Naval War of 1812* (London 1998).

Gates, D., 'The transformation of the Army 1783–1815' in *The Oxford History of the British Army*, ed. D. G. Chandler and I. Beckett (Oxford, 1994), 132–60.

George, D., *London Life in the Eighteenth Century* (London, 1925, repr. 1965).

Gilbert, K. R., *The Portsmouth Blockmaking Machinery* (London, 1965).

Gilboy, E. W., 'Demand as a factor in the Industrial Revolution' in *The Causes of the Industrial Revolution in England*, ed. R. M. Hartwell (Bungay, Suffolk, 1967), 121–38.

Glete, J., *Navies and Nations: Warships, Navies and State-building in Europe and America, 1500–1860* (2 vols., Stockholm, 1992).

   Warfare at Sea, 1500–1650: Maritime Conflicts and the Transformation of Europe (London, 2000).

   War and the State in Early Modern Europe: Spain, the Dutch Republic and Sweden as Fiscal-Military States, 1500–1660 (Abingdon, 2002).

Glover, R., *Britain at Bay: Defence against Bonaparte* (London, 1973).

   Peninsular Preparation. The Reform of the British Army 1795–1809 (Cambridge, 1970, repr. 1988).

Goldenburg, J. A., 'An analysis of shipbuilding sites in *Lloyd's Register* of 1776', *MM* 59(1973), 419–35.

Gough, B. M., 'The Royal Navy and empire' in *OHBE*, V, 327–41.

Gradish, S. F., *The Manning of the British Navy during the Seven Years War* (London, 1980).

Graham, G. S., *Empire of the North Atlantic: The Maritime Struggle for North America* (London, 1958).

   The Politics of Naval Supremacy: Studies in British Maritime Ascendancy (Cambridge, 1965).

Green, G. L., *The Royal Navy and Anglo-Jewry 1740–1820* (Ealing, 1989).

Gwyn, J., *Ashore and Afloat: The British Navy and the Halifax Naval Yard before 1820* (Ottawa, 2004).

Haas, J. M., 'The introduction of taskwork into the Royal dockyards, 1775', *JBS* 8(1969), 44–68.

   'The Royal Dockyards: the earliest visitations and reforms, 1749–1778', *HJ* 13(1970), 191–215.

Hall, C. D., *British Strategy in the Napoleonic War 1803–15* (Manchester, 1992).

'The Royal Navy and the Peninsular War', *MM* 79(1993), 403–18.

Wellington's Navy: Sea Power and the Peninsular War 1807–1814 (London, 2004).

Hancock, D., *Citizens of the World: London Merchants and the Integration of the British Atlantic Community, 1735–1785* (Cambridge, 1995).

Harding, R., *Amphibious Warfare in the Eighteenth Century: The British Expedition to the West Indies 1740–1742* (Woodbridge, 1991).

The Evolution of the Sailing Navy, 1509–1815 (Basingstoke, 1995).

'The expeditions to Quebec 1690 and 1711: the evolution of trans-Atlantic amphibious power' in *Guerres Maritimes 1688–1713* (Vincennes, 1996), 197–212.

Seapower and Naval Warfare 1650–1830 (London, 1999).

'Sailors and gentlemen of parade: some professional and technical problems concerning the conduct of combined operations in the eighteenth century' in *Naval History 1680–1850*, ed. Harding, 127–47.

ed., *Naval History 1680–1850* (Burlington, VT, and Aldershot, 2006).

Harland, K., 'The Royal naval hospital at Minorca, 1711: an example of an admiral's involvement in the expansion of naval medical care', *MM* 94(2008), 36–47.

Harling, P. and Mandler, P., 'From "fiscal-military" state to laissez-faire state, 1760–1850', *JBS* 32(1993), 44–70.

Harris, J. R., *The British Iron Industry* (Basingstoke, 1988).

Hattendorf, J., *England in the War of the Spanish Succession: A Study of the English View and Conduct of Grand Strategy, 1702–1712* (New York and London, 1987).

ed., *The Influence of History on Mahan* (Newport, RI, 1991).

Hellmuth, E., 'Why does corruption matter? Reforms and reform movements in Britain and Germany in the second half of the eighteenth century' in *Reform in Great Britain and Germany 1750–1850*, ed. T. C. W. Blanning and P. Wende (Oxford, 1999), 6–23.

Henriques, U., 'Jeremy Bentham and the machinery of social reform' in *British Government and Administration: Studies Presented to S. B. Chrimes*, ed. H. Hearder and H. R. Loyn (Cardiff, 1974).

Hill, R., *The Prizes of War: The Naval Prize System in the Napoleonic Wars 1793–1815* (Stroud, Glos., 1998).

Hodgskin, T., *An essay on Naval Discipline shewing part of its evil effects on the minds of officers, on the minds of men, and on the community, with an amended system by which pressing may be immediately abolished* (London, 1812).

Hogg, O. F. G., *The Royal Arsenal: Its Background, Origin and Subsequent History* (Oxford, 1983).

Hollander, J. H., 'The work and influence of Ricardo' in *David Ricardo: Critical Assessments*, ed. J. Cunningham Wood (4 vols., Beckenham, Kent, 1985), I, 42–5.

Holmes, G., *The Making of a Great Power: Late Stuart and Early Georgian Britain* (Harlow, Essex, 1993).

Holmes, R., *Redcoat: The British Soldier in the Age of Horse and Musket* (London, 2001).

Hoppit, J., *Risk and Failure in English Business 1700–1800* (Cambridge, 1987).

'Checking the Leviathan, 1688–1832' in *The Political Economy of British Historical Experience, 1688–1914*, ed. Winch and O'Brien, 267–94.

Hornstein, S. R., *The Restoration Navy and English Foreign Trade 1674–1688* (Aldershot, 1991).

Hueckel, G., 'War and the British economy, 1793–1815: a general equilibrium analysis', *Explorations in Economic History* 10(1973), 365–96.

Huston, J. A., *The Sinews of War: Army Logistics 1775–1953* (Washington, 1966).

Hutton, C., 'The force of fired gunpowder and the initial velocities of cannon balls, determined by experiments; from which is also deduced the relation of the initial velocity to the weight of the shot and the quantity of powder', *Philosophical Transactions of the Royal Society* 68(1778), 50–85.

Ingen-housz, J., 'Account of a new kind of inflammable air or gas, together with a new theory of gunpowder', *Philosophical Transactions of the Royal Society* 69(1779), 376–418.

Ingram, E., 'Illusions of victory: the Nile, Copenhagen and Trafalgar revisited', *Military Affairs* 48(1984), 140–3.

Jackson, H. H. and de Beer, C., *Eighteenth Century Gunfounding* (Newton Abbot, 1973).

Jackson, R. V., 'Government expenditure and British economic growth in the eighteenth century: some problems of measurement', *EconHR* 2nd ser. 43(1990), 217–35.

James, P., *Population Malthus: His Life and Times* (London, 1979).

James, W., *The Naval History of Great Britain from the declaration of war by France in 1793 to the accession of George IV* (6 vols., London, 1859).

Jarvis, R. C., 'Army transport and the English constitution with special reference to the Jacobite risings', *Journal of Transport History* 2(1955–6), 101–4.

Jenkins, E. H., *A History of the French Navy* (London, 1973).

John, A. H., 'War and the English economy, 1700–1763', *EconHR* 2nd ser. 7(1954–5), 329–44.

Jones, A. G. E., 'Shipbuilding in Ipswich, 1700–1750', *MM* 43(1957), 294–305.
'Shipbuilding in Ipswich, 1750–1800', *MM* 58(1972), 183–93.

Jordan, G. and Rogers, N., 'Admirals as heroes: patriotism and liberty in Hanoverian England', *JBS* 28(1989), 201–24.

Kemp, B., *King and Commons 1660–1832* (London, 1957).

Kemp, P., *The British Seaman: A Social History of the Lower Deck* (London, 1970).

Kennard, A. N., *Gunfounding and Gunfounders: A Directory of Cannon Founders from Earliest Times to 1850* (London, 1986).

Kennedy, P., *The Rise and Fall of British Naval Mastery* (London, 1976).
The Rise and Fall of the Great Powers: Economic Change and Military Conflict from 1500 to 2000 (London, 1989).

Keppel, S., *Three Brothers at Havana 1762* (Wilton, 1981).

Kirkaldy, A. W., *British Shipping: Its History, Organisation and Importance* (London, 1914).

Knight, R. J. B., 'Sandwich, Middleton and Dockyard appointments', *MM* 57(1971), 175–92
'The Royal Dockyards in England at the time of the American War', unpub. University of London Ph.D. thesis, 1972.

'The introduction of copper sheathing into the Royal Navy 1779–1786', *MM* 59(1973), 299–309.

'The performance of the Royal dockyards in England during the American War of Independence' in *The American Revolution and the Sea* (Basildon, 1974), 139–44.

'Civilians and the navy, 1660–1832' in *Sea Studies: Essays Presented to Basil Greenhill*, ed. P. G. W. Annis (Greenwich, 1983), 63–70.

'The building and maintenance of the British fleet during the Anglo-French Wars 1688–1815' in *Les marines de guerre européennes XVII–XVII siècles*, ed. M. Acerra, J. Merino and J. Meyer (Paris, 1985), 35–50.

'New England forests and British seapower: Albion revised', *American Neptune* 46(1986), 221–9.

'The first fleet – its state and preparation 1786–1787' in *Studies from Terra Australis to Australia* (Canberra, 1989), 121–36, 256–62.

'The Royal Navy's recovery after the early phase of the American Revolutionary War' in *The Aftermath of Defeat: Societies, Armed Forces and the Challenge of Recovery*, ed. G. J. Andreopoulos and H. E. Selesky (New Haven, CT, 1994), 10–25.

'From impressment to task work: strikes and disruption in the Royal dockyards, 1688–1788' in *History of Work and Labour Relations in the Royal Dockyards*, ed. K. Lunn and A. Day (London, 1999), 1–20.

The Pursuit of Victory: The Life and Achievement of Horatio Nelson (London, 2005).

'Politics and trust in victualling the navy, 1793–1815', *MM* 94(2008), 133–49.

Kynaston, D., *The Secretary of State* (Lavenham, Suffolk, 1978).

Lambert, A., *The Last Sailing Battlefleet: Maintaining Naval Mastery 1815–1850* (London, 1991).

'Preparing for the long peace: the reconstruction of the Royal Navy 1815–1830', *MM* 82(1996), 41–54.

The Foundations of Naval History: John Knox Laughton, the Royal Navy and the Historical Profession (London, 1998).

Lavery, B., *The Ship of the Line: The Development of the battlefleet 1650–1850* (London, 1983).

The Arming and Fitting of English Ships of War 1600–1815 (London, 1987).

Nelson's Navy: The Ships, Men and Organisation 1793–1815 (London, 1989).

'The British navy and its bases, 1793–1815' in *Français et anglais en Méditerranée de la Revolution française à l'indépendance de la Grèce (1789–1830)* (Vincennes, 1992).

Le Fevre, P. and Harding, R., eds., *Precursors of Nelson: British Admirals of the Eighteenth Century* (London, 2000).

Lenman, B., *Britain's Colonial Wars 1688–1783* (Harlow, 2001).

Lester, M.,'Vice-Admiral George Murray and the origins of the Bermuda naval base', *MM* 94(2008), 285–97.

Lewis, M., *England's Sea Officers: The Story of the Naval Profession* (London, 1948).

A Social History of the Navy 1793–1815 (London, 1960).

The Navy in Transition 1814–1864: A Social History (London, 1965).

Lieberman, D., 'Economy and polity in Bentham's science of legislation' in *Economy, Polity and Society: British Intellectual History 1750–1950*, ed. S. Collini, R. Whatmore and B. Young (Cambridge, 2000), 107–34.

Little, H. M., 'The Treasury, the Commissariat and the supply of the Combined Army in Germany during the Seven Years' War (1756–1763)', unpub. University of London Ph.D. thesis, 1981.

'The emergence of a Commissariat during the Seven Years War in Germany', *JSAHR* 61(1983), 201–14.

'Thomas Pownall and army supply, 1761–1766', *JSAHR* 65(1987), 92–104.

Lloyd, C., 'New light on the mutiny at the Nore', *MM* 46(1960), 285.

'The introduction of lemon juice as a cure for scurvy', *Bulletin of the History of Medicine* 35(1961), 123–32.

The British Seaman 1200–1860: A Social Survey (London, 1968).

'Victualling of the fleet in the eighteenth and nineteenth centuries' in *Starving Sailors: The Influence of Nutrition upon Naval and Maritime History*, ed. J. Watt, E. J. Freeman and W. F. Bynum (NMM, Greenwich, 1981), 9–15.

London Gazette, 1808.

Lubimenko, I., 'The struggle of the Dutch with the English for the Russian market in the seventeenth century', *TRHS* 4th ser. 7(1924), 27–51.

MacBride, D., *An Historical Account of a New Method of Treating the Scurvy at Sea; containing ten cases which show that this destructive disease may be easily and effectually cured without the aid of fresh vegetable diet* (1764).

MacDonald, J., *Feeding Nelson's Navy: The True Story of Food at Sea in the Georgian Era* (London, 2004).

MacDougall, P., 'The formative years: Malta dockyard 1800–1815', *MM* 76(1990), 205–13.

Mackay, R. F., 'Lord St Vincent's early years (1735–55)', *MM* 76(1990), 51–65.

Mackesy, P., *British Victory in Egypt, 1801: The End of Napoleon's Conquest* (London, 1995).

'Problems of an amphibious power: Britain against France, 1793–1815' in *Naval History 1680–1850*, ed. Harding (Aldershot, 2006), 117–26.

Mahan, A. T., *The Influence of Sea Power upon History 1660–1783* (Boston, MA, 1890, repr. 1965).

Malone, J. J., 'England and the Baltic naval stores trade in the seventeenth and eighteenth centuries', *MM* 58(1972), 375–94.

Maltby, W. S., 'The origins of a global strategy: England from 1558 to 1713' in *The Making of Strategy: Rulers, States and War*, ed. W. Murray, M. Knox and A. Bernstein (Cambridge, 1994), 151–77.

Marshall, J., *Account of the Population in each of Six Thousand of the Principal Towns and Parishes in England and Wales . . . at each of the three periods 1801, 1811 and 1821* (London, 1831).

Marshall, P. J., 'Introduction' to *OHBE*, II, 1–27.

'The first British empire' in *OHBE*, V, 43–53.

The Making and Unmaking of Empires: Britain, India and America c. 1750–1783 (Oxford, 2005).

Masefield, J., *Sea Life in Nelson's Time* (London, 1905, repr. 1972).

Matcham, M. E., *A Forgotten John Russell* (London, 1905).

Mathias, P., 'Swords and ploughshares: the armed forces, medicine and public health in the late eighteenth century' in *War and Economic Development*, ed. J. M. Winter (Cambridge, 1975), 73–90.

'Risk, credit and kinship in early modern enterprise' in *The Early Modern Atlantic Economy*, ed. J. J. McCusker and K. Morgan (Cambridge, 2000), 15–35.

McCord, N., 'The impress service in the north-east of England during the Napoleonic War' in *Pressgangs and Privateers*, ed. T. Barrow (Whitley Bay, 1993), 22–37.

Merrett, L. H., 'A most important undertaking: the building of the Plymouth breakwater', *Transport History* 5(1977), 153–4.

Middleton, R., 'Naval administration in the age of Pitt and Anson, 1755–1763' in *The British Navy and the Use of Naval Power in the Eighteenth Century*, ed. Black and Woodfine, 109–27.

'British naval strategy 1755–62: the Western Squadron', *MM* 75(1989), 349–67.

Mitchell, B.R. and Deane, P., *Abstract of British Historical Statistics* (Cambridge, 1962).

Monaque, R., 'On Board HMS *Alexander* (1796–9)', *MM* 89(2003), 207–12.

Morgan, K., *Bristol and the Atlantic Trade in the Eighteenth Century* (Cambridge, 1993).

'Atlantic trade and British economic growth in the eighteenth century' in *International Trade and British Economic Growth in the Eighteenth Century*, ed. P. Mathias and J. A. Davis (Oxford, 1996), 8–27.

'Business networks in the British export trade to North America, 1750–1800' in *The Early Modern Atlantic Economy*, ed. J. J. McCusker and K. Morgan (Cambridge, 2000), 36–62.

Slavery, Atlantic Trade and the British Economy, 1660–1800 (Cambridge, 2000).

'Mercantilism and the British empire, 1688–1815' in *The Political Economy of British Historical Experience, 1688–1914*, ed. Winch and O'Brien, 165–91.

Morriss, R. A., 'Labour relations in the Royal dockyards, 1801–1805', *MM* 62(1976), 337–46.

'The administration of the Royal dockyards in England during the Revolutionary and Napoleonic Wars, with special reference to the period 1801–1805', unpub. University of London Ph.D. thesis, 1978.

'Samuel Bentham and the management of the royal dockyards, 1796–1807', *BIHR* 54(1981), 226–40.

'St Vincent and reform, 1801–04', *MM* 69(1983), 269–90.

The Royal Dockyards during the Revolutionary and Napoleonic Wars (Leicester, 1983).

Cockburn and the British Navy in Transition: Admiral Sir George Cockburn, 1772–1853 (Exeter, 1997).

'Charles Middleton, Lord Barham' in *Precursors of Nelson*, ed. Le Fevre and Harding, 301–23.

Naval Power and British Culture, 1760–1850: Public Trust and Government Ideology (Aldershot, 2004).

'The supply of casks and staves to the Royal Navy, 1770–1815', *MM* 93(2007), 43–50.

'Colonisation, conquest, and the supply of food and transport: the reorganization of logistics management, 1780–1795', *War in History* 14(2007), 310–24.

Mott, R. A., 'Dry and wet puddling', *Transactions of the Newcomen Society* 49(1977–8), 153–8.

Namier, L., 'The end of the nominal Cabinet' in *Crossroads of Power: Essays on Eighteenth Century England* (London, 1962), 118–23.

Namier, L. and Brooke, J., eds., *The History of Parliament: The House of Commons, 1754–1790* (3 vols., London, 1964).

Naval Chronicle 2(1799), 14(1805).

Neal, L., 'The finance of business during the industrial revolution' in *The Economic History of Britain since 1700*, ed. R. Floud and D. McCloskey (2nd edn., Cambridge, 1994), 151–91.

Norris, J., *Shelburne and Reform* (London, 1963).

O'Brien, P. K., 'The political economy of British taxation, 1660–1815', *EconHR* 2nd ser. 41(1988), 1–32.

'Political pre-conditions for the Industrial Revolution' in *The Industrial Revolution and British Society*, ed. P. K. O'Brien and R. Quinault (Cambridge, 1993), 124–55.

'Central government and the economy, 1688–1815' in *The Economic History of Britain since 1700*, ed. R. Floud and D. McCloskey (2nd edn, Cambridge, 1994), 205–41.

'Fiscal exceptionalism: Great Britain and its European rivals from Civil War to triumph at Trafalgar and Waterloo' in *The Political Economy of British Historical Experience, 1688–1914*, ed. Winch and O'Brien, 245–65.

O'Brien, P. K. and Hunt, P. A., 'The rise of a fiscal state in England, 1485–1815', *HR* 66(1993), 129–76.

O'Rourke, K. H., 'The worldwide economic impact of the French Revolutionary and Napoleonic Wars, 1793–1815', *Journal of Global History* 1(2006), 123–49.

Olson, A. G., *The Radical Duke: The Career and Correspondence of Charles Lennox, Third Duke of Richmond* (Oxford, 1961).

Oman, C., *Wellington's Army 1809–1814* (London, 1913).

Omer, R. and Panting, G., eds., *Working Men Who Got Wet* (Newfoundland, 1980).

Oppenheim, M., *A History of the Administration of the Royal Navy and of merchant shipping in relation to the Navy from 1509 to 1660 with an introduction treating of the preceding period* (London, 1896, repr. Aldershot, 1988).

Palmer, S., *Politics, Shipping and the Repeal of the Navigation Laws* (Manchester, 1990).

Pares, R., 'The manning of the navy in the West Indies, 1702–63', *TRHS* 4th ser. 20(1937), 54–60.

Limited Monarchy in Great Britain in the Eighteenth Century (Hist. Assoc., 1957).

Parker, G., *The Military Revolution: Military Innovation and the Rise of the West, 1500–1800* (Cambridge, 1988, 2nd edn. 1996).

Parkinson, C. N., *Trade in the Eastern Seas 1793–1813* (London, 1966).
  ed., *The Trade Winds: A Study of British Overseas Trade during the French Wars 1793–1815* (London, 1948).
Pearsall, A. W. H., 'Jonas Hanway and naval victualling', *Journal of the Royal Society for the Encouragement of Arts, Manufactures and Commerce* 134(1986), 657–9.
Plumb, J. H., *The Growth of Political Stability in England* (London, 1967, repr. 1980).
Pool, B., 'Some notes on warship building by contract in the eighteenth century', *MM* 49(1963), 105–19.
  Navy Board Contracts 1660–1832: Contract Administration under the Navy Board (London, 1966).
  ed., *The Croker Papers* (London, 1967).
Porter, B. D., *War and the Rise of the State: The Military Foundations of Modern Politics* (New York, 1974).
Pressnell, L. S., *Country Banking in the Industrial Revolution* (Oxford, 1956).
Price, J. M., 'The imperial economy, 1700–1776' in *OHBE*, II, 78–104.
Pringle, C. D. and Knoll, M. J., 'Why Trafalgar was won before it was fought: lessons from resource based theory', *Academy of Management Executive* 11(1997), 73–89.
Pritchard, J., *Louis XV's Navy 1748–1762: A Study of Organisation and Administration* (Kingston and Montreal, 1987).
Prothero, I. J., *Artisans and Politics in Early Nineteenth Century London: John Gast and his Times* (London, 1979).
Quarterly Review 8(1812).
Ramana, D. V., 'Ricardo's environment' in *David Ricardo: Critical Assessments*, ed. J. Cunningham Wood (4 vols., Beckenham, Kent, 1985), I, 196–208.
Ranft, B. M., 'Labour relations in the Royal dockyards in 1739', *MM* 47(1961), 281–91.
Raudzens, G., 'Military revolution or maritime evolution? Military superiorities or transportation advantages as main causes of European colonial conquests to 1788', *The Journal of Military History* 63(1999), 631–41.
Reading, D. K., *The Anglo-Russian Commercial Treaty of 1734* (New Haven, CT, 1938).
Redgrave, T. M. O., 'Wellington's logistical arrangements in the Peninsular War 1809–1814', unpub. University of London Ph.D. thesis, 1979.
Riley, J. C., *International Government Finance and the Amsterdam Capital Market, 1740–1815* (Cambridge, 1980).
  The Seven Years' War and the Old Regime in France: The Economic and Financial Toll (Princeton, NJ, 1986).
Robins, B., *New Principles of Gunnery* (1742).
Rodger, N.A.M., *The Admiralty* (Lavenham, Suffolk, 1979).
  The Wooden World: An Anatomy of the Georgian Navy (London, 1986).
  'The inner life of the navy, 1750–1800: change or decay?' in *Guerres et paix 1660–1815* (Vincennes, 1987), 171–9.
  Naval Records for Genealogists (Public Record Office, London, 1988).

'"A little navy of your own making": Admiral Boscawen and the Cornish connection in the Royal Navy' in *Parameters of British Naval Power 1650–1850*, ed. Duffy, 82–92.

'The continental commitment in the eighteenth century' in *War, Strategy and International Politics: Essays in Honour of Sir Michael Howard*, ed. L. Freedman, P. Hayes and R. O'Neill (Oxford, 1992), 39–55.

The Insatiable Earl: A Life of John Montagu, 4th Earl of Sandwich (London, 1993).

'The mutiny in the *James & Thomas*' in *Pressgangs and Privateers*, ed. T. Barrow (Whitley Bay, 1993), 5–11.

'Sea-power and empire, 1688–1793' in *OHBE*, II, 169–83.

'George, Lord Anson 1697–1762' in *Precursors of Nelson*, ed. Le Fevre and Harding, 176–99.

'Honour and duty at sea, 1660–1815', *HR* 75(2002), 425–7.

The Command of the Ocean: A Naval History of Britain 1649–1815 (London, 2004).

ed., *Articles of War: The Statutes which Governed our Fighting Navies 1661, 1749 and 1886* (Havant, 1982).

Roseveare, H., *The Treasury 1660–1870: The Foundations of Control* (London, 1973).

Rothenberg, G. E., *The Art of Warfare in the Age of Napoleon* (London, 1977).

Rothschild, E., 'The English Kopf' in *The Political Economy of British Historical Experience, 1688–1914*, ed. Winch and O'Brien, 31–60.

Royal Kalendar: or complete and correct annual register for England, Scotland, Ireland and America (printed annually, London, 1767–1814).

Russell, C., 'Richard Watson, gaiters and gunpowder' in *The 1702 Chair of Chemistry at Cambridge: Transformation and Change*, ed. M. D. Archer and C. D. Haley (Cambridge, 2005), 57–83.

Ryan, A. N., 'The defence of British trade in the Baltic, 1807–1813', *EHR* 74(1959), 443–66.

'William III and the Brest fleet in the Nine Years War' in *William III and Louis XIV: Essays 1680–1720 by and for Mark A. Thomson*, ed. R. Hatton and J. S. Bromley (Liverpool, 1968), 49–67.

'The Royal Navy and the blockade of Brest, 1689–1805: theory and practice' in *Les marines de guerre Européennes XVIIe–XVIIIe siècles*, ed. M. Acerra, J. Merino and J. Meyer (Paris, 1985), 175–93.

'An ambassador afloat: Vice-Admiral Sir James Saumarez and the Swedish court, 1808–1812' in *The British Navy and the Use of Naval Power in the Eighteenth Century*, ed. Black and Woodfine, 237–58.

Sainty, J. C., *Treasury Officials 1660–1870* (London, 1972).

Officials of the Secretaries of State 1660–1782 (London, 1973).

Officials of the Board of Trade 1660–1870 (London, 1974).

Admiralty Officials 1660–1870 (London, 1975).

Saunders, A., 'Upnor castle and gunpowder supply to the navy 1801–4', *MM* 91(2005), 160–74.

Saxby, R., 'The blockade of Brest in the French Revolutionary War', *MM* 78(1992), 23–35.

Sheridan, L. A., *Fraud in Equity: A Study in English and Irish Law* (London, 1957).

Sherwig, J. M., *Guineas and Gunpowder: British Foreign Aid in the Wars with France 1793–1815* (Cambridge, MA, 1969).

Sinclair, J., *The History of the Public Revenue of the British Empire* (3 vols., 3rd edn, London, 1804).

Skempton, A. W., *A History of the Steam Dredger, 1797–1830* (London, 1975).

Society for the Improvement of Naval Architecture, *A View of the Naval Force of Great Britain* (1791).

Spengler, J. J., 'Adam Smith's theory of economic growth, parts I–II' in *Adam Smith: Critical Assessments*, ed. J. Cunningham Wood (3 vols., Beckenham, Kent, 1983), III, 110–31.

Spinney, J. D., 'The Hermione Mutiny', *MM* 41(1955), 123.

Stack, D., *Nature and Artifice: The Life and Thoughts of Thomas Hodgskin, 1787–1869* (London, 1998).

Stark, S. J., *Female Tars: Women Aboard Ship in the Age of Sail* (London, 1998).

Steer, D. M., 'The blockade of Brest and the Royal Navy 1793–1805', unpub. University of Liverpool MA thesis, 1971.

Steer, M., 'The blockade of Brest and the victualling of the Western Squadron 1793–1805', *MM* 76(1990), 307–16.

Stokes, E. T., 'Bureaucracy and ideology: Britain and India in the nineteenth century', *Transactions of the Royal Historical Society* 5th ser. 30(1980), 131–56.

Stone, L., ed., *An Imperial State at War: Britain from 1689 to 1815* (London, 1994).

Storrs, C., ed., *The Fiscal-Military State in Eighteenth Century Europe: Essays in Honour of P. G. M. Dickson* (Farnham, 2009).

Stout, N. R., 'Manning the Royal Navy in North America, 1763–1775', *The American Neptune* 23(1963), 174–85.

Syrett, D., *Shipping and the American War 1775–83: A Study of British Transport Organization* (London, 1970).

'The shipowners during the American War, 1776–1783', paper presented to 'The Shipowner in History' symposium, NMM, 1984.

The Royal Navy in European Waters during the American Revolutionary War (Columbia, SC, 1988).

'The Victualling Board charters shipping, 1775–82', *HR* 68(1995), 212–24.

'Christopher Atkinson and the Victualling Board, 1775–82', *HR* 69(1996), 129–42.

'The Victualling Board charters shipping, 1739–1748', *IJMH* 9(1997), 57–67.

'The methodology of British amphibious operations during the Seven Years' and American Wars' in *Naval History 1680–1850*, ed. Harding, 309–20.

'Towards Dettingen: the conveyancing of the British army to Flanders in 1742', *JSAHR* 84(2006), 316–26.

Shipping and Military Power in the Seven Years War: The Sails of Victory (Exeter, 2008).

Talbott, J. E., *The Pen and Ink Sailor: Charles Middleton and the King's Navy, 1778–1813* (London, 1998).

Taylor, J. S., 'Jonas Hanway: Christian mercantilist', *Journal of the Royal Society for the Encouragement of Arts, Manufactures and Commerce* 134(1986), 641–5.

Thomas, J. H., *The East India Company and the Provinces in the Eighteenth Century*, vol. I: *Portsmouth and the East India Company 1700–1815* (Lampeter, Wales, 1999).

Thompson, B., 'New experiments upon gunpowder . . . to which are added an account of a new method of determining the velocities of all kinds of military projectiles', *Philosophical Transactions of the Royal Society* 71(1781), 229–328.

Thompson, E. P., *The Making of the English Working Class* (London, 1963, repub. Pelican 1968).

Tomlinson, H. C., 'The Ordnance Office and the Navy, 1660–1714', *EHR* 90(1975), 19–39.

'Wealden gunfounding: an analysis of its demise in the eighteenth century', *EconHR* 29(1976), 383–400.

Guns and Government: The Ordnance Office under the later Stuarts (London, 1979).

Torrance, J., 'Social class and bureaucratic innovation: the Commissioners for examining the public accounts, 1780–1787', *Past and Present* 78(1978), 56–81.

Tracy, N., *Navies, Deterrence, and American Independence: Britain and Sea Power in the 1760s and 1780s* (Vancouver, BC, 1988).

Manila Ransomed: The British Assault on Manila in the Seven Years War (Exeter, 1995).

Trinder, B., *The Industrial Revolution in Shropshire* (Chichester, 1973, repr. 2000).

Turner, E. H., 'Naval medical service, 1793–1815', *MM* 46(1960), 119–33.

Unger, R., 'Warships, cargo ships and Adam Smith: trade and government in the eighteenth century', *MM* 92(2006), 41–59.

Vale, B., 'The conquest of scurvy in the Royal Navy 1793–1800: a challenge to current orthodoxy', *MM* 94(2008), 160–75.

Van Creveld, M., *Supplying War: Logistics from Wallenstein to Patton* (Cambridge, 1977).

Victoria County History of Kent (3 vols., London, 1908–32).

Ville, S., 'Total factor productivity in the English shipping industry: the north-east coal trade, 1700–1850', *EconHR* 39, 3(1986), 355–70.

English Shipowning during the Industrial Revolution: Michael Henley and Son, London Shipowners, 1770–1830 (Manchester, 1987).

'The growth of specialization in English shipowning, 1750–1850', *EconHR* 46, 4(1993), 702–22.

Wadia, R. A., *The Bombay Dockyard and the Wadia Master Shipbuilders* (Bombay, 1955), 31–9.

Wagstaff, J. M., 'Network analysis and logistics: applied topology' in *General Issues in the Study of Medieval Logistics: Sources, Problems and Methodologies* (Leiden and Boston, 2006), 69–92.

Ware, C., 'The Glorious First of June: The British strategic perspective' in *The Glorious First of June 1794: A Naval Battle and its Aftermath*, ed. M. Duffy and R. Morriss (Exeter, 2001), 25–45.

Waters, D. W., 'Seamen, scientists, historians and strategy', *The British Journal for the History of Science* 13(1980), 207–9, printed as appendix B to the introduction of *The Defeat of the Enemy Attack on Shipping, 1939–1945*, ed. E. J. Grove (NRS, 1997), xl–xlii.

Watson, J. S., *The Reign of George III, 1760–1815* (London, 1960).

Watson, P. K., 'The Commission for victualling the navy, the Commission for sick and wounded seamen and prisoners of war, the Commission for transports, 1702–14', unpub. University of London Ph.D. thesis, 1965.

Watt, J., 'Some consequences of nutritional disorders in eighteenth-century British circumnavigations' in *Starving Sailors: The Influence of Nutrition upon Naval and Maritime History*, ed. J. Watt, E. J. Freeman and W. F. Bynum (NMM, Greenwich, 1981), 51–71.

'Surgery at Trafalgar', *MM* 91(2005), 266–83.

Webb, P., 'The rebuilding and repair of the fleet 1783–93', *BIHR* 50(1977), 194–209.

'Construction, repair and maintenance in the battle fleet of the Royal Navy, 1793–1815' in *The British Navy and the Use of Naval Power in the Eighteenth Century*, ed. Black and Woodfine, 207–19.

'British squadrons in North American waters, 1783–1793', *The Northern Mariner* 5(1995), 19–34.

Webster, A., *The Debate on the Rise of the British Empire* (Manchester, 2006).

Weller, J., 'Wellington's Peninsular War logistics', *JSAHR* 42(1964), 197–202.

West, J., *Gunpowder, Government and War in the Mid-Eighteenth Century* (London, 1991).

Western, J. W., *The English Militia in the Eighteenth Century* (London, 1965).

Wilkinson, C., *The British Navy and the State in the Eighteenth Century* (Woodbridge, Suffolk, 2004).

Willan, T. S., *The English Coasting Trade, 1600–1750* (Manchester, 1938).

Williams, G., '"To make discoveries of countries hitherto unknown": the Admiralty and Pacific exploration in the eighteenth century', *MM* 82(1996), 14–27.

The Great South Sea: English Voyages and Encounters 1570–1750 (Newhaven and London, 1997).

Wilson, K., 'Empire of virtue' in *An Imperial State at War*, ed. Stone, 128–64.

Winch, D. and O'Brien, P. K., eds., *The Political Economy of British Historical Experience, 1688–1914* (Oxford, 2002).

Woodfine, P., 'Ideas of naval power and the conflict with Spain, 1737–1742' in *The British Navy and the Use of Naval Power in the Eighteenth Century*, ed. Black and Woodfine, 71–90.

Wrigley, E. A., 'The supply of raw materials in the Industrial Revolution', *EconHR* 2nd ser., 15(1962), repr. in *The Causes of the Industrial Revolution in England*, ed. R. M. Hartwell (London, 1968), 97–120.

'Urban growth and agricultural change: England and the continent in the early modern period' in *The Eighteenth Century Town: A Reader in English Urban History, 1688–1820*, ed. P. Borsay (Harlow, 1990), 39–82.

'Society and economy in the eighteenth century' in *An Imperial State at War*, ed. Stone, 72–95.

Zahedieh, N., 'London and the colonial consumer in the late seventeenth century', *EconHR* 2nd ser. 47(1994), 239–61.

   'Economy' in *The British Atlantic World, 1500–1800*, ed. D. Armitage and M. J. Braddick (Basingstoke, 2002), 51–68.

Zulueta, J. de, 'Health and military factors in Vernon's failure at Cartagena', *MM* 78(1992), 127–41.

Index